The Conquest of the Russian Arctic

The Conquest *of the* Russian Arctic

Paul R. Josephson

Harvard University Press
Cambridge, Massachusetts & London, England
2014

Second Printing

Library of Congress Cataloging-in-Publication Data

Josephson, Paul R.
The conquest of the Russian Arctic / Paul R. Josephson.
pages cm
Includes bibliographical references and index.
ISBN 978-0-674-72890-5
1. Arctic regions—Politics and government. 2. Arctic regions—Strategic aspects. 3. Soviet Union—Territorial expansion. 4. Soviet Union—Relations—Arctic regions. 5. Arctic regions—Relations—Soviet Union. 6. Arctic regions—Environmental conditions. 7. Natural resources—Arctic regions—History. 8. Environmental degradation—Arctic regions—History. 9. Indigenous peoples—Arctic regions—History. 10. Political persecution—Arctic regions—History. I. Title.
G615.J67 2014
947'.1084—dc23
2013046803

Contents

Illustrations and Tables

Illustrations

Tables

Note on Transliteration

In rendering Russian names, I have adopted a modified version of the Library of Congress system of transliteration, dropping many soft signs and adopting more common "western" spellings for some names and places (Glavsevmorput, *not* Glavsevmorput'; Trotsky, *not* Trotskii; Naryan-Mar, *not* Nar'ian-Mar; Yamal, *not* Iamal; and so on).

The Conquest of the Russian Arctic

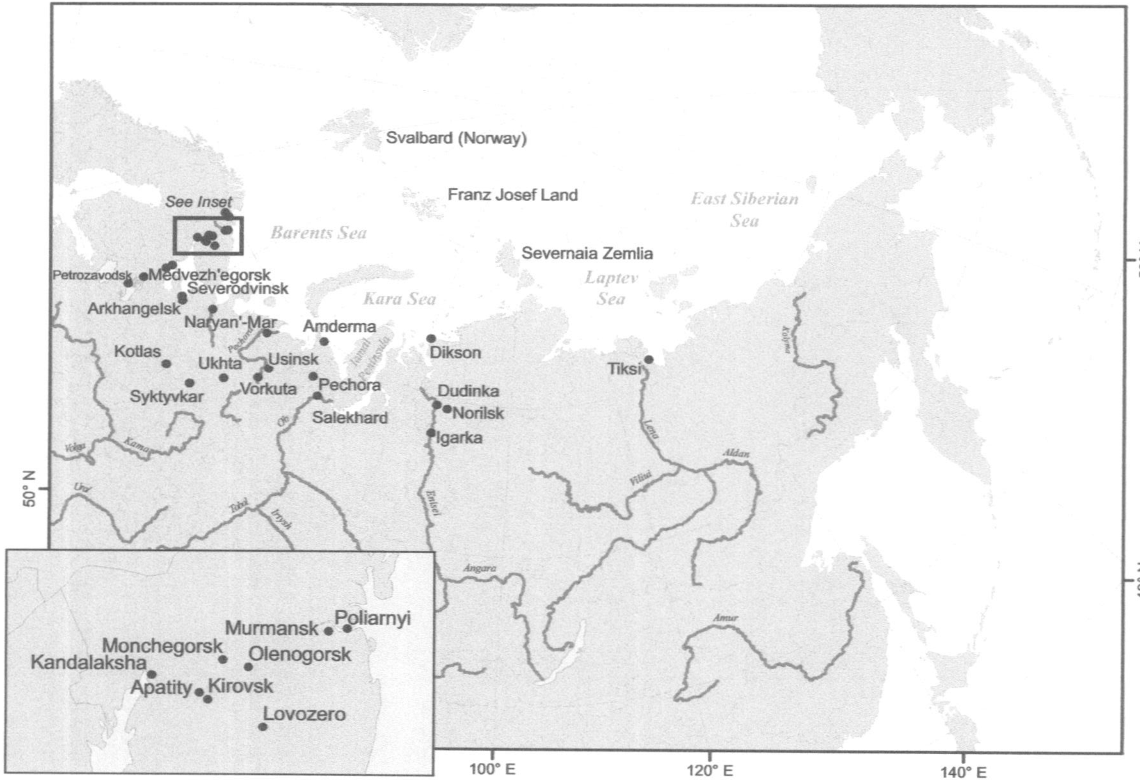

Russia’s western Arctic Circle. (Map by Manny Gimond)

Introduction

The emergency evacuation of the Russian polar drift station NP-40 ("North Pole-40") in May 2013 was a vivid reminder of the century-long presence of Russian explorers and scientists in the Arctic, the risks and uncertainties they faced in charting the landscape, and its growing fragility because of global warming and other human impacts. Stretching roughly halfway around the world, the Russian Arctic covers nine time zones from Norway to the Bering Straits. Roughly one-fifth of the Russian landmass is north of the Arctic Circle. Of fourteen million square kilometers that comprise the entire Arctic region (along with landmass of Canada, the Scandinavian countries, and the United States), Russia's share is roughly 3.5 million square kilometers, one-quarter of the total; Canada has the largest Arctic landmass. This book explores the Russian and Soviet conquest of the Arctic.

Under Soviet power, the Russians undertook a massive effort to industrialize the Arctic, collectivize reindeer herding, modernize small fisheries, establish large-scale forestry operations, overlay taiga and tundra with centralizing technologies of production, especially mining and metallurgical, communication, and transport, and transform nature—and the local and indigenous people within nature—so that the vast region operated according to central plans. The plans that emanated from the Moscow and Leningrad Communist Party apparatus reflected the goals of modernizing those people, controlling and reshaping nature, and extracting natural and mineral resources from the Arctic and subarctic regions of the USSR. The goals indicated Bolshevik certainty that Arctic conquest must be pursued at all costs toward the ends of state political, economic, and military power. The plans reflected Marxist ideology with its emphases on rapid modernization and industrialization. While this ideology shifted its emphases and goals over the nearly seventy-five years of Soviet power, the ends of securing and developing rich resources never abated. One of the major reasons for this was the firm belief that modern science and technology were panaceas for problems that party officials, specialists, workers, and other Soviet citizens faced in inhabiting and taming the desolate north. The path of Soviet Arctic conquest, no matter how heroic—and many heroes contributed to conquest—was also exceedingly costly from the environmental and social points of view. Many of those costs—pollution, haphazard disposal of wastes, high accident rates, coercive policies, destruction of traditional cultures, and so on—will be felt for decades to come.

Soviet explorers and scientists opened northern latitudes to economic development, discovered new lands and geographic features, charted ice drift, founded new understandings in oceanography, geography, geology, and other fields, and created a Northern Sea Route on the foundation of this science—and on such technologies as powerful icebreakers including nuclear ones, well-equipped research fleets, meteorological stations, airplanes, and ice runways. Managers and party officials succeeded in raising vast new industrial enterprises from the tundra. They organized construction brigades to build hydroelectric power stations—some of the first above the Arctic Circle—to power industrialization. They established entire new cities in what they saw as empty space—Monchegorsk for nickel and copper, Severodvinsk for shipbuilding, Medvezhegorsk as a transport nexus and for lumber and aluminum, and many others.

Presidents Dmitrii Medvedev and Vladimir Putin of the Russian Federation have redoubled efforts in the twenty-first century to secure the Arctic region, extract valuable resources from it as a central piece of the nation's economic development policy, and build on Arctic assimilation to remain a superpower by mobilizing state resources—and authority—to do so. As it had for Joseph Stalin, the 5,000-kilometer Northern Sea Route from Murmansk to Vladivostok along the Arctic Circle has assumed mythic scale for today's leaders. In June 2010, Medvedev called for the modernization of both military and civilian shipbuilding to enable Russia to engage in the "recently toughening competition for Arctic resources."[1] On May 12, 2012, Putin issued an executive order about the need to modernize Russia's military-industrial complex. He referred without irony to the Stalinist legacy of building military industry in the 1930s with his instructions for "developing the Navy, first and foremost in the Arctic areas and in Russia's Far East with the aim of protecting the Russian Federation's strategic interests."[2] The Kremlin has issued a series of other proclamations regarding the Northern Sea Route and Arctic oil, gas, and other resources, with occasional reference to environmental and social issues connected with the Arctic. But, as with Soviet power, so in the twenty-first century the Russian government will provide at best only modest protection for fragile Arctic ecosystems.

My focus is three provinces in the Russian northwest (913,000 km^2): the heavily forested Arkhangelsk Province (including the Nenets Autonomous Region that occupies a vast tundra) at 587,000 km^2; the Republic of Karelia, at 181,000 km^2, also rich in forests, lakes, and rivers; and Murmansk Province (essentially the Kola Peninsula, also with significant tundra and some birch forests) at 145,000 km^2. Including Arctic and contiguous subarctic regions, I focus on the Russian northwest because it was the major and longstanding focus of Soviet efforts to settle the north, develop its resources, and "modernize" indigenous people. To this day, the northwest is the most heavily settled, urban, and industrial of all Russian Arctic regions.

While Norwegians, Danes, Germans, Americans, Englishmen, Canadians, and Russians had explored the Arctic for almost two centuries, their efforts remained small scale until the late nineteenth century, as a result of which the Russian Arctic remained superficially investigated. An expansive frozen ocean (although global warming has shrunk the polar ice considerably) deterred exploration as did harsh climate and the weak technological foundations for conquest in shipbuilding and industry. The British and Norwegians had commercial advantages in Arctic regions because of more advanced and larger fleets, including motorized boats well before the Russians.

Technological advances in icebreaking, aeronautics, medicine, and construction accelerated discovery, claim, and development of Arctic regions in the twentieth century. They were fed by the symbolic importance of colonial and imperial control and of human conquest. Visionaries, and not only Soviet visionaries, imagined how modern technology including electricity and cities and factories would arise on the tundra. And not only nations contiguous with polar latitudes but others as well competed for land and resources. Siberian, Eskimo, Inuit, and First Nation peoples were caught in the rush for discovery; ethnographers, folklorists, archaeologists, veterinarians, and others studied them and their animals; planners and officials then pushed them aside.[3]

Under Soviet power the Russian Arctic fell to exploration and settlement. In the 1920s specialists connected with the Academy of Sciences, the Russian Geographical Society, and other groups organized a series of polar expeditions to chart geography, conduct geological surveys, set up experimental stations and observation posts, and begin the process of Arctic systematic assimilation. During the civil war (1918–20) efforts to reestablish trade between Siberia and European Russia also stimulated efforts. The expeditions were important steps in bringing together a community of specialists, getting a sense of the lay of the land, and understanding the limits of technology and finance available to conduct field research. They enabled specialists to realize the richness of resources and the diversity of peoples who lived in Karelia, on the Kola Peninsula, in Arkhangelsk—and far beyond the Ural Mountains in northern Siberia and the Far East. Yet until the rise of Joseph Stalin to power, Soviet Arctic exploration was episodic—and disasters and accidents frequently waylaid the efforts. "Wintering" (spending the winter floating on or trapped unexpectedly in the ice, *zimovka*) challenged men, animals, and supply lines.

Soviet Arctic assimilation involved the sacrifice and uncertainty of exploration in extreme conditions and under great political and economic pressure; the application of modern science and technology to the tasks of study and exploitation of resources, often with utopian certainty that science would enable conquest of those resources; and the surprising urbanization of Arctic regions that accompanied efforts to bring industrial processes to the extraction of ore, timber, gas, and coal from the tundra and taiga. Settlement of polar regions was coercive in the Stalin period; it consumed hundreds of thousands of slave laborers and exiles along with dedicated volunteers. From the Khrushchev era until the fall of the USSR, the state relied on heavy subsidization to encourage settlement. But whether coercive or not, Russian leaders, explorers, party members, and collective farm and factory managers brought an ideology of modernization to the Arctic based on Marxian ideas about the creation of urban, industrial outposts of civilization among local people, including indigenous reindeer herders. They assumed that modernity would always be good for the Arctic environment and the people in it. In all of these ways, the history of the Soviet Arctic reveals a surprising number of continuities with ongoing twenty-first-century Russian programs.

Of course, a sharp break in Arctic research and development accompanied the fall of the Tsarist regime and the Bolshevik seizure of power in 1917. If state power was important to research before the revolution, then its extent grew manyfold under the Soviets. More important, the Soviets attempted to impose a Marxian, urban, and industrial form of the economy on the tundra and taiga—as they did throughout the empire. Marx and Engels provided little direct guidance of what they expected in the future socialist society. From the *Grundrisse, Critique of the Gotha Program, The German Ideology,* and *Capital,* however, it is clear that they anticipated a political and social revolution occurring in an advanced industrial urban society in which the working class would seize control of capital—machines, tools, the modern factory—and create a society of plenty, production, and leisure. Because economic development was the engine of change from feudal to capitalist and to socialist society, some observers have called their doctrine "technologically determinist."[4] Lenin, Stalin, Trotsky, and others amended, emended, and extended these ideas, not the least because the Russian empire lagged significantly behind other powers, in particular England, Germany, and the United States, in level of technology. Lenin, for example, went to great lengths in *The Development of Capitalism in Russia* (1895) to argue that Russia had entered an advanced stage of capitalism and hence was ripe for revolutionary seizure of power. Yet he and the others recognized Russia's lag in industrial development and imagined the United States as the "highest form of culture" for its massive factories, Fordist mass production, and its tractors and steel mills. They wrote about electrification, tractors, hydropower stations, and other symbols of advanced society as the key to the Soviet future; Lenin saw electricity revolutionizing agricultural production. These leaders would weld Bolshevik determination and organization to modern technology and create a modern industrial utopia. They would overcome the supposed "break" between agricultural and urban production endemic to capitalism. The working class would grow, cities would expand, and supply and production bottlenecks would disappear from Central Asia to the Far East, from Siberia to the Arctic. Sleepy, backward villages in the taiga would give way to gloriously operating production facilities; new cities would rise in the tundra.

Stalin and his followers determined to apply these ideas through collectivization of agriculture, rapid industrialization, and revolutions of production throughout the economy. Under Stalin, Soviet leaders vigorously pursued modernization of the entire economy based upon large-scale technological systems of energy production, transportation, shipbuilding, mining, and smelting. They imagined each ecosystem, taiga and tundra, steppe and desert, and lakes and rivers, as an object to be transformed through rapid industrialization initiated from above and pursued without hesitation or uncertainty. Granted, as the transformation of Arctic and subarctic regions indicates, ambitious plans were fraught with inconsistency as political desiderata often overwhelmed scientific concerns, scarcities, and true costs. The Soviets believed, however, that planning was a panacea for any problems they encountered. In this book I shall refer to this ensemble of approaches, ideas, and concerns variously as the Marxist industrial imperative or urban industrial imperative, and in its most violent form the "Stalinist" industrial imperative, and will examine its uneven and at times costly impact on the northern latitudes and the people who inhabited, settled, or were exiled to the region.

Arctic endeavors resembled those elsewhere in the USSR. Leaders set out to remake the nation of new industrial forms and new Soviet men and women. They sought to plan the entire endeavor and master nature and the people in it. The goal was a new civilization; achieving that civilization through industrialization, collectivization, a housing program, and other transformationist projects created extensive difficulties, even if many people embraced the idea of creating a new world. Many of the most loyal citizens believed they had created a new society even as the inefficiencies and perversities of socialist planning led to shortages of goods and services including housing and food; on top of this, hypercentralization of decision making in Moscow-based bureaucracies led to challenges in meeting planned targets in the provinces.[5] The effort to conquer the Arctic reflected this tension between plans and realities. It was at once successful in imposing a Soviet world on taiga and tundra and unsuccessful because of the utopian expectations tied to plans that emanated from Moscow. Climate, distance, low population densities, difficulties in securing adequate resources of capital and labor—all of these things made Arctic endeavors exceptional.

Whether in the Soviet period or in twenty-first-century Russia, this is also a story of tremendous achievements of polar explorers, specialists, and industrial managers, of workers who built and operated the smelters on the Kola Peninsula, the lumber mills of Karelia, and the shipbuilding yards on the White Sea. As leading participants in the second and third International Polar Years (1932–33 and 2007–8, respectively) and the International Geophysical Year (1957–58), Soviet Russian scientists, drawing on national polar science, reshaped a number of fields empirically and theoretically. The Soviets studied the Arctic and then settled it in the form of planned towns and cities that were dedicated to some form of industrial production.

Science and engineering expertise were crucial tools for Arctic conquest. Nascent fields of polar science—geography, geology, metallurgy, and so on—complemented those of engineering—mining, reclamation, and construction. Eventually specialists turned branch institutes of Moscow and Leningrad research centers that had been established in polar cities into independent institutions in the Arctic. In the 1930s, through the efforts of Leningrad Party leader Sergei Kirov and geologist Alexander Fersman, an entire research base arose in the Khibiny tundra in the center of the Kola Peninsula—today's Kola Science Center that in 2010 had eleven institutes with more than 600 employees of whom roughly 350 were scientists. Arctic medical and veterinary science developed more slowly, in fits and starts, to ensure the health of people and establish modern animal husbandry, and the social sciences even more slowly, for the first order of business was industrial and military conquest for state purposes.

Soviet leaders, civil and mechanical engineers, planners, and others sought to transform the entire country, from top to bottom, from industry to agriculture, from north to south, from the Arctic Ocean to Central Asia and the Chinese border, and from Europe to the Pacific Ocean. Thus, the tasks of Arctic conquest were similar to and yet competed with myriad other such tasks of construction, transformation, building, and rebuilding throughout the empire. Local party officials stressed the importance of their regions to the nation's future, of their industry and agriculture to the glorious working class. Given the tremendous challenges of bringing the country's citizens and economy to heel, Arctic projects languished somewhat until the late 1920s. The Soviet economy reached 1913 levels of production only in

1926. Yet, even in this environment of tight resources, scientists and engineers were grateful for the significantly higher and constant level of funding they commanded than in the Tsarist era. If they rejected the radical Bolshevik political agenda of workers' control of the means of production and class war, then they welcomed the government's support for their research agendas.

For their part, party officials were happy to work with scientists toward the ends of Arctic assimilation. They wished to secure Arctic regions militarily. During World War I and the civil war, American, British, and other troops occupied the Russian northwest in the hopes of defeating the Bolsheviks and keeping Russia in the war against Germany. Party leaders feared the traditionally free-thinking residents of Arctic and subarctic regions; after all, they were far from Moscow. Roughly until 1920, a relatively independent government survived in Arkhangelsk to the consternation of the Bolsheviks. The Bolsheviks determined to subjugate these peoples to the center, including ultimately the indigenous reindeer herders and fisherman who inhabited the north, in all some twenty different ethnic groups. Finally, they recognized the economic importance of the Arctic—although how important it would become because of its great chemical, mineral, metallurgical, fossil fuel, and other resources they could not have imagined; in the twenty-first century the Russian Arctic provides much of the world's platinum, nickel, and copper and tons and barrels of fossil fuels.

Vladimir Lenin signed hundreds of proclamations after the Bolsheviks seized power in 1917, many of them dealing with issues of political authority, nationalization of property, and so on. Others concerned education, culture, and the arts. A large number reflected Bolshevik belief that science and technology were keys to constructing the glorious communist future. Within a few years scientists had secured support from the government to establish a series of new research institutes representing all fields of science. The institutes ranged from basic research to applied in their focus including development of natural resources. Yet surveys of the location and extent of natural resources remained rudimentary, especially in the Far North, the Far East, and Siberia. A number of Lenin's proclamations referred specifically to scientific expeditions that set in motion Arctic conquest. Lenin also pushed the New Economic Policy from 1921 to permit small-scale capitalism and encourage economic recovery after the revolution and civil war.

When Stalin rose to power, many of the final obstacles to exploration of the Far North, subjugation of its resources, and transformation of the indigenous people into figures of socialist modernity quickly disappeared. Granted, shortages of manpower, equipment, and financing continued to plague the attempt to impose modern technology on the landscape above the northern latitudes. But through the creation of a powerful new bureaucracy, the Main Administration of the Northern Sea Route (Glavsevmorput), a coordinated effort to assimilate the northern latitudes began. On top of this, specialists in the Institute of Arctic Research (today the Arctic and Antarctic Research Institute, hereafter AARI) expanded expeditionary and research efforts manyfold, launching icebreakers and other ships into the Arctic Basin each navigation season from late July to October, sending researchers to drift on the ice pack for months at a time while taking extensive oceanographic, meteorological, and atmospheric readings, and using aerial reconnaissance and freight to support these programs. They wintered on the ice, occasionally because they were trapped suddenly, not because zimovka was planned. *Ex post facto* the authorities announced many of these unanticipated winterings and trapped ships had in fact been planned events. Research involved increasingly visible public expeditions whose leaders became heroes overnight; more mundane efforts to chart ocean currents and depths and river flow, to catalog chemical and ore deposits, oil, and gas, and to map out terrain; and efforts to settle an inhospitable landscape of fierce winter cold and snow and summer swarms of biting insects. "Settling" in the Stalin period involved both large numbers of volunteers and labor camp ("gulag") prisoners. The volunteers were simple folk seeking jobs or adventure, individuals trying to escape political intrigues in central cities, and devoted party members, including people from the Komsomol (Communist Youth League), while the prisoners were innocent victims of political arrest who were accompanied by large numbers of peasant exiles from agricultural regions.

If the 1920s were a decade of Arctic exploration that floated on the economic recovery of the New Economic Policy and the 1930s were a decade that brought together in northern latitudes a technological revolution and state power, then, after World War II and the difficult rebuilding of the USSR that followed, the 1950s were a decade of redoubled efforts to secure the Arctic through accelerated settlement and urbanization from no more than

500,000 inhabitants in 1930 (with Arkhangelsk alone at 240,000 people) to more than 1.4 million inhabitants in 1959 and more than 2.2 million inhabitants in 1979. Soviet Marxism was an urban, industrial ideology, so logically programs that emanated from Moscow centered on the creation of industrial centers in Arctic and subarctic regions—pulp and paper mills, aluminum, nickel and copper smelters, mines, power stations, shipbuilding facilities, and so on—and new villages, company towns, and cities to support them. Arctic assimilation became much more regular, modern, systematic, and based on science and engineering in the Khrushchev era (1953–64) and beyond. In the Brezhnev period (1964–84), Arctic research, development, and exploitation became unabashedly big science and technology, based on modern machinery and equipment including nuclear-powered icebreakers. Where explorers had had to hunker down in the cold, snow, and winds in tents and wooden huts, now pre-fabricated, insulated "boxes" that served as housing and laboratories were dropped onto the ice. The episodic drift stations of the 1930s became a constant presence, and institutes connected with various resources or branches of central institutes filled the northern latitudes. From Murmansk to the Tiumen region of Siberia, the authorities determined to extract fossil fuels and mineral ore in large-scale operations in part as an important component of Soviet foreign trade.

In the Gorbachev era (1985–91), but especially under Russian President Boris Yeltsin (1991–2000), Arctic assimilation fell on extremely hard times. Budgets dried up, while inflation consumed what was left. Political uncertainty halted research programs. Government subsidies to housing and salaries disappeared, and the last twentieth-century drift stations were pulled from the ice in 1993.

All this has changed in the twenty-first century under Presidents Putin (2000–8, 2012–?) and Medvedev (2008–12). Russian leaders have determined to reestablish a strong military presence in the Arctic and to harness resources, especially oil and gas. They see Arctic programs as crucial to Russia's identity as a superpower, as space and nuclear energy were in the Soviet period (and still are). As part of central state programs, President Putin has called for a rejuvenation of the Northern Sea Route in bureaucracies, programs, terms, and rhetoric that resemble the glory days of the Soviet Glavsevmorput. In the twenty-first century state power, ideology, science and engineering, and break-neck development have come together in ways reminiscent of the 1930s—even if socialist ownership has given way to state capitalism.

Four overlapping tensions played out in the expensive and frustrating drive to conquer the vast Arctic. The first concerned the determination of officials and specialists in Leningrad and Moscow to employ tried-and-true practices and approaches, technologies, and standards, which they exported from the central industrial regions to the Arctic with little modification and applied to the tasks of exploitation of resources. While it seemed to many engineers and planners that a smelter would smelt in the Khibiny tundra in the same way it operated in the "hero" steel city of Magnitogorsk in the southern Urals, in fact managers and specialists on the ground encountered never-ending climatic, geological, and other pressures that hindered production. They also faced unrelenting pressures from central officials to make their science conform to plans. That is, if plans from the center dictated certain targets, then managers of industrial or forestry enterprises were at professional and personal risk to argue that local conditions (river flow, available raw materials, numbers of workers, quality and extent of machinery and equipment) prevented reaching a target.

A second and similar tension existed between utopian visions of a fully "Soviet" Arctic and the realities of polar weather, inadequate investment, and poorly paid and housed workers whose motivation flagged on a daily basis and many of whom turned to alcohol for solace. As in other settings in other countries, the third tension between an industrial ethos and the natural world created constant problems of resource supply and raw material extraction. Given the pressures they faced, it is not surprising that specialists and officials underestimated the fragility of Arctic climes and accelerated environmental degradation wherever they pursued conquest. Finally, the ideology of rapid modernization based on central notions of industrialization of nature—and the people in nature—required the imposition of an unforgiving universal language of science and power that overwhelmed—if not simply rejected—the value of local knowledge, culture, and beliefs.

Yet nature's agency could not be ignored in the taiga and tundra. The challenges of penetrating winds, deep freezes, and permafrost told upon every activity. In their single-minded pursuit of resource extraction officials in warm Moscow offices often failed to heed the distinctiveness of Arctic landscapes, while local party officials and scientific specialists were all too aware of the daily challenges of organizing labor brigades, requisitioning supplies and materiel, keeping machinery operating, and ensuring that workers were constantly focused on the job when they and their families faced poor living conditions. Managers and scientists frequently communicated with Moscow how Arctic realities demanded increased investment of labor and capital, special equipment, more research, and careful consideration of that research. They pointed out that the maximum spring flow of rivers established a firm limit to lumber deliveries to the mills downstream; rivers did not understand the incessant ratcheting up of production targets. They noted how reclamation projects could not possibly continue without delivery of graders, bulldozers, and suction dredges. They argued that without increasing resources for housing, for food, and for hospitals and schools their workers would in no way remain hard-working and energetic. The Arctic cold and summer insects drained them of good health, strength, and enthusiasm, exhortation from party officials in various ideological campaigns notwithstanding. They would have preferred better homes, food, and access to better medical care.

In all these ways, the Arctic experience underscores how the Soviet system, with its centralized planning to ensure maximum efficient use of inputs to produce the maximum output, was in no way supremely rational. In spite of the claims of leaders, economists, and others that Soviet planning avoided the endemic duplication and waste of capitalism, in fact it was irrational in ways that capitalist economists understood. The twentieth century was a century of increased gathering and analysis of economic data and of the advancement of schemes to maximize production and employment in all political systems—authoritarian and democratic—and in all economic systems—capitalist, socialist, and mixed.[6] Everywhere state economic programs failed to achieve maximization and everywhere had unanticipated social and environmental impacts. The self-proclaimed supreme rationality of Soviet planning; the inherently political nature of planners' preferences in the effort to use limited resources to support investment in one sector—or geographic region—at the expense of another; and the excessive human and environmental costs of development, especially in the Arctic, distinguished the Soviet economic system.

Of course, the creation of heavy industry, the advancement of the notion of socialism in one country, the defense of the nation from hostile capitalist encirclement, and the securing of territory—all of these things, whether slogans or real programs, both contributed to and required the creation of a powerful military-industrial state. Toward that end, the authorities hoped to acquire the world's most advanced technology one way or another. Early on, they invited western specialists to assist in installing the machinery at the same time as the country engaged a massive literacy campaign that also involved training engineers. The USSR imported some 300,000 machine tools from 1929 to 1940,[7] and roughly one-quarter of the equipment brought into operation was imported.

British, Swedish, German, and especially American technology and personnel played an important role in Soviet economic history in the 1920s and 1930s. Some of the technology came from outright purchases (tractors, locomotives) that the Soviets often reverse engineered—to learn from disassembly and reassembly how they might manufacture such a thing. Germans vitally supported several branches of industry including such equipment at the Elektrosila factory in Leningrad as turbogenerators and motors (as did the Americans through General Electric Company). The Swedes provided locomotives and equipment for a nascent aluminum smelting industry. England and Finland provided icebreakers to the Tsar that became the first Bolshevik icebreakers. The entire Magnitogorsk steel combine was modeled on the Gary, Indiana, US Steel Bessemer mills, considered among the most modern in the world. It was designed with the help of American engineers and run with their help and using western equipment.[8] Occasionally the Soviets bought entire turnkey plants that gave impetus to the development of indigenous industry. Bolshevik leaders revered American technology, with Trotsky calling it the "highest form of culture." This fascination with the world's most advanced technology, *Amerikanizm* as it came to be called, persisted until Stalin's rise to power. Industrial espionage was also a factor, although we must not exaggerate its importance; rapid industrialization was an achievement based on Soviet know-how, Bolshevik willpower and determination, and a ruthless investment policy to extract capital from the countryside to transform the nation into an industrial superpower in ten short years. All of these things shaped the Arctic experience as well.

Yet by the late 1930s Stalin had settled on a strict policy of economic autarky to ensure self-reliance in the face of "hostile capitalist encirclement," and foreign trade dropped to 1 percent of gross domestic product by 1937.

For the next fifty years, Soviet policy moved between positions of turning westward for critical technology and asserting little need for western technology out of pride and the need for self-reliance. The Soviet shipbuilding, power generation, railway, aviation, and other industries produced the icebreakers and airplanes, transformers and turbines, rail and locomotives required for Arctic conquest.

Regarding its "technological culture," the Arctic experience differed from the broader Soviet experience in several ways. With the exception of Finnish pulp and paper operations in Karelia and original equipment for smelters near the Norwegian border at the town of Nickel, Arctic technology was indigenous technology.[9] Technology transfer existed, but it was transfer from the center to the periphery, from central ministries and design institutes to industries being built in the Arctic involving well-known processes and techniques—milling, ore refining, production of copper, mining of coal. Transfer, when it occurred, involved precisely those central ministries and their research organizations (e.g., Gipronikel for nickel and copper, Gidroproekt for hydroelectricity) that in turn disseminated know-how and technology in a glacially slow and bureaucratic fashion out of the unrealistic fear that industrial secrets might escape the borders. One exception to the role of western technological in the Arctic concerns lend-lease programs of World War II that brought vital armaments and transport equipment from the United States to the USSR. But this transfer involved primarily materiel—armaments, trucks, airplanes, ordinance, boats—and not entire processes, and we may say that it did not change the innovation process in any Arctic industry or sector of the economy.[10]

The desiderata of economic development, technological modernization, and military security had direct impact on people and environment. By the 1930s, under Stalin, a more coercive attitude had replaced a more altruistic view of human potentialities, including those of indigenous people. For example, officials came to see indigenous peoples as not only backward but as inherently hostile to Soviet power and thus to be modernized—as their reindeer were to be collectivized—as quickly as possible. "Cultural bases" of various sorts sprang up in the tundra as tools of modernization of the indigenes, newly created written languages became tools of control, and political consciousness replaced backwardness. Gulag prisoners and peasant exiles were also forced into the tundra, later to be freed, or pardoned, or released, or posthumously rehabilitated. Arctic conquest involved therefore transformation of humans as well as nature. This found response in culture, arts, and literature, as a number of scholars have explored.[11]

The tensions between universal and local knowledge and between modernizers and local people found full reflection in Arctic culture, ideology, and values. In what ways was Arctic conquest significant for the Soviet people from a social and ideological point of view? No less than programs for Arctic conquest in Canada, Denmark and Greenland, Norway, and the United States, those in Soviet Russia served broader cultural purposes of indicating the superiority of the socialist political and economic systems. Explorers were "the first" to achieve some great feat, but they could not have been the first without the support of the people and, of course, of the Communist Party. Their achievements represented the glory of their nation. Soviet ideologues, writers, and journalists no less turned each step of the northward plunge into the tundra into an achievement of great national importance. Newspapers, journals, commemorative postage stamps, lacquer boxes and flags, books, and plays all reveled in these achievements and convinced the citizen of the grandeur of polar conquest.

No less than overconfident Soviet writers, those in the capitalist world also wrote how science, ingenuity, and technology would come together "to enrich man's empire of the world." According to one western author, Arctic conquest was a "brilliant victory over forces [of nature] really worth fighting."[12] If in the 17th and early 18th centuries trade routes were abandoned as too difficult to pursue and unfamiliar, then since Nils Adolf Erik Nordenskjöld who managed to travel the Northern Sea Route in 1878–79, and especially since the first International Polar Year in 1881, scientists learned to solve crucial problems. Nansen and Peary showed the importance of organization, "heroic strength of character," and technological innovation.[13] The modern expedition took "capital into its service in the form of highly perfected technical apparatus. We have carried the machine with us to the Arctic, and through it we take the fullest advantage of our experts."[14] Airplanes and modern icebreakers, especially from Russia, played a crucial role here. Beginning in 1936 the addition of diesel electric engines (and easier transport and storage of fuel) made exploration and reconnaissance regular activities. By the late 1930s "the specialist is a self-contained unit in the modern expedition. We carry him up to the Arctic with all his equipment—his whole laboratory—and give him the same working conditions as at home." Writing in 1938 this author excitedly observed that "it is in Russia that organization has been brought to the highest perfection."[15]

A number of Soviet writers focused on Arctic assimilation in their literature and poetry. Many of them acknowledged the human and other challenges of assimilation, but more often they glorified the achievements and infallibility of leaders who made them possible. The orchestration of the glory of conquest crucially enabled spokesmen for the Soviet state to view the Arctic regions as empty—devoid of culture, devoid of the artifacts of modernity, even devoid of people. In the 1920s as the specialists who went northward into the tundra encountered such indigenous people as the reindeer herding Komi, Nenets, and Saami, they commented on their "backwardness," but they sought to help them preserve their cultures and ethnicity, including by creating written languages for them. They wished to bring the modern sciences of medicine and veterinary medicine to the Arctic to improve the quality of life for indigenous people and other local residents so they would no longer be at the complete mercy of the elements. By the mid-1930s officials had abandoned this approach, for the indigenous people and their reindeer and fish and furs must be tied to the needs of the center.

Through literature, newspapers, journals, film, and other cultural activities, party officials, scientists, engineers, artists, and writers created a common language of Arctic assimilation, although one that avoided mention of the prisoners. Newspapers and other publications appeared from the first days of Arctic exploration and exploitation that reflected heroic deeds. They included *Pravda Severa* (Arkhangelsk, 1917–present), *Poliarnaia Pravda* (Murmansk, 1920–present), *V Boi za Nikel'* (1938–8?), *Monchegorskii Rabochii* (1935–present), and dozens of others that local party committees, industrial enterprises, icebreakers, fish processing factories, and other organizations published. This was joyous celebration at a time of great achievements in society, economy, and culture.

For example, the writer Evgenii Kokovin (1913–1977), a native of Arkhangelsk province and a graduate of a merchant marine school who served as a mechanic onboard ships, celebrated the trials and successes of Arctic conquest and encounters between settlers and indigenous people. Kokovin published his first story, "Deserter," in 1931. In 1940 "Childhood in Solombala" appeared, which dealt with life in the Solombala region of

Arkhangelsk city, home to shipbuilding yards, docks, and a pulp and paper mill. The story was republished in 1947 and gained Kokovin an audience throughout the socialist world. His other stories illustrated life in the north during a period of rapid transformation of industry and social structure and often focused on Nenets reindeer herders. The premise of Kokovin's *Poliarnaia Gvozdnika* (Polar Carnation) is a series of Nenets legends. But these are new legends, for example about the "epic" hero Lenin, and the legends are told like an epic poem, for example *The Tale of Igor's Campaign* (likely dating to the twelfth century). Kokovin's effort to create legends reveals the danger of folklorization, of depicting a culture in a stereotypical ways or seeing local people from the point of view of colonizers who are superior while the local people are primitive. Thus the artificial preservation of traditions in the USSR was seen by many people as something quaint, but never authentic. Kokovin's legends displayed a new power versus the backwardness of indigenes. In the legend "Son of a Trapper," the author described the natural surroundings of New Island (part of Novaia Zemlia). Under Tsarist power foreigners had hunted and carried out surveys for mineral resources without interference. But now island dwellers lived in houses, and on the shoreline of Bear Harbor were two villages, a Russian trading post, and a Nenets collective farm, with a club, boarding school, generator, store, and storehouses of food and goods.[16] In "Bear Cove," the Russian Natasha asks a Nenets youth to show her a teepee. "What teepee? Where would I take you? There haven't been teepees for a long time. All of us, Nentsy, live in houses. Our little village is referred to as a Sedentary Base." She asks, "And what is that noise? "It's an electrical power generator."[17] In a word, the new legends revealed that Soviet power was a positive force of modernization of worldview and a pathway to wondrous labor-saving, efficient, powerful technology.

Population densities across Arctic and subarctic regions increased during Soviet history as leaders sought to secure the lands and exploit vast natural and mineral resources. They encouraged settlement through economic development and forced prisoners and exiles to move north and east, and not only during the Stalin period. Taking a page out of the Tsarist practice of exiling intellectuals, the Brezhnev-era attorney general's office (Procuracy) never hesitated to strive to silence independent, critical minds. The Nobel Laureate Joseph Brodsky was arrested in February 1964, convicted of being a "social parasite," and exiled for five years at hard labor to Norenskaia in Arkhangelsk province. He was sent first by train to Konosha, a major railroad junction for Arkhangelsk, Karelia, Vologda, Leningrad, Vorkuta, and Syktvkar whose lines were frequently built by gulag prisoners—and were used to send prisoners further into the bleak tundra.[18] From Konosh Brodsky made his way to Norenskaia. Brodsky's exile included a stay at a mental institution where his "treatment" consisted of being wrapped in cold, wet sheets. Brodsky's sentence was commuted in 1965 after protests from Jean-Paul Sartre, Anna Akhmatova, and Dmitry Shostakovich.

Born in Leningrad in 1940, Brodsky dropped out of school at fifteen, having failed to get into a naval academy; he wanted to work on submarines. He took a job as a lathe operator at the Arsenal Factory and moved about at other odd jobs that kept him fed and housed as he became more and more involved in writing about politics, religion, and morality. Brodsky claimed he began to despise Lenin already in the first grade. During the Khrushchev-era "Thaw," Brodsky began to write poetry and translated a number of poets into Russian (including Czeslaw Milosz, W. H. Auden, and the Beatles' "Yellow Submarine"). In 1972 the Brezhnev regime exiled Brodsky. He moved to the United States and became a U.S. citizen, a professor at Mt. Holyoke College, and U.S. Poet Laureate in 1991–92.[19]

In the rural cold of Arkhangelsk, Brodsky wrote a poem, "Tractors at Dawn," unlike most of his other works on the simple life in the countryside.[20] He later claimed his happiest moments were during exile when he was not bothered by the complexities of Soviet life.

Tractors wake up with roosters
Roosters wake up with tractors,
With the engines and plowshares,
Chopping the silence as with axes,
And in the morning fog knee-high
Humming, they line up along the front
Silence falls apart like a log,
On both sides of the horizon.
The stoves start to burn. The smoke curls up straight.
Birds bend over chicks.
The wood, like a giant sawmill,

Cuts the clouds with teeth.
The sun rises, and stares blindly,
And mows the sleepy shacks with its rays.
The tractors ascend like birds to the sky
And lift the fields to the sun with their ploughs.
This is a working morning. Morning of the People!
A morning of labor with an ancient smile.
Nature looks at people as into a great river,
And gets up, glassing itself, from sleep with the village.

Even Brodsky found something positive in the gradual industrialization of Arctic regions and the transformation of life under the impact of machines.

Conquest of the Soviet Arctic resembled that for other circumpolar nations. Costly and dangerous industrial operations enabled Canada, Denmark, the United States, and Norway to develop oil, gas, mineral, and other resources. The United States and Canada joined the Soviet Union in destroying fragile ecosystems through the establishment of military bases and radar stations and through haphazard disposal of drums of oil, PCBs, and other chemicals. Each of them desecrated Inuit lands and destroyed Inuit customs. Over time, each nation succeeded in developing technologies appropriate to Arctic conquest—from icebreakers to diesel vehicles, from construction techniques to mining operations. In each case, business, industry, and government provided subsidies and other incentives to encourage workers northward.

Yet the Soviet experience differed in a number of ways. The first was the role of the Stalinist system in shaping expectations about and approaches toward Arctic conquest. Stalinism was both a polity and a crash economic development program. As rational as planners contended their plans were, significant disjunctions existed between goals established in the center and realities of daily life in factories, mines, collective and state farms, fishing enterprises, shipbuilding yards, and so on throughout the entire Glavsevmorput enterprise, not the least of which was the presence of hundreds of thousands of slave laborers. The second was the relative poverty of Soviet operations. In spite of healthy subsidies for housing, salaries, and the like, daily life was hard. Food, housing, transport, stores, medical care and other social overhead capital lagged significantly behind the needs of northern residents—and behind other regions and especially Moscow, which took more than their share of resources. Because a central planning system was required to provide for these things, it was also responsible for the low quality of life. Third, the technology of Arctic conquest was not fully up to the task. Until the 1930s explorers, researchers, and sailors of the merchant marine and of the navy had to use outdated and underpowered ships dating to before World War I and for which it was difficult to get parts. Regarding motorboats for inland waterway transport, locomotives and rails, tractors and bulldozers, telephone and telegraph poles, roads, and so on—the quantity and quality of equipment never met demand. If, in the 1960s and 1970s, nuclear-powered icebreakers and freighters begin to ply the Northern Sea Route, then elsewhere—at farms, in factories, and along the road and railways—even telephones, lightbulbs, paved roads, and generators were rarely encountered. The result was a labor-intensive, frustrating, expensive, and environmentally costly endeavor.

Party officials, Nenets herders, geologists, factory workers, fishermen, peasant exiles, gulag prisoners, captains, and sailors crawled out from under the covers, stoked stoves in the frigid early morning with temperatures at –30°C or –40°C, left their deerskin tents, started tractors, operated sawmills, pushed papers, built factories and cities, erected power, telephone, and telegraph lines, laid rail, bulldozed roads, felled trees, operated smelters, explored the Arctic Ocean, discovered islands, charted currents, and sought to create a great Northern Sea Route. Together they created an urban, industrial Arctic that in the twenty-first century faces redoubled efforts at strategic resource development.

1

Charting the Arctic Landscape

In winter, the persistent Arctic night; in the summer, the unending Arctic sun. Under those two extremes lay a landscape rich in natural and mineral resources, but with inadequate infrastructure to develop them. Beginning in the second half of the nineteenth century, Russian explorers turned their attention increasingly to the polar regions of the empire to chart them. In some cases they followed expeditions of Norwegian, Danish, and British explorers and businessmen who recognized the vast natural and mineral wealth of the northern regions, especially fish, but also lumber, graphite, and some diamonds and gold. The Russians followed not because of lack of interest or any hesitancy associated with fear, but because of the lack of support of their professional scientific societies, disbelief that such modern technologies as icebreakers might enable exploration, and insufficient government funding.

With the turn of the twentieth century and the grudging recognition among Tsarist bureaucrats of the economic and military significance of the region from Arkhangelsk on the White Sea shore to Novaia Zemlia between 70° and 78° north latitude that serves as the "border" between the Barents and Kara seas, explorers and scientists began to make their way into the treacherous Arctic Ocean with its short navigation season from late June to mid-autumn. They sought a Northern Sea Route around Norway, past the Kola Peninsula, through the Barents Sea, perhaps near the Novaia Zemlia island archipelago and through the Kara Gates or south to Vaigach Island and through the Iugorskii Strait to the Kara Sea.[1] Steamers skirted the north Siberian coast through the Laptev Sea, the East Siberian Sea, and the Bering Strait to the Pacific Ocean.

Gradually, captains and explorers learned about the Arctic basin navigation season, ice floes, cyclones, and currents that made passage hazardous. They learned about the rich geology of the region, of coal, iron ore, and later oil and gas. They identified extensive fisheries that Norwegian sailors had already managed to exploit with motorboats, while the local Pomor fisherman struggled to fish with sailboats. Often out of foolish haste they belatedly realized the dangers associated with Arctic exploration; they knew little of water chemistry, ice regime, and weather patterns, and they did not have the icebreakers and research vessels appropriate to explore the Arctic seas. Until discovery of extensive fossil fuel, mineral, and other resources in the twentieth century, fishing was the most important activity of local residents and international visitors.

But geopolitics further stimulated Arctic exploration. Ultimately, Tsar Nicholas II determined to compete with European, Chinese, and Japanese interests in the Far East. This accelerated the development of the Northern Sea Route, both for its own sake and to assist in and accompany the construction of the Trans-Siberian Railway. For reasons of national self-determination and national defense and growing recognition of the extent of natural resources, the episodic nature of exploration and discovery of the Russian Arctic ended in the 1920s. Augmented by the Bolshevik certainty that explorers could conquer all obstacles, an atmosphere dedicated to promoting new Soviet heroes, and determined leadership under Joseph Stalin, Arctic assimilation accelerated in the 1930s. There would be countless victims of the Arctic, of the weather, the ice, the backward technology, and also of the Stalinist prison labor camps and purges. But welcoming significant government support, polar explorers charted the great wealth of the region, ethnographers studied the indigenous people of the region, and political operatives set out to reorder the region in the name of modernity.

Early Russian Efforts in the Arctic

In one of the first major Russian expeditions in the mid-1880s, the poet and amateur naturalist Konstantin Sluchevskii traveled more than 11,000 kilometers, 8,000 by ship, through the Russian north, accompanying the Grand Duke and brother of Tsar Nicholas II, Vladimir Alexandrovich. One of his goals was to get a sense of daily life of local residents. In their lengthy travels they visited a whaling factory, opened in 1883 on the Ura Bay northwest of today's Murmansk on the Kola Peninsula. Founded in the 1860s, by this time Ura had 200 residents in the summer months, most of whom were male. The men sailed in fifteen undecked whaling boats and while ashore tended seventy-four head of cattle, sixty sheep, 309 reindeer, and one horse. Meanwhile the woman folk apparently remained in Kem on the White Sea coast. In addition to modest whale harvesting the men caught sharks for their oil and fat. Sluchevskii found little of interest in the factory, but learned about the Pomor people—Russian and Norwegian settlers whose life centered on the sea, which they feared for its unpredictability and which they had personified in their dialect. Hiking with some difficulty some 200 meters up the Murman fjord, Sluchevskii and the Grand Duke viewed the landscape—"endless tundra, swamp, low-growing forest, completely devoid of human habitation and roads, if you don't consider those shacks in numbers of one or two, dispersed at great distances."[2]

As those explorers who followed him, Sluchevskii saw great promise in development of northern resources, but also tremendous challenges. The British and Norwegians fished extensively in the region, not so the Russians, and the most successful local residents were Finns. Sluchevskii noted a large number of bays with potential to serve as a permanent sheltered harbor for the Russian merchant marine and navy. But the main point was how the region remained so undefended and the fact that the Ekaterina harbor (now Murmansk, a city of 250,000 people) and the entire Arctic northwest were so desolate, with the people at great risk from the elements and ignored by the tsars in their poverty and in poor health. The Tsarist government made small steps to end the isolation with limited and indirect support with the construction of a railroad across the Kola Peninsula during World War I, in part built with slave labor, to permit military supplies to reach Petrograd and Moscow from the north.[3] This has been a long pattern of Russian history for the government actively to promote technological change only in a moment of crisis—usually some military defeat and always with a self-conscious recognition among thinking statesmen about how backward Russia was in comparison with European powers—in fact, this has persisted under the Tsars and the Soviets and in the Russian Federation under Presidents Medvedev and Putin.

By the mid-nineteenth century sea captains knew that the extension of the Gulf Stream (the North Atlantic drift) made it all the way to the Barents Sea and even to the shores of Novaia Zemlia, a phenomenon first predicted in 1848 on the basis of the comparatively high air temperatures observed in Arctic regions. In 1870 the Russian Alexander Middendorff explored Novaia Zemlia and the Taimyr Peninsula, followed by Fridtjof Nansen, who explored the Siberian Arctic on the "Fram" from 1893–96, and Roald Amundsen in the small sloop "Gjøa," equipped with a thirteen-horsepower engine, who was the first to traverse Canada's northwest passage in a three-year journey beginning in 1902, and all of whom contributed to understanding of Arctic conditions, climate, and resources.[4]

The Gulf Stream keeps Murmansk, some 200 kilometers north of the Arctic Circle, ice free year-round. In 1871 the Ministry of Finance formed a commission to look at the construction of a port to support the development of the rich Barents Sea fisheries and establish a military presence. After a survey, the commission determined to equip Kildin Island roughly at the entrance to the Murmansk fjord. The commission also considered a site 350 kilometers to the southeast on the Iokanskie Islands protected by Cape Sviatyi Nos—a very small naval base was established here in 1915.[5] The cape divides the Barents and White seas, and a beautiful lighthouse sits above the water, rich with marine wildlife. The Pomors had a proverb about the rich fisheries near Sviatyi Nos: "Wherever fish swim, they do not miss Sviatyi Nos." The Soviet Union later established a submarine base here.

There were several stimuli to more extensive exploration of the Northern Sea Route at the turn of the century. One was the economic wealth of Siberia. Between 1893 and 1899 several British companies struggled to establish regular commercial voyages to the Ob and Enisei rivers to engage in trade for lumber, gold, and other products. But they eventually tired of fighting the short navigation season. Moving ice floes that varied in thickness from two to six meters hampered sailing at the best of times. The absence of year-round meteorological stations precluded any regular weather communications.[6] An 1893 expedition to bring rails and other equipment from England indicated how inadequate knowledge and equipment prevented normal transport. On July 18, a crew of six officers, a doctor, and forty-three sailors from Kronstadt Naval Base departed Dumbarton for the Enisei River with fourteen months of food supplies. The Tsarist government decided not to finance the trip, but the Committee for the Siberian Railroad secured support from the Englishman Francis Leyborne-Popham who organized a convoy under the direction of Captain Joseph Wiggins. Wiggins had made a series of cargo trips supported by British syndicates.[7] Wiggins, with 1,500 tons of rail destined for the Trans-Siberian Railway and several vessels, joined the Russians at Varde on August 5, 1893. Wiggins's knowledge of the Kara Sea, in spite of the fact he had traveled to the Enisei and Ob at least eight times, was primitive; he did not even have charts based on his previous voyages. The expedition ran into all sorts of bad luck including an "orgy of storms" and inadequate conditions on arrival. Six barges at Golchika on the upper reaches of the Enisei met the ships, all in poor condition and springing leaks; three of them sank immediately with nearly 3,000 rails on them, although apparently most of the rails were recovered the following year. The Committee of the Siberian Railroad nevertheless declared the expedition successful.[8]

Not only economic interests stimulated Arctic research; concern about the precarious life of the Pomor also played a role. In 1898 the oceanographer Nikolai Knipovich initiated the "Expedition for Fisheries Research off the Coasts of Murman." He commenced the research in response to a freak storm that devastated Pomor fishing communities and the failure of the Tsarist government to respond with emergency support; of course, governments have learned glacially throughout the twentieth century how to respond to natural and other disasters. As a first step, Knipovich carried out survey work from a small sailing vessel, *Pomor,* to study meteorological and oceanographic conditions. But in 1899 Knipovich took delivery of a specially designed research vessel, the *Andrey Pervozvannyi,* and in that summer the ship completed her first season of field work. During the next decade the vessel carried out an impressive range of surveys of the fisheries and the oceanography of the Barents Sea and acted as a valuable training ground for an impressive group of young scientists. They sailed primarily in the southern Barents Sea to just south of 76°, took 2,000 biological and 1,500 hydrological observations, and gained better understand of the annual oscillation in the amount of heat of the Gulf Stream and its influence on atmospheric conditions and animal life.[9] These expeditions contributed to the rise of Arctic trawling industry, although Russian industry remained extremely modest.

Between 1908 and 1913 Russian fisherman harvested in all 512 tons while England and Germany fished more than 86,000 tons.[10] Eventually, the Murmansk Scientific-Industrial Expedition received more than one million rubles from the Tsarist government.

Another important facility to generate scientific data on which to expand fisheries was the Murmansk Biological Station, at first located on the Solovetski (or Solovki) Islands in the middle of the White Sea. Founded in 1881 at the initiative of Nikolai Petrovich Vagner, known better by his literary pseudonym, Kota-Murlyki, the station was moved to Ekaterina Harbor in 1899; it operates to this day in another incarnation. Because of insufficient funds, the construction of a building and purchase of laboratory equipment lagged, and it opened officially only in 1904. Vagner (1829–1907) was a zoologist and entomologist, but gained fame for his fairy tales and short stories including *The Tales of Kot-Murlyka* (1908). The Arctic had yet to capture the national imagination as it would under the Soviets, in part because of low literacy rates, in part because the Tsarist government did not tout the achievements of Arctic explorers, let alone sponsor them, and in part because educated society thought of the Arctic as largely barren.[11]

Of course, plying the Arctic Ocean required more than intrepid escapades. In order to achieve success, new equipment was needed, especially icebreakers. A naval visionary, Admiral Stepan Makarov, gained funding from Minister of Finance Sergei Witte to build a 6,000-ton experimental ship, the *Ermak,* to demonstrate the feasibility of icebreakers. The *Ermak* was the first of several Russian icebreakers built in Newcastle. It set sail in 1899 to Spitsbergen Island and later approached the North Pole.[12] Witte, who doggedly pushed the Trans-Siberian Railway as crucial to Russia's modernization, believed the icebreaker would enable Russian to open trade in the Pacific Ocean, compete with England and Japan, and serve as an example of the way that infrastructure generally was the key to Russia's economic future.[13] Yet Makarov's achievements did little to dispel continued skepticism. First, several figures opposed further developments because they saw the Arctic as providing natural protection against aggressors in its ice fields.[14] Second, not only was the government bureaucracy unable to support polar research consistently and sufficiently, it could not even organize simple matters in the Arctic. On one occasion Petersburg bureaucrats instructed the governor of the Enisei region to order indigenous people, the "Samoeds," to "extend their wanderings farther to the north so as to meet the expedition, deliver its mail and also pick up any outgoing mail" during its wintering on the Taimyr Peninsula. This order reflected why ignorance of indigenous people required study to remedy. Officials in the capital had no understanding of the fact that the "natives" spent the summers along the Arctic Ocean, then returned to the forests about 300 kilometers from the seacoast in the winter for shelter from constant storms and blizzards for themselves and their reindeer. Delivering the mail would upset this time-proven way of life.[15]

Admiral Nikolai Chikhachev, minister of the navy, hoped to see the railway built quickly because of Russia's growing military and economic interests in the Far East and the fact that the railroad would more rapidly and cheaply ship military supplies to Vladivostok. He proposed that the Committee for the Siberian Railroad employ the Kara Sea for transportation of railway cargo that would be purchased in England and shipped to construction sites upstream on the Ob and Enisei rivers as Captain Wiggins had attempted. The absence of telegraph made it difficult to communicate with the ships, to ensure timely deliveries, and to coordinate labor and capital, and the rivers needed to be dredged in many places.[16] The proposal foundered because of opposition from officials in Witte's ministry of finance both because of their protectionist views and because the navigability of the Kara Sea remained an open question. Ultimately, the Committee for the Siberian Railroad budgeted only 300,000 rubles of its total budget of 21.9 million rubles (1.5 percent) for the development of the Northern Sea Route.[17] In contrast, the Soviet Union pushed transport, communication, and scientific technologies northward with great determination.

The Russo-Japanese War (1904–5) demonstrated the importance of a Northern Sea Route, in addition to a Siberian railroad, because Russia had to send its naval units to Japan by way of the Cape of Good Hope, while the Trans-Siberian Railway was "overloaded" to the breaking point with soldiers and materiel and fully finished only in 1916. Siberia itself was then afflicted by shortages of grain, sugar, and kerosene. Rolling stock and track had to be transported from European Russia to Siberia. On top of this, the distance from Moscow to Arkhangelsk by rail and by sea to Tiski at the Lena River delta totals 5,337 kilometers, while from Moscow and Irkutsk by rail thence by river and sea was 9,421 kilometers and involved several transfers of cargo. On March 28, 1905, Minister of Transportation M. I. Khilkov chaired a meeting to discuss how to ship freight through the Kara Sea to the Enisei River. Russian shipyards were unable to build the ships quickly enough, so the government turned to Germany and Holland to build them with small draft, high degree of maneuverability, strong engines, and boilers that could handle wood or coal and capable of using salt or fresh water. The Russians acquired only four tugboats and two steam and nine non-steam lighters, the latter freighted to the mouth of the Enisei by German seagoing tugs leased by the Russian government. The government also leased four large English merchant steamers, but only one of these ships reached the Enisei. The *Ermak* icebreaker and *Pakhtusov* polar vessel led the fleet north; it waited two weeks at the Iugorskii Strait for ice floes to give way. Excluding the anchoring, the trip from Murmansk to the Enisei took nine days and convinced several individuals of the value of a seat route to supplement the railroad.[18]

Siberian economic growth in the early twentieth century slowly stimulated the development of the Northern Sea Route, with the expansion of the grain production a crucial stimulus, itself stimulated by the Trans-Siberian Railway and the arrival of roughly one million settlers from the European parts of the empire from 1894 to 1902 and reaching 320,000 settlers annually in the years 1906–9 primarily to the southern regions of Siberia and on the steppes. With something like the Homestead Act (1862) in the United States, Prime Minister Stolypin encouraged peasants to settle plots of land in Siberia as a way to destroy their persistent communal institutions and remove them from the influence of radicals. Siberian farms were more likely also to have horses and agricultural machinery than the ones of European peasants because of the wealth generated by export—perhaps one-fifth or one-sixth of production was exported.

To facilitate transport of agricultural goods from Siberia, officials ordered the building of the Perm-Kotlas railway line (from Perm in the western foothills of the western Ural Mountains to Kotlas, a lumber town on the Northern Dvina River) at great speed and expense (occasionally work already completed was abandoned for another route). But because the line was poorly equipped, grain growers hesitated to use it, and shipping along the Northern Dvina River between Kotlas and Arkhangelsk was so inadequate that grain often wintered at the river. Ultimately, the Trans-Siberian Railway had such low capacity that it was overwhelmed by grain shipments—grain often rotted in the open air as it waited to be shipped. All of these difficulties "destroyed all illusions concerning the low cost of railroad shipment" and led to renewed interest in a Northern Sea Route. Owners of mines spoke about the potential for exports of asbestos, mica, and graphite to Europe and the receipt of machinery and equipment as payment. Timber interests expressed the view that a Northern Sea Route was indispensable for their industry's development. Parliamentary representatives, merchants, and public and industrial groups in support of development of trade flooded the government with letters and telegrams. Yet no Russian commercial voyages took place through the Kara Sea to the Ob and Enisei rivers between 1902 and 1912.[19] The Northern Sea Route would have to wait for extensive exploration and determined government interest.

Foolish Courage

Having learned of the inadequacy of Siberian transport during the Russo-Japanese War, the Tsarist government grudgingly supported Arctic research. In 1908 the Council of Ministers determined to give an 80,000-ruble subsidy to any vessel making an annual trip from Russian Pacific ports to Nizhnekolymsk in the Kolyma River estuary. But of course this had no bearing on the problem of European-Siberian transport. In May 1911 it allotted funds for construction of radiotelegraph stations. The first three stations, located on Vaigach Island, at the Iugorskii Strait, and at Mare Salo on the Yamal Peninsula, opened in 1914.[20] The ministers eventually authorized the organization of a scientific expedition for the overland exploration of the northern coasts of Siberia from the Lena River to the Bering Strait. Yet insufficient funding left the success of expeditions up to the ability of independent explorers to master difficult conditions. Some survived through good fortune while others perished.

Among the expeditions, one succeeded in traveling the entire sea route from east to west under V. A. Vilkitskii.[21] Vilkitskii (1886–1961), who graduated from the Naval Academy in Petersburg, was later in charge of the Main Hydrographic Administration. In 1894, 1895, and 1896 he conducted three hydrographic voyages to demonstrate that the Enisei was clear of shoals and accessible to ocean-going vessels for more than 1,500 kilometers upstream from its mouth. Unfortunately, the shallow Ob River permitted no ships with a draft of more than twelve feet. Next Vilkitsii oversaw preparation of charts of the Ob and Enisei gulfs. Finally, in 1898 the Council of Ministers established the small-scale Arctic Ocean Hydrographic Expedition. Yet Vilkitskii and other members of the expedition could not rely solely on financial support from the government, seeking instead "voluntary donations." The members of the expedition considered such questions as the feasibility of the Northern Sea Route, the kind and number of vessels needed for shipping and exploration, organization of supply along the route, and the location of sites for observation stations.[22]

Vilkitskii led hydrographic expeditions of the Arctic Ocean (1910–15) with two specially constructed icebreakers, the *Taimyr* and the *Vaigach*, each at 1,200 tons displacement. The ships explored the Arctic coasts and in 1914 sailed from the Pacific to the Atlantic, wintering en route. In 1914–15, the expedition succeeded in completing the first voyage from Vladivostok to Arkhangelsk, repeating that of Nils Adolf Erik Nordenskjöld in 1878–79 in the opposite direction. Vilkitskii charted the islands of Severnaia Zemlia and discovered other islands (e.g., Vilkitskii). He presented the results of the expedition to members of parliament, the council of the admiralty, and other groups. The Russian Geographical Society awarded him the Medal of Constantine and the council of the French Geographical Society the gold medal "La Rockette." During World War I Vilkitskii commanded a ship for the Baltic Fleet. In 1918 he was named director of the Soviet hydrographic expedition of the west Siberian region of the Arctic Ocean. He journeyed to Arkhangelsk where he joined the monarchist Whites and threw support behind the Allied intervention. He led an operation to evacuate the troops of the northern army from Arkhangelsk in February 1920, simultaneously commanding the icebreaker *Kosma Minin* and directing other ships with soldiers and refugees on board. They sailed to Norway, entering a fjord near Tromsø. Vilkitskii emigrated to London in 1920, although at Bolshevik invitation he returned to lead Soviet commercial expeditions. He later was employed as a hydrographer in the Belgian Congo and died in Brussels.[23]

The polar explorer Vladimir Rusanov joined an underground revolutionary circle at Kiev University, the "Workers' Union," in 1896, participated in student demonstrations, was arrested, and was exiled to Orel where he was re-arrested for his work as a union organizer. In prison he read Fridtjof Nansen's *In Night and Ice* (1897) in which Nansen described the 1893–96 voyage of the *Fram*.[24] Upon release Rusanov returned to revolutionary activity, was arrested again, and was sent to Ust'-Sysolsk in the Vologda region for two years. Being required as part of his punishment to analyze regional statistics, he conducted a series of field trips to the vast, almost unknown Pechora region. When he finished his term, he was not allowed to return to a central city, so he went to Paris where he continued studies at the Sorbonne in geology and decided to write his doctoral dissertation on Novaia Zemlia. He returned to Arkhangelsk in 1907 where he learned that local officials were willing to support his research, which they saw as a way to counter an expanding Norwegian presence. He carried out a journey to the Matochkin Strait in Novaia Zemlia with a student-geologist of Kharkov University, L. A. Molchanov, and with Nenets hunters sailed to the Kara Sea and back. In 1908 the French selected Rusanov as a geologist for their expedition to Novaia Zemlia; he traveled much of the length of the archipelago. In 1909 he traversed the interior of Novaia Zemlia. He recognized that the archipelago could serve as a base to store coal and other supplies for the Northern Sea Route. He began to ponder in print whether a Northern Sea Route was possible not south from Arkhangelsk through the Iugorskii Strait and the Kara Gates, but northerly around Novaia Zemlia. In 1912 he disappeared without a trace as part of an expedition on the *Hercules* near Cape of Desire. In 1934 parts of the boat were discovered that had washed up on the shore of the Taimyr Peninsula.

Georgii Brusilov also drifted to a frozen death. In 1910–11 he took part in the Vilkitskii expedition on the *Taimyr* and *Vaigach*. In 1912 he led an expedition of the sail- and motor-powered schooner *St. Anna* to navigate the Arctic route from the Atlantic to the Pacific oceans. Passing through Iugorskii Strait to the Kara Sea in mid-September, the ship became icebound near the west bank of the Yamal Peninsula then drifted northward. In spring 1914, when the schooner drifted to the north of Franz Josef Land, part of the crew abandoned ship and headed southward on foot along the drifting ice. Two men were rescued by Georgii Sedov's expedition on the *Sviatoi Foka* ("Saint Martyr Foka," hereafter *Sv. Foka*), while the others disappeared without a trace.

One of the reasons for these failed expeditions was a rudimentary understanding of Arctic conditions. Another was that such emotions as patriotism and heroism overwhelmed those of common sense and self-awareness. The 1912–14 expedition of Georgii Iakovlevich Sedov, an officer of the Main Hydrographic Administration, and his crew showed the limits of technology and the disadvantages of foolish courage over circumspect preparedness.[25] Sedov organized an expedition to the North Pole in 1912, because of a late start to the voyage wintered in 1913 and 1914, and died during a futile attempt to reach the North Pole with sled dogs. Perhaps foreshadowing the challenges and human losses associated with future Arctic expeditions, Sedov's name belongs to an archipelago, an island, a cape, a peak, two peninsulas—and a ship that drifted for 800 days in Arctic ice at Stalin's great displeasure.

Sedov was the son of an Azov fisherman and until fourteen was illiterate. After he completed a three-year church school, he left home. The twenty-one-year-old Sedov earned a diploma as a navigator for the merchant marine, took a commission as a warrant officer in the Admiralty, and was sent to a hydrographic expedition in the Arctic Ocean. He continued to move up through the service, in 1908–10 working on a Caspian expedition in Kolyma where he charted the delta of the Kolyma River and Novaia Zemlia where he produced maps of Krestovaia Bay and Ol'ginskii Poselok ("Olga's Village").

Perhaps at the urgings of the Arkhangelsk governor and the Orthodox Church and in an effort to demonstrate to the Norwegians that Novaia Zemlia belonged to Russia, small groups of Russians had established such settlements as "Olga." There were but four families at Olga, and only one had lived on Novaia Zemlia earlier. This family had come to make a living, while the others apparently hoped to find happiness in this unfamiliar, barren place. The settlers were young: the Usovs without children, the Dolgoborodovs with a six-year-old girl, the Fomins, and the Konechnyis with two children lived here at the end of the earth.

In 1912 Sedov advanced the idea for the first Russian expedition to the North Pole. Perhaps not knowing that Robert Peary had claimed the Pole in 1909 or more concerned about the Norwegian presence in the Arctic, Sedov said, "Amundsen desires no matter what to place the honor on Norway for the discovery of the North Pole. He wants to go in 1913, and we will go in this year and prove to the entire world that Russians are capable of this achievement." The Navy gave him a two-year leave of absence while transferring his commission from that of captain in the Admiralty to senior lieutenant in the Navy. However, a committee of the Main Hydrographical Administration sharply criticized the plans for the expedition. Its members pointed out that while Sedov had worked in the north, he did not know the winter Arctic nor have any experience sailing in drift ice, and in striving to "beat" Amundsen, Sedov fixed the date for sailing as July 1, which clearly did not permit adequate time to prepare. Sedov presented revised plans by the end of May that raised other questions, and if the Tsar had initially promised 10,000 rubles for the expedition, then the government now refused to release the funds. A committee of newspaper men, seeing a business opportunity, raised funds, and by July 10 Sedov had succeeded in chartering the *Sv. Foka* which set sail very late in the season on August 27. In addition to Sedov, Vladimir Vize, a geographer, Mikhail Pavlov, a member of the Petrograd Society of Admirers of Nature and of the Society for the Study of the Russian North, and the artist Nikolai Pinegin would join a crew of nineteen others. Pinegin, who had lived on Novaia Zemlia for three months in 1910, wrote a diary of the journey, recently republished, that recounts the fateful mistakes of Sedov. The plan was to put a detachment of the crew on Franz Josef Land and to winter in housing brought on board a ship. They would explore Franz Josef Land with sled dogs and plan the assault on the North Pole, while others returned to Arkhangelsk for more supplies. But the late start and an early onset of ice meant difficulties from the beginning, on top of which they brought inadequate rations and, at the last moment, the Naval Ministry refused to release a radio operator for the expedition. But Sedov, who felt overconfident because of his Kolyma research, did not think to wait.[26]

August in the Arctic is no longer a summer month, but already autumn, and brings unpredictable winds and wet snows. When Sedov and his crew set out on August 15, the White Sea was calm and welcoming. By August 19, as the *Sv. Foka* entered the throat of the White Sea, it met a strong northern wind, on top of which the *Sv. Foka* was overloaded with supplies, and the wind pushed the heavy vessel backward. Sedov had dreamed of soon seeing Franz Josef Land, but instead they had to hold somewhere in the White Sea where they were buffeted by waves and took on water. The storm continued; on some days in the strong wind they barely moved a mile over several hours. As they approached Novaia Zemlia they were enveloped in a fog and hit thick ice floes. They struggled, tacked, and finally founded protection in Krestovaia Bay. Then ice set in. By early September it was clear to Sedov that he would have to winter earlier—and elsewhere—than planned, having consumed most of the coal they had brought on board the old *Sv. Foka*.[27] In June 1913 there was a chance to save the expedition, when the assistant machinist and three sailors managed on a sloop to reach Matochkin Strait and take a passing steamship to Arkhangelsk.

When the *Sv. Foka* did not return, businessmen and journalists in Arkhangelsk wondered how to send out a rescue party; the ship had no radio. An aerial search was unsuccessful. But Sedov was determined to sail to Franz Josef Land and set out for the North Pole. Yet the ice held the ship all summer. On September 6 an eastern wind came up that carried it away from the shore. The ship's officers urged Sedov to abandon the expedition; he refused. During the second zimovka in Tikhaia Bay they burned boards, desks, and old boxes to stay warm. The polar night descended. Conditions grew worse, ice penetrated the ship, little food remained, and nearly all men took ill with scurvy. In spite of all this, Sedov and two sailors embarked on a futile effort to reach the North Pole by sleds. He set out on February 15, 1914, with G. Linnik and A. Pustoshnyi. Sedov wrote in his diary, "We saw above the mountains the first dear, mother sun. Ah, how it is so beautiful and good!" But it was so cold the mercury could fall no lower. Sedov died shortly thereafter, and the two sailors buried him in canvas bags with skis as a cross. The sailors made it back with fourteen dogs, surviving on frozen pork fat and water thawed from their breath. After five days there was no more food or fuel, but they made it back to the ship. The men gathered, read Sedov's diary, and talked about his last days. They learned how he had hidden his wasting, could not walk, and suffered. Perhaps they discussed Sedov's unfounded self-confidence, his lack of common sense, or their suffering owing to inadequate knowledge of weather patterns, ice regime, currents, and miserable preparedness. The modern technology of icebreakers would enable future expeditions to have greater success, although each technological advance led to more risk taking.[28]

The First Icebreakers

At first, primarily for reasons of cost, many officials and businesspeople hoped that ordinary vessels might be able to handle Arctic conditions. But at the very least these vessels had to have reinforced hulls and be accompanied by auxiliary vessels to lend assistance if need be. Polar stations, an expensive undertaking, were also an important consideration. Hence the capital, insurance, and other fixed costs would be higher for freighting in polar regions than along existing but exceedingly long and expensive southern hemisphere routes. The short navigation season made it difficult to cut costs through repeated or regular voyages. High cargo volumes and rapid turnaround remained an impossibility until the Soviet era.[29] The icebreaker was the solution, in the Soviet-era *Arktika* class reaching 25,000 tons displacement and 150 meters in length and powered by two 171-megawatt nuclear reactors.

The *Ermak* was perhaps the world's first true icebreaker, with a strengthened hull shaped to ride over and crush pack ice. Admiral Makarov supervised the construction and launch of the *Ermak* at the Armstrong Whitworth Shipyards in Newcastle, England, and then commanded in two Arctic expeditions in 1899 and 1901. Makarov followed Nansen's conclusion that ships had significant advantages over dogs and sledges if they could be made strong enough to resist the pressure of polar ice; such a ship could be carried by currents when ice-bound. Of course, more expensive than sleds and dogs, the icebreaker had advantages of safety, capacity, power, and speed. Makarov devoted significant effort to convincing officials and officers alike of its utility. He pointed out that it would clear ice from the Gulf of Finland to St. Petersburg in the winter and in the summer facilitate navigation to the Kara Sea with access to Siberian rivers and their goods. No sooner had the *Ermak* been launched to sail to St. Petersburg than it encountered 160 miles of ice in the Baltic Sea and freed thirteen steamers caught in ice near Revel, then freed another forty some ships going to St. Petersburg before sailing on to Newcastle to take on coal for the trip to the Kara Sea.

In thick ice the *Ermak* might take four hours to go two miles instead of the usual speed of 2.5 miles per hour. The captain and navigator learned how to break ice, at which angles, and so on. Like the *Fram* (only 200 hp) before it, the *Ermak* presented the opportunity to study the nature of polar ice and how to use powerful cranes and wenches in the process. They often hauled huge chunks of ice on board to measure its temperature, strength, and specific gravity. The strongest ice was glacier ice, while the weakest ice was from ice floes. The staff also conducted hydrological observations.[30] The voyagers took great joy in bear shooting; Makarov wrote, "There is such an abundance of polar bears that one cannot avoid having a shot at them. Fresh traces of the white bear are seen on almost every other ice-floe." The so-called sportsmen pursued bears and cubs to "finish them off" at close range.[31]

Many of the first Russian icebreakers were built at the Newcastle shipyards during World War I under Evgenii Zamiatin.[32] Zamiatin served his nation although he was twice imprisoned by the Tsarist government and exiled for his leftist political views. Known mostly for his dystopian novel *We* (My, 1923) that anticipated all of Aldous Huxley's criticisms of authoritarian industrial society in *Brave New World* (1930), Zamiatin was an engineer by training. He commenced his writing and completed his engineering education simultaneously. In 1916 and 1917 Zamiatin spent eighteen months at the Armstrong Whitworth shipyards. On his return to Russia on the eve of the revolution Zamiatin turned fully to writing, in addition to *We* publishing a number of satires in fairy tale form that were increasingly critical of Bolshevism. Zamiatin found it more difficult to publish than to build icebreakers. A member of the Leningrad branch of the Writers' Union, he was arrested in 1919 and 1922 and released, but eventually expelled from the Union along with Boris Pilnyak, the chairman of the Moscow branch of the Union, in 1929 for publishing "anti-Soviet" works abroad.[33]

Zamiatin saw his posting in England as keeping a widow open onto Europe, likening it to the achievement of Peter the Great 200 years earlier. The Baltic Sea had been closed by the Germans during World War I; in theory icebreakers would enable the White Sea and Arkhangelsk port to be open to war and supply ships. Russia needed more than the *Ermak*. Zamyatin himself said later that Russia did not begin building large icebreakers until 1935, when the Soviet shipyards embarked on construction of 11,000 ton-Stalin class vessels. But the Armstrong shipyard icebreakers, several of which engaged the allies under Soviet flags in 1918–19 near Arkhangelsk, were ahead of their time.

The *Sviatogor* was launched in February 1917. Because of its possible use

Table 1.1 Early Russian/Soviet icebreakers of British manufacture

Name	Soviet name	Gross tons	When completed
Sviatogor	*Krasin*	4,902	February 1917
St. Alexander Nevsky	*Lenin*	3,375	June 1917
Koz'ma Minin		2,432	November 1916
Kniaz Pozharsky		2,432	December 1916
Ilya Muromets		1,651	December 1915
Dobrynia Nikitich		1,664	1916

during the civil war and allied intervention in Russia in 1918–19, the *Sviatogor* was scuttled in the White Sea by the Royal Navy, but later raised, used in the White Sea, then brought to the Orkney Islands for use as a minesweeper. The British returned *Sviatogor* to the USSR under a trade agreement negotiated by the Old Bolshevik and engineer Leonid Krasin, the commissar of foreign trade.[34] The *Sviatogor* was renamed the *Krasin* in his honor (see Table 1.1).

The *Krasin* was instrumental in rescuing the *Italy* balloon expedition of Italian General Umberto Nobile in the Arctic in 1927. Journalists on board the *Krasin* filed daily reports about the heroic effort that thrilled the Soviet reader. On July 12 the *Krasin* rescued the survivors of the *Italy*, with General Nobile himself having been pulled from the ice by the Swedish pilot Einar Lundborg. Upon returning to Leningrad on August 5, 1928, some 250,000 people met the great icebreaker, according to *Izvestiia*. The ship and its staff earned proclamations, orders, and medals.[35] In 1933 the *Krasin* first reached the northern shores of Novaia Zemlia.[36] But greater things awaited the *Krasin*. It rescued other ships and their passengers and, notably, in May 1938 freed the *Sadko*, *Malygin*, and *Sedov*, which had been forced to winter because of poor planning—or rather because Glavsevmorput (the Main Administration of the Northern Sea Route, established in 1931; see Chapter 2) was forced to pursue the sea route with still inadequate technology and incomplete weather forecasts.

The Northern Expeditions of the Early 1920s

If military and political competition for the Arctic were relatively muted during World War I and the war forced an immediate interregnum in Arctic research, then in the 1920s, and especially after the Russian Revolution, national designs on the polar cap grew significantly. One reason was the end-of-the-war Allied intervention in the Russian northwest as France, England, and the United States supplied Russia through Arkhangelsk and then intervened in the war after the Bolsheviks seized power. White opponents of the Reds also opposed the Bolsheviks in the Arctic and in Siberia.[37]

But while the efforts of scientists to expand exploration at the end of the war and after the collapse of the Tsarist regime accelerated, political uncertainties, inflation, and civil war interfered with attempts systematically to develop the Northern Sea Route. Yet the collapse of the Provisional Government (March–October 1917), the rise of the Bolshevik regime after the October coup, and the resulting chaos paradoxically gave impetus to exploration or the Northern Sea Route. Civil war broke out between Reds—the Bolsheviks and their sympathizers—and Whites—monarchists and anyone else who opposed the Bolsheviks—in summer 1918 and lasted until 1920. The Bolsheviks recognized immediately their precarious military situation and were determined to secure the Arctic, even if they lacked the troops to do so. On top of this, commerce between cities and the countryside was disrupted, food production plummeted, and starvation and epidemics spread. In an effort to bring grain and other foodstuffs from Siberia to the European parts of the nation, the government—both the Red and the local White governments—supported a series of expeditions and ship caravans along the Arctic shore through the Kara Sea to Siberian river delta ports where finished goods might be exchanged for food.

Scientists, explorers, and naturalists realized that, in spite of the efforts of Sedov, Makarov, Vilkitskii, and others, they had yet to determine how to sail safely. In recent decades, explorers and specialists had made headway in charting the shoreline of the Kola Peninsula, Novaia Zemlia, and river deltas along the Arctic Ocean. They knew more about fisheries and mammals of the White and Barents seas and their economic potential. They speculated that interior expanses of tundra must be rich in minerals and fossil fuels. They suggested establishing coal and other fuel depots and food stores in the Arctic basin. They recognized the potential of transforming rivers into transportation routes to foster exchange of Siberian and European goods and raw materials. They had some sense of the challenges faced in their hope of establishing a Northern Sea Route from Arkhangelsk to Vladivostok. Yet the foundering efforts to regularize shipping in the Arctic showed in stark relief all of the challenges of Arctic assimilation: a weak technological foundation, insufficient scientific knowledge, and inadequate government support.

Established upon the abdication of Tsar Nicholas II in February 1917, the Provisional Government struggled to stay in World War I against growing opposition and anarchy until the Bolsheviks seized power in October in a nearly bloodless coup. Although the Provisional Government moved from crisis to crisis in 1917, scientists, educators and others attempted to continue normal work in societies and universities. Even during the difficult days of the new republic, and the more difficult days of fuel and food shortages in Petrograd, where many specialists resided, the leading geographers, cartographers, hydrographers, geologists, and biologists engaged nascent bureaucracies to secure funding to explore the north. In October 1917, members of the Academy of Sciences' Committee for the Study of the Natural Productive Forces *(Kommissiia po Izucheniiu Estestvennykh Proizvoditel'nykh Sil)*[38] met to establish a Northern Department within KEPS to focus on study of the natural wealth of the north. They defined "the north" as Arkhangelsk, Vologda, Olonetskaia, Viatskaia, Permskaia, and Siberian guberniia (provinces). The department's first chairman was well-known biogeochemist Vladimir Vernadskii, whose concept of the noosphere contributed greatly to the modern idea of the ecosystem. But when Vernadskii moved to Ukraine to head up efforts to establish an Academy of Sciences of Ukraine, the Russian Academy president A. P. Karpinskii and his deputy, A. F Shidlovskii, took over. At the October meeting the participants heard V. D. Nikolskii lecture on the wealth of the Murmansk Region; V. K. Brazhnikov on the fisheries of the Murmansk shore; G. A. Kliuge, a short history on the role of the Murmansk Biological Station; and A. I. Auzana on cartography of the north. Other talks concerned climatic and hydrological data, on the study of permafrost soils of the north, and on dirt roads of the north.[39]

After discussion, the participants charged L. L. Breitfus and an assistant with producing a detailed map of resources, which they turned to only a year later in November 1918.[40] Another goal of the committee was to take advantage of the Murmansk railroad, built in 1916 during the war to facilitate delivery of materiel for the war effort, through study of the regions contiguous to the railroad including those on which the Finns and others had claims (e.g., Pechenga, with its schools, brick factory, electrical service, and so on).[41] Leonid Breitfus, who had assisted Knipovich on the 1898 Murman expedition, headed the hydrometeorological service of the Main Hydrographic Administration, established four stations in the Kara Sea, and organized airplanes for searching the Arctic for Sedov, Brusilov, and Rusanov. Involved in several expeditions after the revolution, in 1921 Breitfus emigrated to Germany where he died in 1950.

The Bolshevik October Revolution ended any further activities in KEPS's Northern Department until May 1918 when scientists modestly determined to produce an inventory of those individuals who worked in fields of study connected with polar regions to avoid duplication of research including provincial scholars and coordinate activities with such existing groups as the Geological Society, Main Hydrographic Administration, and Military-Topographic Department of the General Staff. The scientific agenda centered on the effort to provide basic biological, hydrographic, geological, anthropological, and ethnographic data on the Far North.

Allied intervention resulted in great risks for the Bolsheviks and chaos in Arkhangelsk harbor; the allies scuttled and sank ships that had to be removed from shipping lines. The ships were concentrated in four different regions: sixteen at the throat of the White Sea, five others somewhere in the White Sea, six ships near Kildin Island at the entrance to the Murmansk fjord, and ten others scattered throughout the Arctic Ocean. A fact-finding mission to England and Norway focused on determining the cargo and value of those vessels. A harbormaster estimated that, in spite of the absence of bills of lading or other legal documents, the sunken ships had a value of twenty-five or thirty million rubles, but there is no evidence of recovery of cargoes. When raised, some of the ships were ordered to shipbuilding yards to be refurbished. While the war and civil war dragged on, Murmansk and Arkhangelsk essentially could not support a merchant marine, and dry docks for new ships or repair fell into disrepair themselves. The Bolsheviks had envisaged a northern navy of fifty ships, but they had no spare parts, which had to be acquired from abroad with hard currency, nor an indigenous industry.[42]

On another front, however, there was immediate success, if only on paper. Less than three months after seizing power, the Bolshevik Council of Ministers budgeted 100,000 rubles for a Northern Expedition to protect the small steps achieved in Arkhangelsk Province in the previous years and make recommendations for military, trade, and industry. A few days later the council allocated another five million rubles to bring order to the Arkhangelsk port.[43] Among the many organizations created after the revolution to resurrect production in a specific sector of the economy was the White Sea Administration of Fish and Fur Industry. In part with these funds, the administration slowly accumulated ships and equipment, although coal remained in short supply. It began to acquire vessels from abroad, as well as to raise sunken ships. Another problem was that sailors were few in number and inexperienced and had limited access to housing and food.[44]

An additional push to the development of the Northern Sea Route came from business interests in Siberia. On November 14, 1918, representatives of business and trade of the Enisei River region called for the provisional Siberian government to address the problem of the Northern Sea Route with all deliberate speed. British capitalists also eyed Siberian resources with interest and advanced a project to open shipping along the Ob River. All of this led to preparation for an Arctic expedition from the fall of 1918. Alexander Kolchak, a naval officer and polar explorer who led the White Army until his defeat, capture, and execution by the Red Army in January 1920, formed a Committee of the Northern Sea Route in April 1919. He ordered the requisitioning of all available Russian ships and icebreakers and also ships of the interventionists, with Vilkitskii to lead the expedition. It departed on August 15, too near the end of the polar shipping season to achieve much of anything.[45]

If they could understand the navigation season and acquire seaworthy vessels, captains could sail Siberian rivers as excellent freighting routes into the Arctic basin: the Ob, Taz, Enisei, and Pisina rivers flowed into the Kara Sea; the Khatanga, Anabara, Olenek, Lena, and Iana into the Laptev Sea; the Indigirka and Kolyma into the Eastern Siberian Sea. The Ob, Enisei, and Lena had a series of navigable tributaries, with the 4,000-kilometer-long Enisei fed by the Upper, Podkamennaia, and Lower Tunguska rivers. Until 1878, not one steamship had appeared on the Lena River, and before the

Russian Revolution the only one was the "Lena" that made its way under Swedish explorer Nordenskjöld. The total Siberian river system might reach 27,000 kilometers of navigable steamship waters. The problem was that the deltas—and much more—were frozen for nine to ten months a year. Yet the advantages of the Northern Sea Route—much shorter distances between Europe and Asia and significantly lower costs per ton of transport on the seas versus across Siberia on land—meant a concerted effort to achieve this end.

Vilkitskii commanded three groups of ships.[46] The 1918 expedition revealed the great challenges of opening the Northern Sea Route with inadequate planning or equipment. In the Iugorskii Strait the expedition encountered foreign ships and difficult ice floes that damaged the propellers of the *Kildin* and *Pakhtusov*. Barges sank, and several "cowardly" captains returned to Norway. A few ships made it to Belyi Island by late August, from which they sailed into the gulf of the Ob River to await a caravan of freighters and barges coming downstream. They picked up nearly 500,000 poods (one pood = about thirty-six pounds) of bread as well as 28,000 poods of copper from Ural factories. But the passions of the civil war played out when the river workers refused to participate in this "White Guard" expedition, and there were not enough longshoremen. While boats waited in Tobolsk on the Irtysh River upstream from the confluence with the Ob, panic broke out; the men were poorly dressed, an epidemic began, and nearly 70 percent of the expedition got sick. When the barges arrived, loading and unloading moved at a snail's pace, and much of the freight was returned upstream to Kolchak's troops in Tomsk. The ships made it to Arkhangelsk on September 28, and one ship reached Liverpool at the end of October, with a cargo of 450 tons of vegetable oil, 100 tons of seed, 100 tons of wool, 355 reindeer hides, seven barrels of furs, and other items.[47]

Future expeditions were no more successful. One reason was tense international relations. On May 29, 1923, British Foreign Secretary Lord Curzon sent a note to the Soviet government protesting its continued propaganda activities in Britain in breach of the Anglo-Russian Trade Agreement. In such a situation exporters were not permitted to sail, and by the time relations had improved in late summer, it was too late to organize shipping, although several Norwegian, German, and British ships managed to transport modest amounts of goods through the Iugorskii Strait. The ships arrived in late July at the Enisei delta to deliver machinery and equipment for a soap works, a carton factory, and other finished goods and to pick up wool and graphite, returning to Arkhangelsk by September 13. If in 1920 ships in the Kara Sea sailed forty-eight days, and in 1921 for forty-two days, and in 1922 only thirty-seven days, then in 1923—they were in the Kara Sea in all for twenty-three days. In 1924 the international situation improved, as did the financial situation of the USSR[48] and shipping.

In spite of these difficulties, the revolutionary years sparked a rebirth of interest in projects that had languished in the Tsarist era to expand large-scale transportation infrastructure through the far reaches of the empire. One project involved the construction of some kind of transfer canal between Siberia and Europe, perhaps through the upper reaches of the Pechora River that was proximate to the Kama River Basin. The project had been debated for 100 years—and among some Russian engineers continues to be discussed in the twenty-first century.[49] Another project focused on a "Great Northern Railroad" whose supporters presented a project to the Petrograd Division of the Supreme Economic Council in February 1919. Bolshevik officialdom, interested in resurrecting transport, still rejected the project because, as proposed, it would have required payments in huge tracts of contiguous land to the private developers, and resources would be better focused on improving the "ruinous condition" of Russia's existing railroads.[50]

More to the point, a crisis of food and trade triggered Arctic shipping projects. Two days before the Red Army took Arkhangelsk in February 1920, a special food commission of the northern front held a meeting of individual specialists and representatives of various bureaucracies to discuss the economic promise of scientific research. They referred to possibilities for mining and industry, rational organization of fish and fur-bearing mammal industries, and also reindeer husbandry. The meeting resolved that the region's physical-geographic peculiarities, the nature of its economic life and low technical level, extremely low population densities, and poorly developed cultural institutions made the need for immediate action questionable. Yet the region's international significance, importance of its grain exports, and the wealth of down, fur, and other products indicated great promise. After a series of reports the commission voted to create a committee on the scientific and industrial assimilation of the Arctic. This organ would focus "on the furthest rational development of economic and industrial life of the region in the general interests of the Republic, carrying out all practical measures through local economic organs with the help of constant technical and scientific assistance to these organs." This led to the creation of the Northern Scientific-Industrial Expedition of the Supreme Economic Council,[51] later known as the Northern Expedition (Sevekspeditsiia). Two new Bolshevik bureaucracies that were responsible for research and development, Glavnauka (the Administration for Science) and NTO (the Scientific-Technical Department of the Supreme Economic Council) offered support for basic Arctic science.

Often lost in Soviet hagiography that exaggerates his involvement in so many crucial activities of the early Bolshevik state are areas where in fact Lenin's direct participation was crucial. Lenin clearly recognized the value of science and technology for the struggling republic, including Arctic science, and he supported specialists in several endeavors. He believed that scientists were naturally "materialists" and hence in an important philosophical sense were allies of Bolshevik programs. He signed decrees to support the expansion of normal research. He presided at meetings of various bureaucracies that considered how to apply science to the task of resurrecting the economy. The scientific community, buffeted by loss of funding, the interruption of research and teaching, and international isolation because of the war, revolution, and civil war, sought funding wherever it could. Lenin approved the formation of Glavnauka and NTO to manage and fund scientific institutes. Through their support scientists reestablished contacts with the international scientific community, purchased equipment and chemicals, formed new research centers, and embarked on expeditions, including to Arctic regions. In a word, Lenin saw the importance of strengthening the scientific foundation of a modern economy, including charting of resources, for the future of the Soviet state.

Sevekspeditsiia, a forerunner of Glavsevmorput, operated from 1920 until the mid-1920s. Its major results were four Kara Sea rescue operations to reestablish economic and political ties between Siberia and the central industrial region of the country and thus fulfilled the Bolshevik determination to tie the city and the countryside closer together. Its goals included establishment of regular trade between Europe and Siberia, with finished goods being exchanged for Siberia raw materials. It secured grain and bread from Siberia to fight a growing famine in the Volga that killed two million peasants; the American Relief Administration, organized under future president Herbert Hoover, operated in Russia from 1919 to 1923 and saved millions of people from starvation.[52] Sevekspeditsiia fostered scientific research in the Arctic basin; conducted surveys of islands and the coast; contributed to the development of such resources as forests, fish, reindeer, salt, oil, and peat; pushed the expansion of marine shipping; organized salting, canning, drying and smoking, refrigeration, and small processing facilities; and underwrote the manufacture of fish and mammal byproducts. Although running on a shoestring budget and small in terms of personnel, Sevekspeditsiia involved a huge program of research, supply, and economic activity from Arkhangelsk to the Enisei delta and attracted such leading specialists as Aleksander Fersman, Nikolai Knipovich, and Rudolph Samoilovich in its efforts. In all these ways, Sevekspeditsiia's many operations foreshadowed the vast empire of Glavsevmorput, Stalin's administration for Arctic conquest, in the 1930s.

In those days, before the advent of the Stalinist system in whose bureaucracies money moved at the speed of a Pamir mountain glacier, weighted down by dozens of forms and receipts, each signed and stamped by officials, bureaucrats, and secretaries, the blessed individuals who secured funds from a nascent Bolshevik bureaucracy might see them appear suddenly. In the spring of 1920 Samoilovich and two other scientists went to the State Bank on Tverskaia Ulitsa (now Gorky Street) to claim two bags stuffed with fifty million rubles, released to them according to a resolution of the presidium of Supreme Economic Council to organize Sevekspeditsiia. Sevekspeditsiia was based in four cities: Petrograd, Moscow, Arkhangelsk, and Vologda. In Petrograd it occupied offices at the Geological Museum of the Russian Academy for its academic council and publication activities. In Moscow Sevekspeditsiia established what appears to have been a lobbying office to deal with the Supreme Economic Council's Scientific-Technological Department and other government institutions at 7 Mamonovskii Alley. In Arkhangelsk, the explorers established quarters at 35 Pinezhskaia Street focused on matters of practical planning, and in Vologda at 9 Struggle Square they opened an agency to acquire the machinery, equipment, and instruments.[53]

While the frightful economic conditions of the Bolsheviks' "War Communism" (commandeering of supplies, confiscation of grain, militarization of labor, plummeting production) hindered its first efforts, Sevekspeditsiia grew to coordinate all scientific research and industrial trade in the Arctic and Siberia. Already in the summer of 1920 the expedition worked the shores of the White Sea, the Kola Peninsula, the Murmansk shores, the Iugorskii Strait, and the basin of the Pechora River. The ships worked extensively in the short summer season from July to September from Iugorskii Strait to the Kara Gates and to the Yamal Peninsula. It commenced activities very late in its first year because it was nearly impossible in conditions of civil war to secure supplies, food, fuel, ships, and barges in spite of resolutions of government. By 1921, some twenty-three detachments with a total of 400 men carried out work under its auspices in mining-geology, geographic, biological, ethnographic, and economic-statistical regions of activity and thereby supported the recovery of the economy. As soon as allied intervention had ended, Sevekspeditsiia acquired the *Del'fin* trawler, the *Sharlotta* and *Nadezhda* schooners, and the unfinished *Persei* that would become a research vessel.

The first Kara expedition, in the absence of a money economy, was organized in 1921 on the basis of exchange and barter of goods; it was large scale and broad profile with industrial goods from Europe being exchanged for raw materials of Siberia, including grain. In April 1921 the Council on Labor and Defense approved 1.1 million gold rubles for emergency purchase of fishing equipment to facilitate fishing along the Murman coast. By May 1921 the local authorities succeeded in securing six Norwegian ships for Oblastryb, the provincial government fishing trust, two of them motorized and one a motor-sail boat, that were loaded with 15,000 tons of food. (The authorities also hoped soon to have the naval power to push foreign fishing vessels out of White Sea waters, but this would come later.) In any event, fresh and salted fish would help the residents deal better with severe shortages of meat products.[54] The first expedition was limited by insufficient capacity of barges and confusion over how to organize the entire enterprise. On top of this, government control of the fishing industry led to small catches because fishermen worried about confiscation of catch and low prices, and they struggled in the absence of supplies, bait, nets, and so on.[55]

The second Kara expedition in 1922 commenced with the principal task of shipment of food supplies from Siberia where millions of tons of grain sat that might be exported to the northwest to fight spreading famine. The struggle against famine was so critical that Felix Dzerzhinski, head of the secret police, the Cheka, journeyed to Siberia to organize freight by rail and steamer; he learned that the Enesei fleet was ill-equipped to handle regular shipments. Using the power of the Cheka, Dzerzhinski ordered four steam tugs and lighters along with sixteen barges with a capacity of roughly 400,000 poods (65,000 tons) from the White Sea to the Enesei River. The hope was to export graphite, lumber, asbestos, and other products in exchange for finished European goods. All trade was to be without customs stamps. The *Lenin* icebreaker (formerly the *Alexander Nevsky*) would lead the caravan. Shipping would be dangerous given the fact that there were only four radio stations along the route that worked twenty-four hours per day—at the Iugorskii Strait, Vaigach Island, Mare-Sale on the Yamal Peninsula, and Dikson Island. On August 1 the ships left England, loaded with 492,000 poods of machines, metal, medicines, and so on, stopped at Tromsø for supplies, and three weeks later entered the Kara Sea. Because of overloaded barges and few sailors, they arrived at Novyi Port only on September 20. The wooden barges leaked, goods spoiled, and they delivered 600,000 poods of bread and grain to Arkhangelsk, but returned to London in early October with 356,000 poods of leather, fur, and other products.[56]

In 1922 through 1924 Sevekspeditsiia carried out two more expeditions that included hydrobiological research on Novaia Zemlia in connection with the establishment of a radio station and geophysical laboratory at the Matochkin Strait. The northern end of the Novaia Zemlia archipelago had at this time hardly been explored; Vladimir Rusonov's efforts in 1908–13 produced mostly information on the general geological profile, while the eastern shore remained uncharted in spite of what researchers sensed were great geological resources.[57] In 1924, under Samoilovich, a ten-person expedition identified copper deposits and the possibility of coal on Pen'kovaia Zemlia. Sevekspeditsiia also sponsored the four-year researches of Fersman in the Khibiny Mountains near the future cities of Apatity and Kirovsk; Vittenburg and others along the Murman shore; S. L. Zernov and V. K. Soldatov in the Pechora River basin; exploration of the Upper Pechora basin in search of coal that also uncovered graphite, asbestos, lead, and iron ore; hydrobiological work in the White Sea; work on Murman fisheries; and studies of an economic and ethnographic character involving reindeer nomads of Arctic regions.[58] Sevekspeditsiia researchers offered a five-year plan of activities for the future: continued exploration of Novaia Zemlia under Samoilovich; surveys in the Pechora Basin and Northern Urals in search for coal; hydrobiological and fisheries research; and ethnographic studies, in all involving seventy-eight people including forty-five scientific workers and the rest support staff.[59] In 1924 Sevekspeditsiia became the Institute of the Study of the North of the Supreme Economic Council, later simply the Institute of the North and in 1931 AARI.

The leading personnel of Sevekspeditsiia were largely scientists trained in the Tsarist period. With the revolution, overnight, they had become "bourgeois specialists": Chairman of Sevekpeditsiia Samoilovich, with Academy of Sciences President Karpinskii as chair of the academic council, Fersman, Knipovich, and others, as well as playwright Maxim Gorky, a long-time associate of Lenin who in the 1930s put his imprimatur on Stalin's first slave labor project, the White Sea-Baltic Canal. Other members were representatives of the Russian Geographical Society, the Pulkovo Observatory, the Standing Polar Commission, KEPS, and the Murmansk Biological Station.[60] The fact that these scientists threw their lot behind the Bolsheviks means that many of them remained patriots to Russia if not to the Bolsheviks, that they welcomed more regular financial support, and that the Bolshevik leader, Vladimir Lenin, willingly supported their efforts.

Lenin and Arctic Assimilation

When tensions rose among funding and administrative bureaucracies, Sevekspeditsiia benefited from the acquaintance of Knipovich and Lenin. Knipovich journeyed to Moscow to meet with Lenin. Lenin had fallen ill, so Knipovich consulted with his wife, Nadezhda Krupskaia, told her about ongoing research on polar fisheries that were in danger because of threats from Glavryba (the Maine Administration of Fisheries), whose directors sought to confiscate the expedition's *Del'fin*, a trawler, and control research funds intended for applied ichthyology to expand their fleet. Krupskaia asked him to summarize these points in a letter to Lenin. Lenin later noted in the margin of that letter than Knipovich was "not only a scientific force of the first rank, but also without a doubt an honest man." He asked Knipovich to report to him directly from time to time with practical recommendations about the organization of scientific and economic matters and ultimately threw his support behind Knipovich and a new marine research institute.[61]

On March 30, 1921, Lenin signed a resolution of the STO (Council on Labor and Defense) to protect the fisheries that had fallen into a kind of anarchy of poaching and overfishing during the civil war. On April 6, the STO "militarized" the fisheries of the Russian Republic, including in the White Sea, to prevent poaching and increase production. Shortly thereafter, Lenin authorized 1.1 million gold rubles to buy equipment to assist in the recovery of fisheries through the Glavryba and in May signed another proclamation—in essence a declaration of the extent of the territorial waters to protect them and coastal fisheries.[62]

By the mid-1920s, the authorities had learned to use modern science and technology to reinforce Soviet claims to Arctic islands, regions, and waters—similar in the way, if not in the audacity, with which parliamentarian and polar explorer Artur Chilingarov used a bathysphere in summer 2007 during the third International Polar Year to "prove" Russia's ownership of a vast region of the Arctic Circle. On November 15, 1922, for example, the deputy commissar of enlightenment, N. P. Gorbunov, endorsed a project to strengthen Soviet ownership of Novaia Zemlia to counter the growing presence of Norwegian fishing vessels through the establishment of a permanent radio station on Mys Zhelaniia (Cape Desire) to enable a fledgling hydrographic service to gather information on ice conditions, weather, and other valuable scientific data.[63] A month later at a meeting of the Hydrological Administration, such specialists as Vladimir Vize, active in Arctic research until his death in 1954, and N. A. Kulik, who had traveled the Pechora River basin, endorsed the scientific "colonization" of Novaia Zemlia through the expansion of expeditionary forces.[64]

In January 1923 plans for the station moved slowly with the support of the Navy, the Commissariat of Foreign Affairs, the Council of Ministers, and the State Planning Administration.[65] On January 9, 1923, the deputy chair of the Russian Republic's Revolutionary Military Council and minister of internal affairs, Efraim Sklianskii, a close associate of Trotsky later removed by Stalin from all military posts and "exiled" to the United States as head of

Amtorg, an organization dedicated to acquiring American technology through purchase or industrial espionage, wrote to the Russian Geographical Society. He urged its members to develop this project given its scientific value and the political significance for the Northern Sea Route.[66]

With expeditionary and government support, Soviet fishing vessels supplanted Norwegian, British, and German ones, albeit very slowly. In the period from 1921–26, Soviet ships harvested 42,000 tons of fish, while the German and English catch grew to 140,000 tons. The Soviet trawling fleet in the Barents Sea consisted of only twelve trawlers in 1919, but fifty-four in 1933, although they lagged in power and capacity compared with other nations, and spare parts remained a problem. Yet scientists predicted with confidence that the catch would increase in the 1930s for the Barents Sea alone tenfold to some 20,000 tons on the basis of modern technology.[67]

In 1926 Sevekspeditsiia and several other committees entered the jurisdiction of the Commissariat of Trade and Foreign Trade, and in 1928 what remained of it became the newly organized North Siberia State Stock Society of Industry and Trade, or Komseveroput. As a transport, economic, and scientific organization, its goal, like those of its predecessors, was the economic development of northern Siberia including trade, the flowering of industry, in particular that connected with forestry, fish, and mammals and mineral wealth, and the equipping of the Northern Sea Route. As became more and more characteristic for Soviet organizations in the Stalin era that grew seemingly of their own accord under the pressures of economic centralization and political control, Komseveroput developed—or acquired—a graphite industry on the Kureika River; several ports; thirty-five fish processing and cannery operations; vegetable and milk operations; and by 1929 lumber exports that flowed from a Igarka, a new Soviet port. Later Komseveroput added to its portfolio reindeer and animal husbandry at a 500-hectare facility just above the Arctic Circle. Its ships gained extensive river freighting experience. Beginning also in 1929 the company employed "Komsevmorput-1, -2, and -3" airplanes for ice reconnaissance through a polar aviation administration; it sponsored expeditions of E. I. Igolkin and Iu. M Petranda along the Piasina River that enabled deliveries for the newly established Norilsk Metallurgical Combine. By 1932 the enterprises of Komseveroput employed more than 35,000 people, certainly many of them in the gulag labor camps. As a last major endeavor, Komseveroput organized the expedition of the

Sibiriakov in 1932 that tied together parts of the Arctic and underlined the crucial importance of a Northern Sea Route that would be realized through Glavsevmorput, formed December 17, 1932, while Komseveroput was liquidated shortly thereafter.[68] In its ability to acquire more and more capital from logging operations to reindeer farms to mines, put laborers to work with distant concern about their well-being, and single-mindedly apply science and technology, Komsevmorput, like Glavsevmorput after it, was like a big business, and Stalinist Arctic science was big business and big science.

The director of Komseveroput, Boris Vasil'evich Lavrov, was a Soviet workaholic. According to a journalist who met Lavrov in 1933 for an interview, the director thought only of the conquest of the polar ice. (The journalist was a special correspondent for *Vodnoi Transport* and a stoker on the *Krasin.*) He met Lavrov in Dikson. Lavrov had arrived by plane from Igarka to meet frozen guests—captains, sailors, and government representatives who had arrived aboard steamships from Western Europe and were interested in Siberian lumber, fish, and reindeer. Dikson, an island town in the Enesei delta, became an important stopping and freighting point along the Northern Sea Route and also the site of a crucial radio and hydrometeorological center established in 1936.[69]

According to the young journalist, Lavrov was a hands-on kind of administrator, not someone likely to stay in an office, and willing to make difficult decisions when nearby party officials and secret police agents were poised to pounce on any indication of "wrecking." The indigenous people of Yakutia recognized Lavrov's capable leadership; the Nenets and Dolgans who lived on the Taimyr Peninsula called him the "Bolshevik Boss." The Boss had great obligations in the "socialist reconstruction" of the Far North especially after the promulgation of the First Five-Year Plan. The huge Krasnoiarsk region, at some 900,000 square miles, and even larger Yakutia (the Sakha Republic) at 1.2 million square miles, together 30 percent of Russia's total land mass, but in 2012 with four million people most of whom inhabited southern regions, were home to ore, fossil fields, rivers, and forests. Lavrov presided over the development of these resources, bringing "the sounds of the saws and steam ships and the rev of airplane engines" to the tundra, from Omsk to the Ob delta and from Krasnoiarsk to the Enisei Peninsula and farther into the Kara Sea.[70]

Like many young men in the late nineteenth century, Lavrov (1886–1942) was attracted to Marxism because of the intransigence of the Tsarist regime in the face of pressure for necessary reforms and joined leftist groups as a student. Lavrov received training at the Rybinsk Seminary and Petersburg University and participated in demonstrations against the government during the 1905 revolution. He was arrested and exiled to the north. Upon his release he joined the Bolsheviks. During the civil war, Lavrov carried out underground work against the interventionists and White Guardists in Odessa on the Black Sea and was again arrested in 1919 by the Whites. After the civil war, Lavrov occupied a series of positions involved in the acquisition and distribution of scarce resources: he was a food commissar in Viatka, headed the Central Asian division of Voentorg, an economic organization for military officers, was trade representative in Afghanistan, and was not quite fifty years old when sent to the northern latitudes to organize Komseveroput. He himself worked the float on the Angara and at the docks in Igarka unloading lumber on *subbotniki;* subbotniki were extra workdays, usually on the weekends, that initially tapped the revolutionary enthusiasm of workers who were "building socialism," but eventually became an uninspired state-sponsored effort to get free labor to clean streets and pick up yards. Lavrov impressed his staff on one occasion when, on a return trip from Tiksi, powerful ice halted the caravan not far from the Vilkitskii Strait. But rather than focus solely on local concerns, Lavrov released the *Krasin* icebreaker from local duties so it could steam to the Leningrad port before the deep onset of winter, and he determined to winter with three transport ships (the *Volodarsk, Stalin,* and *Pravda*) at the deserted Samuil Islands (later the Komsomolskaia Pravda Islands). It was he who pushed for winter aerial ice regime reconnaissance into a modern era to avoid unnecessary zimovka.[71]

With the second five-year plan, the government redoubled efforts to establish a regular Northern Sea Route from the White Sea to the Bering Strait. This required ensuring safe passage along the Arctic shore in Krasnoiarsk and Yakutia, regions essentially without roads, with rivers open only half the year for transport of supply, their deltas frozen ten months out of the year, and only one radio station on Cape Cheliuskin along the entire distance from Dikson to Tiksi to ensure communication about weather, ice regime, depths soundings, and so on. Lavrov was determined to bring a bit of the urban to the Arctic Circle. He ordered a brick factory built and then raised four-story buildings on permafrost. For Lavrov this was not a utopian thought; he believed that Igarka, as a contemporary polar city, must have stone houses.[72] According to one source, a British MP in the Enesei region wrote on his return home, "Igarka has united Siberia with the seas of the world, and now the Enisei flows 1,000s of miles further . . . to Leningrad, Hamburg, Rotterdam, London and New York." That is, through Komseveroput the world's great trading ports became more proximate to the Siberian shores with increasing freight making its way through the ice on regular trips from the Ob and Enesei deltas.[73]

With the creation of Glavsevmorput, Lavrov became member of the collegium of Glavsemorput and headed its administration to develop the economy and culture of people of the north. He next organized the massive geological trust Nordvikstroi that worked on the Khatangskii Peninsula—and opened a coal mine. But like so many others who had given their lives to Arctic assimilation, Lavrov fell to the Terror. He was arrested on August 9, 1939, and accused of anti-Soviet activity—participation in a right-Trotskyite organization—confessed to working with a series of other individuals who were also carted off, and he himself probably was shot or died in camps in 1942. After the twentieth Party congress the party recognized his contributions to Arctic development by naming a steamship *Boris Lavrov* to operate in and out of Tiksi. But in modern Igarka, which he founded and built, his name does adorn any street or building.[74]

Indigenous People in the Russian Arctic before the Revolution

Many of the early explorers paid so much attention to the ice floes and cold, to schools of fish and lumber, to ore and transport, and after the Russian Revolution to the Reds and Whites that they lost sight of the people who inhabited northern latitudes, the Komi, Saami, Nenets, and others, the twenty or so indigenous groups, many of whom were nomadic reindeer herders, plus the Pomor, the coastal people who made their precarious livings fishing and trapping. Many of the expeditions created to assimilate the Arctic had little place for these people except as curiosities and sources of horses or food. They—or at least their reindeer—would become crucial to the five-year plans to feed and clothe urban workers. But to backtrack a bit, whom did the early explorers encounter on their expeditions?[75]

The Nenets, Saami, Komi, and others and their lands, their reindeer herding activities, like mineral wealth and other resources of their region, were not systematically exploited, let alone studied until the early twentieth century. Nor did the brave—and often unfortunate—travelers to the tundra have a real sense about climate or terrain or how the people managed to survive in it. They discovered a bleak existence spread out over a huge Arctic desert of tired, hungry, and dirty people who suffered from a variety of medical maladies and lived on the edge. The most sympathetic of the explorers eventually opened the eyes of the educated Russian public to another world.

S. V. Kertselli (1869–1935), a veterinarian who served as the Arkhangelsk province veterinary inspector, after the Russian Revolution was a member of the Committee for Assistance to the Minority Nationalities of the Central Executive Committee (the Committee of the North), and then was director of the Institute of Reindeer Husbandry,[76] spent twelve months over the summers of 1908 and 1909 "cut off from the real world," exploring the Bolshezemelskaia Tundra located today in the Nenets Autonomous Region, traveling more than 1,000 kilometers on foot.[77] Kertselli and two colleagues surveyed the tundra at the request of the Veterinary Inspectorate of the Tsarist Ministry of the Interior to address the periodic wasting diseases of reindeer that had attracted the attention of the government. In the first summer he focused his attention on the reindeer herders; in the second he began more systematically to gather meteorological data at the request of the government with an eye toward evaluating the possibility of the "colonization" of the tundra. He remained skeptical of colonization on two counts, first the poverty of the infrastructure and second the terrifying climate. Kertselli referred to the first stage of his journey from Arkhangelsk, a bustling provincial mill town on the shore of the White Sea, to the Pechora River as "tormented wanderings," a trip made worse by the absence of weigh stations and private horses along the route; before 1911 the Tsarist postal service had not considered it necessary to secure horses for its deliveries in out-of-the-way places. It took Kertselli three weeks in 1908 and ten days in 1909 to travel 780 versts (one verst is a little over one kilometer) to

Ust-Ilimsk, arriving in mid-March.[78] From Ust-Ilimsk Kertselli determined to travel to the villages of Kolva and Mokhcha where herders had gathered with thousands of head of reindeer that he could examine and inoculate. As his trip progressed he faced the danger of sudden ice floes on rivers and streams.[79] He was relegated to riding unfamiliar and uncomfortable reindeer across the remote taiga; he had absolutely no contact with the outside world. "It was as quiet as a cemetery. All around there was absolutely no one . . . here was something entirely new [to me that I had] never before experienced; here were complete emptiness and I."[80]

Toward the end of the first expedition in late September 1908, Kertselli and company nearly died when they moved along the Pechora River in a rowboat suddenly to be struck by a blinding blizzard and ice storm. They were forced to take shelter on an island through the night; they awoke bound in ice. To stay a moment longer threatened them with freezing. The oars repeatedly froze in the ice so that every ten minutes or so it was necessary to bring an axe to release them.[81] They made their way in the absence of charts by seat-of-the-pants orientation, compass, and advice from passing reindeer herders.[82]

To make travel easier they thinned their supplies and made their tent or *chum* (closer to a teepee than a tent) smaller by half, having also gotten rid of the second covering so that there was hardly any protection from weather, and in which one could only lie down, not sit, for their heads were in the smoke of the fire. For the entire trip, expecting support from reindeer herders, they carried with them only thirty pounds of dried bread crusts, twenty pounds of sugar, seven pounds of tea, twenty pounds of millet and rice, twenty pounds of butter, and two packages of dried greens. They dispensed with several jars for specimen collection of plants and insects. Thence they marched from small "village" to small village, each consisting of no more than two houses, and often lived "in tedious inaction, waiting for the arrival of workers with reindeer." Ultimately they moved to the task of vaccinating the animals, but only several hundred, for the reindeer did not approach the *chum*. Rather the men had to go out in difficult conditions, again on foot or on reindeer, into the vast, swampy, hillocky tundra, sometimes covering ten or even twenty kilometers. Over time their own reindeer began to fall from wasting syndrome, and they worried about being left standing in the tundra.[83] It took Kertselli virtually the entire summer to gain the confidence of the reindeer herders. Toward the evening they sought out a dry spot, which was not always possible even given the exceptional knowledge of the tundra of their hosts. In the summer, water seemed to rise up out of even dry ground and flow everywhere.[84]

Kertselli admitted the difficulties in estimating how many reindeer the nomads had and concluded that official estimates in the office of the governor of the province had no foundation. Given 250 to 300 teepees in the Bolshezemelskaia tundra, with 1,000 to 2,000 head among the larger herders, he determined 350,000 to 400,000 reindeer. When the Timanskaia and Malozemelskaia were added, this meant at least a half million reindeer, or nearly twice as many as according to official statistics.[85] The massive size of herds was crucial given the epidemics that decimated them—Kertselli's reason for being in the tundra—as well as hunters who poached and climatic conditions. If the snow was too deep or there was a thaw followed by deep frost that killed most plants, the herds were forced to dig for lichen and starved by the thousands.[86] (Soviet officials determined to carry out "passportization" of reindeer to be certain of their numbers and as a hedge against cheating by the herders, whom they did not trust.)

Kertselli's commentary on the possibility of colonization revealed great skepticism toward the economic promise of the taiga. He remarked that the pine "forest," such as it was, was of quite low quality, with trees in different stages of dying. He estimated that perhaps 15 to 20 percent of trees were shrunken, and on the northern-most islands around 35 to 50 percent of them were, like the reindeer, wasting. The fact that peat ran to a depth of two meters and likely deeper explained the moisture of the land in seemingly dry spots. The further north one moved, the more the tundra turned into swamp and lakes, creating an endless horizon and an unchanging landscape. A special expedition of the forestry bureaucracy in 1908 and 1909 that worked in Pechora district indicated even in its southern parts that the quantity and quality of the forest was inadequate for economic development.[87]

In the second year Kertselli organized a systematic series of meteorological observations, although with "inadequate instruments," both to evaluate the potential for "colonization" and to understand the impact of climate on the Nentsy. The researchers measured temperature and barometric pressure three times a day, cloud cover, wind speed, and direction. The average temperature for May hovered slightly above freezing before rising to 5.5°C in the first half of June, to 6.8°C in the second half, and into double digits for July and the first half of August. By mid-September, the average temperature had dropped to just above freezing again; there were also several freezes in the summer and snow even in July, although it melted quickly.[88] The night temperatures fell rapidly even in the summer months. The sun's rays were so weak that you could look directly at it with an unprotected eye. Another discouraging factor was the absence of clear days, but constant, raw winds, low temperatures, soaking rains, and rain with snow in the second half of the summer. All this made "the climate of the tundra not only unpleasant but unhealthy. It is not surprising therefore that a significant number of reindeer herders, who because of their lifestyle must spend a significant amount of time under the open sky, suffer from rheumatism, the symptoms of which are observed often even in young fellows."[89] Although he worked in the tundra until his death in 1935, I have not found any mention by Kerstelli of the impact of colonization under the Soviets on health of the indigenous people.

Other inhabitants of the tundra made their living through hunting and fishing, although the latter activity was much less important. These people traded with reindeer herders and also stored the herders' goods for the upcoming winter including sleds, dried bread crusts, flour, winter clothing, and heavy winter parkas. Storage was important in generating a small amount of income with payment often in meat and hides. Hunting in recent years had lost importance as the price and quantity of white partridges declined. The hunters had slaughtered birds on the upper reaches of rivers where habitat was more forested, with the feathers used for pens and down and the meat for food. In the tundra Kertselli did not see one mallard and only a few ducks, but noted that hunters found weasels and foxes, which were not plentiful but whose fur was valuable. The money economy was less important than trade of animal products for kerosene, soap, gunpowder, tea, metal instruments, and yeast. Fishing, a major interest of Soviet economic organizations from the 1930s and fairly productive if profligate by the 1950s, served largely only personal needs; Kertselli assumed a low fish population.[90]

Weak and unpromising agricultural development also discouraged colonization. Animal husbandry was limited to milk products because of the low—and declining—quality of meadows and therefore the small choice of grasses. In the tiny settlements of the tundra he counted in all 123 farm animals with twenty-two horses and thirty-four cows. Meadows could not be improved because of shortages of labor and complexities of reclamation—although in the postwar period, Soviet specialists certainly tried to make nature cooperate, in some cases with slave labor. The meadows were of little interest in the summer when inhabitants of the tundra struggled to catch up on chores—buying and selling winter products and provisions, fixing houses and outbuildings, fishing, getting traps ready, hunting goose and duck. Local folks had success only in growing potatoes and knew from experience nothing else would grow. As they explained to Kertselli, "Tried it, planted it, and the frost comes."[91] Based on admittedly superficial observation of the tundra, the swampiness, and the peaty, low quality of the soil, Kertselli doubted there was any basis to hope that reclamation or melioration measures would work: there was simply too much need for fertilizer—manure—to permit farming or expansion of meadows.[92]

These observations—the nature of life on the tundra, the climate, the remoteness, the seeming impossibility of commerce—convinced Kertselli of the impossibility of the rapid colonization of the tundra. The Soviets would dispute these conclusions in words and actions, subjugating the Arctic with a vigor seen nowhere else in the world and incorporating the previously invisible Nenets, Saami, Pomor, and other local people into their vision.

Surprisingly, when the Bolsheviks took power they applied a progressive ethnic minorities policy that Joseph Stalin developed in 1914. At the request of Lenin, Stalin, who abandoned this policy in the 1930s to force the minorities to modernize, to industrialize, to collectivize their reindeer herds, to become Soviet citizens tied to distant markets and subjected to Marxist ideologies of class struggle, wrote "Marxism and the National Question" (1913), which supported the right of self-determination and promised the aid of the new socialist regime in preserving national difference, language, and culture. Toward these ends, in sharp distinction with the Tsarist regime that had ignored the destitution of the minorities and in essence belittled their cultures, the Bolsheviks supported ethnography, linguistic, and anthropological research to understand and preserve those ethnicities. Sevekspeditsiia was involved in these earlier efforts, as was the Commissariat of Minority Nationalities, Narkomnats.

Narkomnats (1917–24) grew out of Sibnats of the Siberian Revolutionary Committee, the leading government body in the eastern RSFSR from 1919–25. Members of Narkomnats believed in the need to assist "those people over there" (*tuzemtsy*—Iakut, Buriat, Ostian, Altai, Tatar, Kirgiz, Chuvash, and others, some twenty groups of people) to advance from a precarious cultural, economic situation. They had been ignored or mistreated by the Tsarist government, were tied to fishing and hunting, yet were at constant risk of poverty and starvation. Narkomnats organized the Committee of the North under P. G. Smidovich. By 1934 Soviet officials had established nine national districts, seventy-seven national regions, and more than 500 national councils to support the "little people."[93]

Peter Germogenovich Smidovich (1874–1935) was active in nationality politics in the USSR in the 1920s and supported the establishment of a Jewish autonomous region with Yiddish as its official language through work on Komzet (an organization purged in 1937 and then liquidated in 1938).[94] From his student years Smidovich was interested in conservation and studied with some of the pioneers of the environmental movement of Russia at Moscow University. He joined the Bolshevik government after the revolution. During his tour as chair of Committee of the North Smidovich oversaw the study of the daily life and needs of the minority nationalities, organization of cooperative organizations, medical and public health service, strengthening of the local economy, defense of their rights, and regulation of trade.

One of the achievements of the Committee of the North was the creation of a higher educational institution precisely for the small peoples, the Institute of the People of the North. The institute established a council to create written languages and textbooks, published *Works* in ethnography, history, economics, linguistics, and folklore, and printed the first publications in Nenets, Enetsk, Nganasansk, Saami, Khanty, and Mansiiskii languages. By the late 1920s the Institute for the Study of the North had acquired a modest icebreaker and shipping fleet—or at least the ability to pursue research and trade during the regular navigation season. This enabled the authorities to score a number of propaganda coups in the world of Arctic exploration. In 1928 the *Krasin, Malygin,* and *G. Sedov* saved survivors of the *Italy* hot air balloon expedition. In 1929 the *Sedov, Malygin,* and *Lomonsov* carried out oceanographic work on unstudied areas of the northern reaches of the Kara Sea and Franz Josef Land and also sent four geological expeditions to establish on Franz Josef Land the northernmost geophysical laboratory. The 1930 expedition on the *G. Sedov* in the Kara Sea discovered several new islands including Shmidt Island in the Severnaia Zemlia archipelago (consisting of Bolshevik, October Revolution, Komsomol, and other islands), the northernmost archipelago in the world, first observed in 1913 but explored only in the 1930s above 81 degrees. In November 1930 the Institute was transferred from VSNKh into the Committee of Scientific and Higher Educational Institutions of the Central Executive Committee and renamed the All-Union Arctic Institute whose director was Otto Shmidt. Unfortunately, in the mid-1930s the nationality policy of the USSR changed radically with the institute losing all of its publication functions. In 1936 it was liquidated and its activities subsumed into Glavsevmorput.

The well-known book and Akira Kurowsawa film of 1975, *Dersu Uzala*, explores the conflict between old and new, traditional and modern, local, indigenous, and national that would play out in the Arctic Circle. Building on his lifelong travels in the Far East, Vladimir Klavdievich Arseniev (1872–1930) turned to fiction late in his short life to depict the vast tracts of seemingly unspoiled nature in riveting accounts. Arseniev, a well-known traveler, geographer, and ethnographer, grew up in Petersburg, dreamed of travel from an early age, and found inspiration from the geographer P. P. Semenov Tian-Shan and other naturalists. Graduating from military school, Lieutenant Arseniev was stationed in the Far East, arriving in Vladivostok in 1900, and began extensive study of the region. He joined the Society for the Study of the Amur region and over the next years actively pursued zoological, botanical, and ethnographic research to support the society.

During the Russian-Japanese War (1904–5) Arseniev led a scientific expedition to the Sikhote-Alin Mountains, the shoreline of the Gulf of Olga, and the headwaters of the Ussuri and Big Ussurka River. His field trips, which continued after the war, provided firsthand insights for *Dersu Uzala*. Arseniev wrote a large number of books that described the surroundings. He produced botanical, geological, zoological, meteorological, and ethnographic notebooks that included detailed advice on how to survive in the wild, take care of soldiers and others in your charge, keep your matches dry, and so on. After the revolution he taught in universities in Vladivostok. In

1926–27, at the request of the Bolshevik government and Academy of Sciences, Arseniev undertook scientific study to support economic development in the Far East, in part to evaluate the utility of building railways to secure the region. In 1930 he took ill on expedition, returned quickly home, and died in Vladivostok.[95] This likely saved him from the labor camps, for he was not a supporter of Stalinism. Arseniev's widow, Margarita Nikolaevna Arsenieva, was arrested in 1934 and again in 1937 after being accused of being a member of an underground organization of spies and saboteurs allegedly headed by her late husband. She was sentenced to death, while Arseniev's daughter, Natalia, was arrested in April 1941 and sent to the gulag.[96] Many other polar explorers perished.

In *Dersu Uzala* (1923) Arseniev describes the Ussuri region in a story that follows the life of Dersu Uzala, a native Goldi who becomes the author's guide and soon dear friend. The author, a captain in the employ of the Tsarist army, is charting the significant animal, fish, forest, and other resources of Ussuria with a detachment of troops. Without Dersu's help, the captain and his men would hardly have understood the forest and might have perished. Thus, Dersu's "local knowledge" is more powerful than the science the Captain brings to the task of subjugating undeveloped resources in the Far East. Like other indigenous people, the Goldi saw their lifestyle threatened by settlement of Russians and Chinese and by the growing tensions between Japan and Russia to control the Far East. The Goldis' territory shrank as they were pushed into new areas of settlement by the Russian, Japanese, and Chinese colonial powers—to areas with inadequate fish or game. The Goldis and others lost not only territory, but population and way of life.

Prelude to Exploitation

As the 1920s ended, explorers and scientists had succeeding in imposing a kind of regularity over northern climes. They understood better the vast expanse of Arctic resources and possibilities. The New Economic Policy, an easing of state control over all but the "commanding heights" of the economy that permitted small-scale entrepreneurship, facilitated recovery from War Communism by 1926. A kind of cultural renaissance unfolded in many fields of art, music, and culture. In this environment, scientific research and development in many fields expanded, including in areas of oceanography, hydrography, geophysics, and chemistry vital to exploration of the Arctic.

The primary focus of party officials and government bureaucrats had been on the economy to pull the new socialist republic out of the war years. They recognized that economic development of Arctic regions would embellish the nation's claim to vast polar resources and secure them militarily. The first expeditions were tied to resurrecting trade between Siberia and the European USSR. By the time NEP was in full flower, they sought more generally to chart and exploit resources along the Northern Sea Route, and they supported the establishment of new ports, trade centers, and meteorological and hydrographic observation points toward that end.

Scientists and polar explorers had excitedly reported on the extent of resources in the polar regions, and not only lumber and coal, but rare metals, various ores, and even possibilities to exploit fish and animal resources to feed the nation's growing working class. They had established the groundwork for much more systematic study of the high latitudes including the roughly 5,000-kilometer shoreline contiguous to the Northern Sea Route to the central Arctic basin. While harsh climate prevailed, they determined that the richness of fossil fuel, mineral, and other deposits justified exploitation. They published detailed maps and charts especially of the Kola Peninsula, the tundra along the White Sea, and Siberia river deltas. The government supported their efforts in a series of new research institutes, expeditionary forces, and publications. The scientists recognized, however, that their findings were preliminary and superficial given the vast expanse of the Arctic, a seemingly empty space filled with nomadic people.

Further evaluation of the extent of Arctic riches and their exploitation—and further marginalization of reindeer herders and nomads—lagged until scientists and explorers could overcome technological backwardness and truly establish systematic and regular expeditions. Their reliance on older, underpowered ships and the inability to get spare parts for them, the shortage of icebreakers, the inability of the nation to manufacture its own icebreakers and modern research vessels, and the lack of full commitment on the part of the regime to costly and uncertain Arctic exploration handicapped their efforts. With the rise to power of Joseph Stalin this changed dramatically, for Stalinism was not only an authoritarian polity, but an economic policy centered on rapid industrialization. A firm policy emanated from Moscow and Leningrad from 1930 to transform the Arctic into a planned economic entity, subject to the political decisions of center, with science and technology serving central dictates and with slave labor contributing to the imposition of modernity on the environment.

2

Neither Cod nor Coal

Some get pies and crumpets, others get fingers and lumps.

—Attributed to JOSEPH STALIN

Soviet leaders sought to industrialize production across the Arctic Circle. They invested extensively in smelters, mills, and mines. They used icebreakers and airplanes to identify and tap resources that were effective both symbolically in demonstrating state power and physically in enabling economic growth. They supported the rapid expansion of the scientific enterprise in the industrialization effort. Yet they faced endless problems of technological lag and were forced to rely heavily on ideological exhortation and gulag slave labor in the massive undertaking.

When Stalin rose to power in the late 1920s he and the Communist Party leadership introduced significant changes in virtually all areas of Soviet life from the political system and economy to science, education and culture, and the family. The major purpose was to transform a still predominantly agrarian society into an industrial power. Through five-year plans the Party abandoned any semblance of capitalist organization of production that persisted in the New Economic Policy of the 1920s, and it accelerated the creation of heavy industry with a focus on metallurgy, mining, construction, railroads, and other transport, with housing, health care, and other sectors of the economy starved for investment. A second goal was to collectivize and modernize agriculture, in part to provide workers—in many cases former peasants—for the burgeoning factories. Marxist leaders believed that class struggle shaped both the domestic and international political arena, that only a socialist society based on socialist industry could withstand the impending assault from capitalist powers that surrounded the USSR, and that the peasant, who they rightly believed was hostile to Bolshevism, had to be forced to abandon small-scale family farming for large-scale mechanized agriculture. Identifying a thin layer of the peasantry as exploiters, the *kulaks,* they carried out violent, even murderous de-kulakization in the countryside. The result was essentially war on the countryside in which millions of peasants were exiled and millions died of starvation in Ukraine and Kazakhstan.

On the cultural front, the Stalinists promoted class war, too. This meant the transformation of relatively autonomous scientific institutions into tools of state power; the introduction of a rigorously ideological component into research and higher education that stressed class origin of students and specialists; development of curricula and foci that emphasized allegiance to the Party and vigilance against perceived enemies; organization of the arts, music, theater, and other media as powerful means to shape the Soviet mindset and reality; and even the redefinition of the family as something more traditional so that it might be one of the few sources of stability in society. In all of these ways, Stalin's revolution from above may be considered the completion of the Bolshevik seizure of power in 1917.

The Stalinist revolution had a long-lasting impact on the Arctic as a physical space, as a region to be fully incorporated within state economic and military programs, and as an ideological construct. This was to be expected given that central policy makers pushed rapid industrialization and collectivization into polar regions as well. They viewed the north both as a border to be secured and a source of rich natural and mineral resources that must be tamed in short order. They considered the north to be nearly devoid of people, abandoning earlier Soviet notions about the need to protect the cultures of nomads, seeing them as backward and in need of modernization; they were something like polar peasants who, like those in agricultural regions, had to be coerced to participate in the five-year plans.

Party officials and managers established scores of massive smelting, transportation, construction, power production and animal husbandry and fishery operations. Party organizations spread into each village, town, and city unit, into each mining and forestry operation, and into the tundra through cultural bases to attract the nomads to communism. They filled the Arctic with new people—gulag prisoners; devoted, militant communist volunteers; exiled kulaks; and others who were dedicated to the tasks of building industry and infrastructure across the taiga and tundra. They brought to bear redoubled efforts in scientific research and development to explore, understand, and subjugate resources and to establish a regularly operating Northern Sea Route. Eventually Arctic conquest acquired ideological along with economic significance—along with such so-called hero projects of the five-year plans as the White Sea-Baltic Canal, the Magnitogorsk iron and steel mills, the Dniepr River hydroelectric power station, and dozens of other dams, mills, canals, factories, and entirely new cities built in short order in the 1930s. Writers, film producers, party officials, workers, managers, and others celebrated Arctic conquest—including a series of aviation feats—proudly and publicly, while hiding or ignoring the great human and environmental costs of the endeavors.

Toward the ends of resource exploitation, scientists and engineers joined in a series of new or transformed institutes and bureaucracies dedicated to Arctic conquest. Soviet polar explorers systematically sailed and dangerously drifted into higher and higher latitudes. They discovered both more islands and other geological phenomena and more uncertainty. The industrialization effort included significant investments into expeditions, radio stations, icebreakers, and the like. The scientific foundation was anchored in the Institute for the Study of the North that became the All-Union Arctic Research Institute (today the Arctic and Antarctic Research Institute, AARI), while the Main Administration of the Northern Sea Route (hereafter Glavsevmorput) acquired responsibility to organize, administer, extract, smelt, transport, and explore the entire vast region from Murmansk to Kamchatka and to modernize the people within it.

Like other nations that used science and technology to reorder the social and natural world for state purposes without feeling encumbered by the past, the USSR embraced high modernism.[1] The Soviet effort in the Arctic set the stage for other high modernist plans "simultaneously [to] overcome the hostility of a northern environment and catapult native northerners into conditions of modern living," especially during the Cold War.[2] From the point of view of a high modernist society, three components of Arctic development and security were lacking in the USSR in the 1920s: government and economic organizations and institutions, modern science and technology (icebreakers, airplanes, roads, telephones, machinery, and equipment), and people dedicated to assimilation of resources and security of borders. In the Stalin era, Soviet leaders provided these things in burgeoning bureaucracies and large-scale projects; in systematic exploration; and in voluntary and coercive programs to settle northern latitudes. The secret police gulag empire of canals, railroads, hydropower stations, mines, and smelters completed this Arctic empire.

One of the striking aspects of modern technological systems is the extensive momentum they acquire. Whether socialist, capitalist, or mixed systems, large bureaucracies become adept at finding new projects and securing funding to justify their existence after their original purpose has been fulfilled.[3] This technological momentum is logical, given the experience, personnel, and machinery and equipment they have accumulated, the reluctance to lose sight of a mission, the unwillingness to fire colleagues who may have become friends, and the ability to frame new tasks. Yet it may be that in the Soviet context the momentum was ineluctably greater than in capitalism systems. As they developed, large-scale Soviet organizations acquired responsibility not only for the specific projects and outputs stipulated in plans, but for housing and feeding their employees. They lobbied for and received funds in their budgets to build housing stock, hospitals, stores, and schools. On top of this, in the USSR there could be no unemployment—in sharp economic and ideological distinction to the western democracies where the Great Depression saw millions of men and women hungry, without work, standing destitute in long lines for soup and bread. When men finished jobs to build dams along the Columbia River or for the Tennessee Valley Authority, they were done, period, and had to move to new jobs wherever they might find them. The Dust Bowl sent hundreds of thousands of families to California in search of a job, as John Steinbeck described in *The Grapes of Wrath*. In the USSR, there would be new projects to keep the workers busy building socialism, who in the evenings might relax in barracks and dormitories, or the lucky ones in apartment buildings, contentedly reading *Pravda* and learning of Stalin's most recent achievements.

Soviet organizations grew inexorably, stoking cities with workers to fulfill industrial ends, moving through the countryside to transform it into a socialist wonderland of smoke-belching factories, beehives of mines, clockworks of rivers with dams and irrigation networks, extensive transportation "magistrals" that extended into the taiga and tundra, and massive teakettles to produce electricity and heat. When organizations had finished one project or grew too large or when the Moscow visionaries determined to pursue another heroic project, they hived off thousands of workers to create new construction, mining, transport, and other organizations that themselves were transformed into massive hero projects. The heroic projects ranged from the far-fetched to the absurd, from the hubristic to the outrageous, but through coercion, propaganda, exhortation, the use of gulag labor, the *propiska* (a residency permit required of all workers to control their movements), and the internal passport, those Moscow visionaries were able to push these projects, whether financially foolish, environmentally bankrupt, and socially disruptive or not, and the workers and equipment for them to the ends of the Soviet earth.

Glavsevmorput—the Administration of the Northern Sea Route—rivaled such other Soviet technobureaucracies with tens of thousands of employees and millions of rubles of capital as Gidroproekt, the hydroelectric and riverine engineering firm that, on a foundation of gulag labor, rebuilt the Volga, Angara, and other rivers; Minsredmash (the Ministry of Middle Machine Building), the nuclear enterprise responsible directly and indirectly for hydrogen bombs, nuclear submarines, reactors, peaceful nuclear explosions, and Chernobyl that emanated from a dozen closed military cities; or today's Gazprom that produces 15 percent of the world's and 78 percent of Russian gas output. Glavsevmorput's employees sailed icebreakers, herded reindeer, felled timber, organized aerial feats across the North Pole, created written languages for illiterate indigenous people, produced sausages, and wintered on drift ice.

The Main Administration of the Northern Sea Route

In November 1930 in light of ever-expanding research programs and numbers of personnel, the All-Union Arctic Institute opened under Otto Iulievich Shmidt, at the same time as energetic measures to develop a regularly operating sea route had commenced. In fact, if in 1920 ten steamships plowed the Kara Sea from the Ob River delta to Arkhangelsk with more than 10,000 tons of bread, then by 1930 Kara operations had reached 150,000 tons of freight with the help of icebreakers; aerial reconnaissance of ice; radio stations; and multidisciplinary studies of the hydrometeorological factors. This made shipping less an expedition and more a normal sea transport operation.

Underlying the point, in 1932 the crew of the *Sibiriakov* carried out the first entire Northern Sea Route in one navigation. Hence the possibility of annual navigation seasons was no longer an "illusion" as Nansen had warned, but a reality.[4]

Unlike many of his staff in Glavsevmorput, Shmidt managed to avoid the pitfalls of Stalinism. Shmidt (1891–1956), who was born in Mogilev, Belarus, attended schools in Mogilev, Odessa, and Kiev before studying mathematics at Kiev University. After the abdication of the Tsar, Shmidt worked in Kiev and in Petrograd in the effort to secure food deliveries for starving cities and transferred to Moscow to continue this work, at the same time becoming a mathematics professor at Moscow State University. From 1921 to 1924 he headed the State Publishing House and worked on the *Great Soviet Encyclopedia*, becoming its editor in 1925. Around this time Shmidt fell ill from tuberculosis, sought recovery in the Tyrol Mountains of Austria, turned to mountain climbing (later leading expeditions in the Pamir Mountains), and acquired a fascination with ice and snow. In 1929 he led an expedition to Franz Josef Land. From this point on, on icebreakers and airplanes, Shmidt explored the north and supported expansion of this Arctic empire as director of Glavsevmorput, itself an empire of the Arctic; in some circles he was called the Ice Commissar.[5]

The creation of Glavsevmorput by proclamation on December 17, 1932, marked the confluence of several factors. One was the approaching second International Polar Year and the intention for Soviet scientists to occupy a leading position in the effort. The USSR would be "the first among all nations of the world," with researchers on the *Knipovich* the first to sail north of Franz Josef Land and those on the *Persei* carrying out systematic study of Greenland Sea. Other expeditions in the Kara and Laptev seas and in Chukotia-Anadyrsk Region produced significant scientific results.[6] A second factor was the increasingly centralized administration of all aspects of Soviet life from industry to education, from cultural endeavors to science.

Glavsevmorput grew rapidly to encompass the entire Northern Sea Route and much of the territory contiguous with it. According to a December 1932 government resolution, it operated from the White Sea to the Bering Strait. It was to equip the entire route, put it into working order, and guarantee its safe operation. The bureaucracy rapidly became a kind of super-commissariat by expanding its activities, bailiwicks, foci, interests, and concerns far

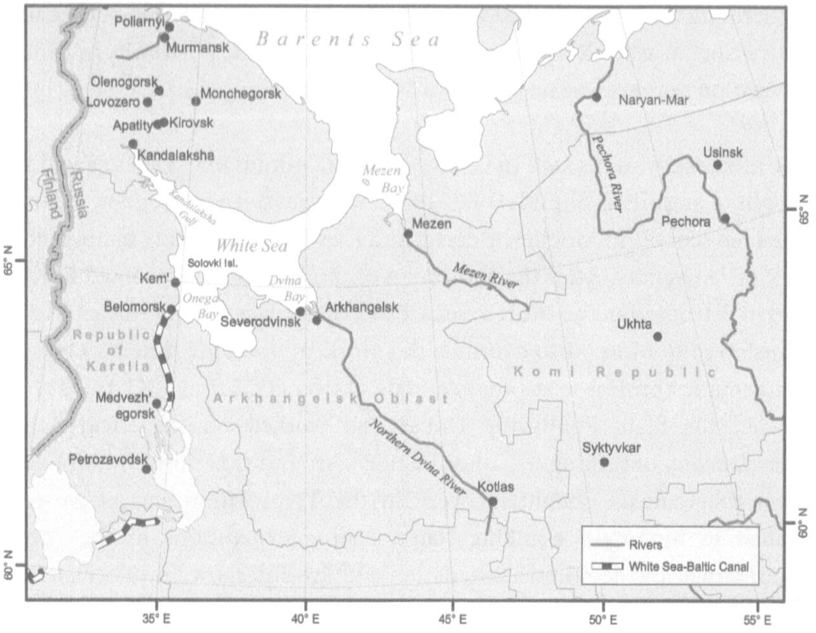

The White Sea. (Map by Manny Gimond)

beyond shipping to exploration, geographic, oceanographic, meteorological, and other research and development programs; organization of mines and smelters; geological prospecting; organization of fisheries; and shepherding of indigenous people who shepherded their reindeer. Glavsevmorput was an imperial agency as well, seeking to lay claim on all islands, ridges, and resources in the Arctic; the official tasks of Glavsevmorput for 1935 included establishing firm Soviet control over Wrangel and Franz Josef Land. When in 1924 explorers raised the Soviet flag on Wrangel Island and when Shmidt piloted the *Sedov* to build a station on Franz Josef Land before the Norwegians, it was to demonstrate Soviet claims. Now Glavsevmorput would secure ore and coal, tin, platinum, gold, molybdenum, graphite, lead, salt, and oil to substantiate the claims on these islands.

To administer its empire, Glavsevmorput administrations and departments covered every conceivable object and ecosystem: sea and river transport with identification and construction of ports, docks, and shipbuilding facilities as needed; polar aviation with a network of airports, airlines, and runways; polar and radio stations and hydrographic services; agriculture of the Far North; cultural and daily life of the people of the Far North; industry including organization of factories; trade; finance and planning; and, of course, a political department responsible for ensuring Bolshevik approaches to the assimilation of the north. Glavsevmorput organized a series of research and educational institutions: AARI; the Institute of the Economy of the North in Moscow; and the Hydrographic Institute. It ran a series of mining and prospecting organizations and trusts for deposits of coal, oil, salt, ore, and so on from Amderma to Nordvik to Chukhotka to the Arikugol Trust in Spitsbergen.[7] Glavsevmorput supervised the delivery of oil, motor oil, kerosene, gasoline, lubricants, and coal throughout the region.[8] It built its own cities like Igarka on the Enisei River with a huge timber combine that swallowed up and spit out Siberian lumber. Glavsevmorput opened three fish processing factories. It pulled Norilsk ore, Nordvinsk oil, salt, and other minerals, Amderma fluorspar, and Vaigach Island zinc out of the frozen ground.[9] The Council of People's Commissars confirmed the expansion of Glavsevmorput's already vast responsibilities in a resolution of June 22, 1936: to use the massive empire for the final assimilation of the Northern Sea Route from the Barents Sea to the Bering Sea.[10]

One of the features of technological momentum is the sense that people mean little to the organization at hand. They are cogs in a machine to be fitted and replaced as need be by a faceless, omniscient administration far away. We shall meet several of these people later in this and other chapters. But Glavsevmorput's massive responsibilities obscured its flesh and blood. It was an empire, sitting on top of Stalin's desktop and providing Moscow with the vast resources of the Arctic with icy hands.

The successes of Glavsevmorput hid significant infrastructural and personnel problems. Through coercion and exhortation and by force of will, shipping improved dramatically in the 1930s—in spite of technological lag, shortfalls in repairs and parts, and personnel problems. In 1935 freight grew from 14,000 to 65,000 tons—an approximately 400 percent increase. But one Glavsevmorput administrator, Kuz'min, warned, "Our world renowned icebreakers and icebreaking steamships should not be in the condition they are now," nor operated as they were: The *Lomonosov* went aground, and while under anchor the *Malygin* was pushed ashore by drift ice. Some of the problems were not of Glavsevmorput's making. The reindeer farms produced millions of rubles of goods, but couldn't get them the last few kilometers to the docks. Kuz'min called for the embrace of Stakhanovite methods to improve performance of shipping.[11] Stakhanovism was a special kind of exhortation of workers to achieve higher output targets by force of will in the absence of salary incentives or modern technology.

In 1936 Glavsevmorput shipped 271,100 tons of freight. This was only a slight overfulfillment of the plan, but significantly greater than in 1935; captains completed fourteen through trips along the entire route versus only four the year before. Two river caravans successfully ran the Indigirka and Iana and tributaries of the Enisei, while a new port shaped up at Igarka. These results had been "achieved during very difficult ice conditions, thanks to shock Stakhanovite work of . . . sailors, pilots and winterers of polar stations." One Glavsevmorput official acknowledged serious failings: Ships anchored in port too long while loading and unloading, and passenger transport was poorly organized, with river travel an especially sore point, especially in the Omsk region and Iakutia. Another problem was a high frequency of accidents, although an official did not give details. He called for "decisive measures" to modernize navigation and blanket the region with aerial photography and geodesy including observation of ice floes. The administration sought to overcome lags and aging icebreakers, with two new icebreakers coming into service both in 1937 and 1938 and with the river fleet slowly being augmented with metal-hulled boats.[12]

Glavsevmorput acquired a series of steamships, freighters, cutters, motorboats, lighters, and icebreakers toward the end of creating a trade route from Arkhangelsk and Murmansk through the Arctic Ocean to Dudinka and ultimately to Vladivostok. It was no easy matter to purchase ships even during the industrialization boom of the 1930s. At first the administration relied on older, refitted vessels. Even with these determined and devoted icebreakers, the shipping season lasted only from late June or July to October no matter how often party officials insisted otherwise. By mid-October cyclones had appeared, wet snow fell, and ships became icebound as early as November 1. The treacherous waters of the Barents and Kara seas created their own dangers. In the best cases, captains had carried out preliminary soundings to identify rocks, reefs, and other submerged hazards. Yet even the experienced captain was often at the mercy of good luck. Dozens of ships were lost, sinking after striking unknown objects or crushed in the ice. Scores of sailors perished in the depths, and some even froze to death on ships. Given the hostile environment in which they sailed, explored, and delivered goods, it is shocking to read in ship manifests how often captains sailed with insufficient supplies of food and fuel. A series of investigations of lost ships or serious accidents in the Glavsevmorput archives reveals the risks that captains and crews took in the name of the glory of the Soviet Union and the glory of their profession.

In the Soviet reporting system, Party officials balanced clear achievements with measured acknowledgement of continued shortfalls. The assimilation of the Northern Sea Route opened "the broad road for the furthest flowering of the Soviet north." It facilitated regular freight on internal river routes. A political officer of Glavsevmorput noted, "The party, government and entire Soviet country have given us unlimited support"—billions of rubles with which they organized a series of polar stations and radio centers, built the new city of Igarka, and trained an army of qualified polar workers. But to turn the route into a regularly operating "magistral" they required modern freighters and an icebreaking fleet. According to a Communist Party "aktiv," "Socialist industry gives us such an icebreaking fleet, and the primary requirement of workers of Sevmorput is to keep it in the proper order."[13] Not only was wear and tear a problem, but wrecking had reared its fur-draped head. For example, while repairs and capital construction began in 1934 at the Ship Repair Factory in Murmansk, the factory would not reopen until 1940. Also every spring the same "feverish" orgy of ship repair commenced, but because Glavsevmorput lacked its own ship repair facilities, other factories were sufficiently overwhelmed so as to meet few orders on time.[14]

For Glavsevmorput, as for many other economic organizations under Stalin, nature was an obstacle no less than spies and wreckers. A major challenge to regular shipping was the difficulty in stoking ships with coal, oil, gas, or diesel. This is one of the reasons that the Soviets sought coal deposits along the Northern Sea Route. That way, they could create fuel depots at various harbors and shore points. According to Party officials, however, Glavsevmorput had spent millions of rubles on geological work and the depots barely functioned; this indicated to them sabotage, not the challenges of locating ore, nor shipping in Arctic conditions with a still inadequate fleet. They found sufficient reason to blame nature itself. One official proclaimed, "In all of our work it does not follow to forget the scientific research character of Arctic navigation. The Northern Sea Route—it is not a route in the Black or Mediterranean Sea. Our path differs from every other one by the presence of ice cover. This is a serious opponent which it follows daily to encounter and fight. We may conquer the ice only in the instance if we study correctly and scientifically fight with it." He continued, "We are obliged persistently to study our enemy—the ice cover." Glavsevmorput had begun to do this with good naval and aerial reconnaissance, the work of the Arctic Institute, Hydrographic Administration, and Gidrometsluzhba.[15]

Slowly but surely, both through the sheer determination of committed employees and as required by the almighty plan, the ice was broken: Kara Sea navigation had reached 100 days, although further east the season was significantly shorter. In New Port shipping from upstream on the Ob River had been extended from sixteen days in 1927 to forty in 1929 and to fifty-four in 1934. Icebreakers stationed across the north by 1936 were responsible to lead convoys in each of four zones, although the experiences of the *Cheliuskin* and *G. Sedov*, both of which were suddenly trapped in the ice, showed that great uncertainties remained.[16]

Daily Life and Stakhanovism in Northern Latitudes

Glavsevmorput gave short shrift to its responsibilities to distribute a variety of goods and services among its many minions because it focused instead on industry. Glavsevmorput determined to improve trade "on new foundations" as "the most important stimulus to increasing the productivity of labor." Unfortunately, Glavsevmorput did not match goods, services, or prices with local desires and needs. Its much-maligned Arktiksnab supply organization failed each step of the way to ensure acquisition and distribution of goods and services. Glavsevmorput requested the assignment of approximately fifty to sixty party members and 250 to 300 Komsomol members to facilitate deliveries of supplies in the Arctic. But it had simply grown too large and could not meet all of its growing obligations. It tried to administer dozens of small geographic units with dozens of small boats to carry goods along every small tributary into great rivers to deltas that emptied into the Arctic Ocean and then to Murmansk and Arkhangelsk and to central cities. It maintained a number of stores and trading posts. In larger towns that drew hunters and fisherman (Bulun, Anadyr, Nordvik, Khatanga, Novyi Port), Glavsevmorput opened a network of dining halls to feed its workers, which in many cases were sooner creaky windblown holes in the tundra; from 1937 its officials strove to increase the number of bakeries to 200 and offer even white bread—which most inhabitants considered unhealthful compared to black bread—in larger villages. At larger polar stations where there were no trade posts, they organized general stores on ships and boats, which, however, in no way replaced free rations for sailors or those at polar outposts.[17] The Arctic empire simply could not accomplish its ever-expanding tasks.

One of the remarkable institutions of Soviet life for its ubiquity, colorfulness, range of concerns, and democratic accessibility to nearly all writers on all topics was the wall (or bulletin board) newspaper *(stengazeta).* The bulletin board newspaper usually appeared weekly or biweekly. Staff, usually volunteers who had other responsibilities within a given establishment, worked with the editor, a person with good political credentials, to ensure that themes, topics, and articles reflected broader sociopolitical concerns: discussion of a recent Communist Party meeting or decision; fulfillment of the plan (or not); upcoming elections; cultural achievements; and birthdays of leading figures. Because the authorities feared the publication of regular newspapers by small or medium establishments as potential carriers of anti-Soviet sentiments and because in many cases the absence of printing presses prevented publication, bulletin board newspapers became widespread. These newspapers reveal that while the official narrative of Glavsevmorput was heroic and successful, on ship, shore, and expedition grave concerns and human hardships prevailed.

The *Arctic Star,* the stengazeta of a Glavsevmorput territorial administration I could not precisely identify, but likely that of the Arkhangelsk district, offered delightful commentary on the quality of life and concerns of workers. It was largely a forum for complaints of acceptable sorts, although it also reported on the successes of the missions. The bulletin board frequently referred to the extreme hardships of life. One article called the failure to provide fuel and other supplies "criminal generosity." The hardships befell both land and sea, ships and schools, hospitals and clinics. Arktiksnab, the regional organization responsible for supply of everyday needs, could not provide smoked goods, flour, and eggs, made delivery of goods after they had spoiled, and behaved as if no herring could be found at sea.[18] Another article, "Abandoned Little Island," revealed how modest living conditions might be. Here the "dormitory" in question was "just one small room almost unfit for habitation, our previous joiner's workshop." From five to nine people lived in the cramped place.[19] The bulletin board could be merciless to those underserving even these hardships. The efforts of Comrades Zhuravlev and Varzin to gain an apartment they thought they deserved provoked ridicule. Zhuravlev did not merit housing because he was "an incurable alcoholic," while Varzin was a rolling stone who voluntarily left work on the *Molotov* to gain employment as a mechanic with no promise of an apartment from his bosses. He then delivered an ultimatum to be awarded precisely a spot in line for the next available apartment and no sooner than that moved on to another job. But the author of the article noted that "the fact remains that we can satisfy the interests of workers no more than 30%."[20]

Clearly, not all was strawberries and whipped cream, nor cod and coal, nor overfulfillment of plan and the acquisition of the comforts of a tundra home. According to one article, "Workers and service personnel, because of the good graces of the suppliers, again, as in last year, apparently, remain without wood. The stengazeta pointed this out already in March, but no one at all bothered to conclude a contract with the lumber department or with forestry organizations. Only recently was the commandant ordered to locate wood within five days . . . but [again] it followed to wait, and no wood was split for us, and therefore the only thing the commandant succeeded in gaining was a response from the forestry folks that 'there is no wood. . . .'" The anonymous author(s) continued, "In spite of such lamentable results among the workers and service personnel, however, the hope for delivery of wood has not been lost."[21]

If goods were in short supply, political education was not. In the effort to ensure the proper consciousness among all employees, party officials sought to strengthen the system of discussion circles with lecturers on various subjects, "red corners" of current literature and the classics of Marxism-Leninism-Stalinism, and consultations. On some ships these efforts met with success. Within Glavsevmorput there were twenty regularly published and 100 bulletin board newspapers, and eight ships regularly published newspapers during season. As one activist noted, "Print media—this is our vital party business. The newspaper is the spirit of the party organization." One such newspaper, *Skvoz' L'dy* ("Through the Ice"), was published on the *Ermak* under editor Gertner who earned a pin as an "Honorary Polar Explorer." Unfortunately, most ocean-going newspapers were not as good. In 1938 the newspaper on the *Stalin* came out irregularly and was, according to an internal observer, boring, dry, and illiterate.[22]

Officials were convinced that the growth of the party apparatus and increased political education would solve a large number of problems. (The Party apparatus generally grew in numbers over the decades, but with periodic purges to ensure only the most devoted members remained; during the Stalin era the purges assumed murderous proportions.) Comrade Kuftin, the political officer on the *Lenin*, proudly noted the rapid growth of the party organization on the ship. If on January 1, 1939, there were seven members and six candidate party members, within half a year they had ten members and seventeen candidates, a doubling of membership, and also twenty-seven Komsomol (Communist Youth League) members. On top of this, the library was well equipped. The holdings included two sets of the collected works of Lenin, more than thirty sets of the three-volume *Lenin and Stalin*, and twenty copies of the *Short Course of the History of the Party*, and everyone on board had a pamphlet about the eighteenth party congress. Twenty-five party members, eleven Komsomol activists, and fifty other individuals actively studied the history of the party.[23] Of course, more newspapers and books did little to change the condition of the ice and shortages of spare parts.

In the absence of the stimulus of the capitalist profit motive, because of lack of experience among many laborers who had only recently left the farm, owing to shortages of machinery and equipment, because of lag in technology even if icebreakers and airplanes served vital political, economic, and ideological functions, and because it was difficult to motivate tired, cold, hungry workers, the authorities turned to various competitions to exhort workers to higher productivity norms. One of the major systems to improve performance of Soviet industry and agriculture in the 1930s was Stakhanovism. Alexei Stakhanov, a Ukrainian miner, apparently overfulfilled his target for coal mining fourteen times in one shift in 1935 and then doubled his output again a few months later. (His comrades, content with modest effort for inadequate pay, must have hated him.) Stakhanov even appeared on the cover of *Time* in 1935. The authorities held Stakhanov up as an example of what could be accomplished through the proper organization of labor and the proper attitude toward one's work. Using exhortations, establishing higher norms, rewarding workers with modest titles and little else, Party officials employed Stakhanovism to encourage devotion of workers to the task at hand as miserly salaries and shortages of consumer goods did little to promote a work ethic. Stakhanovism moved from the mines of Ukraine into factories of the central industrial heartland and then north into the forests, rivers, lakes, seas, and tundra among loggers, fishermen, builders, sailors, longshoremen, and other glorious workers.

In many regions of the economy, party officials, managers, and workers faced unrelenting pressure to increase production in the face of old equipment or equipment in need of repair but no spare parts to be found. Remember that industrial production had reached 1913 levels only in 1926 and that the economy, compared with that in Western Europe and the United States, lagged considerably in terms of most indices, for example, capital investment per worker, speed and reliability of machines, and so on. The machine tool industry was nineteenth century. The Soviets needed pumps, compressors, turbines, locomotives, and the ability to produce them. Hence the system of competitions, awards, financial stimuli, and honors that was created under Stalin to encourage overfulfillment of targets was powerless in the face of technological backwardness. Brutish exhortation had some impact, but often was of an episodic nature, coming at the end of various planning periods (the end of the month, the quarter, the year, and so on).

The introduction of Stakhanovism, shock work, and other stimuli in the fisheries, forestry, and transport industries of the Far North indicates the difficulties in motivating underpaid and overworked laborers. The major problem concerned the old, glorious icebreakers intended to keep the Northern Sea Route open and to lead freighters laden with supplies to the outreaches of civilization, then to bring them back laden with raw materials from the north. Most of the aged vessels dated to the Tsarist era although now carried Soviet names or names of Russian prerevolutionary and Soviet heroes and explorers: *Sibiriakov, Malygin, Sedov, Lenin, Sadko, Litke, Lomonosov,* and *Stalin.* New names did not mean spare parts or powerful engines, so Stakhanovism was pursued.

The age of ships, nature of the workforce, and natural challenges imposed by the northern climate itself strained the ingenuity of captains and their icebreakers. At a November 1935 meeting of leading shock workers, Captain Pechuro of the *Lenin* said he had come to rely on his many Stakhanovites. They had achieved a great deal "without any kind of damage of mechanisms which he had usually [seen] in previous years." But he worried that his seventeen-year-old icebreaker might not keep up with the demands of industrialization and had lost some of its power, even after repairs carried out the preceding year at the Baltic Shipbuilding Factory in Leningrad. Lag on the part of other vessels exacerbated the problem. Pechuro reported, "We escorted barges of wheat from the Ob [River] to Arkhangelsk. We dragged small barges along that were hardly sea worthy and on which it was necessary to put people so that they could bail water . . . This was at that difficult time when they put together the first Kara Expedition when we lost two of the best steam ships that we had." One 3,000-ton barge went down near Belyi Island. It took on water and was gone in five minutes. In another case, the underpowered *Icebreaker No. 8* could not finish a trip with three barges and one tugboat to the Enisei River from Arkhangelsk. It was so small that it couldn't carry sufficient reserves of fuel: only five tons of coal. Pechuro complained, "We need better barges so that it is not necessary to subject icebreakers to such dangers, loading them down with gasoline, oil, and passengers because in the event of unhappiness, we lose them all. We have great icebreakers and can use them for a variety of tasks, but technology has moved ahead—the beautiful electric pumps on the "*Malygin*" are already old and inadequate."[24] Some mechanics were simply wizards at keeping the engines running in the face of shortages of parts. One Ivanov, a machinist on the *Rusanov,* was a "master of a cultured relationship with his machine"; he knew it so well that he could hear its calls for tender repairs even when he slept.[25]

What in fact was Stakhanovism? The nationwide fervor for the movement obscured the fact that Stakhanovism for a miner (how many tons of coal in what seams of what hardness with which tools in what time?) might not hold for a lumberjack, fisherman, dock worker, or sailor. Was Stakhanovism not simply the arbitrary imposition of higher norms imposed from above? What role could individual initiative play? One Comrade Volodarskii testified that the Stakhanovite movement on icebreakers was like a fog of misunderstanding. In one case, while docked in Murmansk in January–February as part of the Stakhnovite movement, the captain instructed his sailors to swab the deck and make things ship shape. This wasn't Stakhanovism, but busywork, or perhaps the authorities believed that Stakhanovism would overcome the seasonality of labor. But Volodarskii asked rhetorically, "What is a Stakhanovite movement on a ship in conditions of cold watch and anchored in port?"[26]

Dukhovskoi, senior assistant of the captain of the *Malygin* icebreaker, who claimed that repair records and rates had improved because of the Stakhanovite movement, acknowledged that the situation nevertheless remained abysmal because of a shortage of materials. How could one be a Stakhanovite without the materials needed for repairs?[27] That was, in fact, the crux of the matter. Stakhanovism was intended to overcome shortfalls of materials, backwardness in technology, years of lack of investment in most sectors of the economy, and workers who needed to have their qualifications improved, whether they desired this or not.

Lags in repair threatened the entire operation. They delayed departures, endangered the crew, and of course put the vessel at risk. But Glavsevmorput was unable to secure timely repairs or spare parts.[28] Captain Korel'skii of the *Malygin* noted that a large number of ships had begun repairs in one season that carried over into the second, third, and even fourth. He blamed the managers of ship repair dry docks. It seems that factories earned more for emergency repairs than for scheduled capital repairs. When Captain Korel'skii took it upon himself to order repairs to rebuild the deck of the *Malygin*, he discovered "no materials, no nails, screws, paint brushes . . . and other seasonal material . . . we need to get ready not in spring but earlier."[29]

Increases in productivity because of Stakhanovism hid serious problems. Comrade Kuz'min, a director in the administration of Glavsevmorput, found reason to criticize operational, safety, and employment records, no matter the newspaper reports to the contrary that praised Stakhanovism effusively. In 1936 nine accidents cost a total of 97,000 rubles, only eight or nine steamships were operational, they sailed a total of fifty-nine days versus ninety-nine planned, and Kuz'min had to fire eleven workers, while eight quit and twenty-one were reprimanded. The upcoming navigation season to Tiksi and back depended upon getting all ships in order figuratively and literally[30]—which was unlikely given these statistics.

Making matters worse, brutal work conditions, poor pay, and even worse housing led even the most devoted Communists to shirk responsibility from time to time. Sailors were a constant source of concern for Glavsevmorput. They might attack their work with vigor when at sea, but not at shore. To make matters worse, sailors worried about their families' living situation on shore when they themselves were at sea. All of this led to poor job performance. Said one administrator, "If we can recondition six apartments with our shock workers, then still many of them live in poor conditions, have big families and live on the edge. The worker will keep himself together on the ship if he knows that his family is well taken care of at home."[31]

Accidents were an endemic problem related to all of these other factors: poor repairs, old vessels, workers without incentives and with relatively low qualifications and insufficient knowledge about Arctic geography, weather, and transport. The captain of the *Lenin,* Pechuro, complained that sailors lacked the kind of discipline needed to avoid accidents. At sea, all was in order, but on land discipline failed. "And administrative rebukes don't give results. [The sailors] fall under the influence of their comrades." Pechuro called for "stern measures." He acknowledged improvement as labor turnover had dropped significantly since 1928. By the mid-1930s, most sailors had been working more than five years. He knew that "in the majority of cases the reason for disaster of a ship is turnover of employees." The disaster of *Icebreaker No. 9* occurred because they didn't know piping.[32] Another official, too, stressed the need to raise sailors' qualifications because "the slightest wavering or carelessness or negligence will have great consequences when at sea."[33]

The party, Komsomol, and trade unions hadn't figured out how to liquidate truancy and tardiness, nor how to improve qualifications, so they turned to Stakhanovism. Yet how could they solve truancy when miserable housing and poor work conditions prevailed? On the *Ermak* they introduced a capitalist artifact, a poorly functioning "time card" system. On its basis, apparently, in the preceding five months, twenty sailors had been fired and fifty-eight received reprimands. Labor discipline had also slipped on the *Malygin*. The third navigation officer, Aurov, left ship during his watch, instead went into town, returned drunk with two friends, and got into a fight. The police came, and the next day he was fired. While in Arkhangelsk port, the second mechanic of the *Sadko*, Shaukhat, played hooky for two days, while the second navigation officer showed up only drunk and was fired for truancy and tardiness.[34]

Ultimately, Stakhanovism involved any activity to improve productivity through better organization and propaganda. The experience of Comrade Tokar, the assistant political officer (Pompolit) of the *Malygin*, gave some sense of the efforts to raise the "cultural" level of workers and passengers on board. During the time of the work of the "historic eighteenth Congress of the Party" in 1939, the *Malygin* was engaged in a sea mammal campaign in the throat of the White Sea. On board, in addition to the crew, was a group of scientific workers and roughly 100 manufacturers, to whom the authorities had "affixed" an agitator communist or Komsomol member. Toward the end of political education, Tokar reported that they organized chess and checker tournament in which eighteen crew members participated, set up an ideological circle for twelve men, and held lectures on such themes as first aid, how to deal with flu, and the dangers of alcohol. Finally, they showed films jointly with the crews of the *Sadko*, *Rusanov*, and *Malygin*, all of whom fished in same area.[35] Another achievement was the creation of an "oral literary newspaper," already in its second year, whose members offered humorous verse *(chastushki)* to the accompaniment of a small string orchestra. Said Comrade Tokar, "Of course, [the chastushki] were far from perfect, but our people were not exacting [in their standards] . . ."[36] One such chastushki made light of the stoker Davydov:

We have Vasil Davydov—
A great stoker
And his whole shift is like Vasilii
And the steam is always at the right level.
The other stokers,
Catch up with him,
To avoid hold-ups
When the steam is lost and we stand still.[37]

Understanding Arctic Climes and Regularities

As noted in Chapter 1, more regular research on the Arctic basin commenced in the 1920s. Even in the face of growing state support, it relied on the determination of individual naturalists to go forward until the establishment of a strong institutional foundation in the 1930s. Even that institutional foundation faced formidable financial obstacles, and it was difficult to attract specialists with higher degrees to a long-term commitment in Arctic-based institutions. Bit by bit naturalists expanded the focus of their research to seas and oceans from land-based research by cultivating government interest.

One of the greatest frustrations for scientists, officials, and workers alike was the lack of systematic scientific knowledge about the north, its resources, its people, and meteorological conditions that were the key to subjugation of the Arctic. On the eve of Stalin's self-proclaimed "Great Break" (1929–31) with the moderate approaches of the past, Soviet scientists had begun to piece together polar data. They did not know in any detail the composition of the forests, the annual or seasonal flow of rivers, or the population and makeup of indigenous peoples, their lifestyles, or reindeer herding techniques. Nor had they established beyond general outlines the nature of seasonal fluctuations and regional differences in climate to determine how late they might sail the Barents Sea, when they might expect to begin lumbering activities, how to prepare for the spring float, and how much lumber a stream or river might be expected to carry downstream.

Great lacunae in collection and analysis of observations hindered the promulgation of economic development programs at every step and permitted political considerations to intrude into science and the economy in full force. Planners always ratcheted up cubic meters of harvest and delivery; correspondence in provincial archives reveals that quite a few local forestry managers protested increases in annual harvest and float targets. The rivers on which they depended simply were unable to carry more than a finite amount of lumber, depending precisely upon maximum flow during spring freshets, not upon planners' dictates and party officials' whims.

Many of the specialists had established reputations during the Tsarist era—Aleksandr Fersman and Otto Shmidt prime among them. They were joined by individuals of only modest training who had been "advanced" from a series of new educational and research and development institutions that were a major feature of Soviet industrialization; personnel lists from a variety of northern research ventures reveal very few specialists with completed higher educational degrees occupied positions of responsibility. On foot and on horse, they headed north, at first in small groups with limited supplies, to chart resources. They assembled the required permissions from the authorities to requisition the necessary machinery and equipment, manpower, boats, ships, and horses to steam out of the throat of the White Sea to the Barents, Kara, and Laptev seas.[38] From 1935 to 1938, with a total budget of roughly 250 million rubles, Glavsevmorput supported a series of expeditions in a diverse region of the Arctic in search of various resources. These included the *Eniseiskaia 1935–38* that centered on prospecting for oil and coal deposits that received 3,290,700 rubles, *Chukotka 1935–38* and the search for tin and gold with 120 participants at a cost of 3,918,000 rubles, and the study of fluorite, in all sixteen different expeditions.[39]

As they sailed from Arkhangelsk and Murmansk to the east through the Barents, Kara, Laptev, Eastern Siberian, and Chukotka seas, explorers, sailors, and scientists studied tides and currents, gathered information on water chemistry, temperature, and ice regime, encountered and charted familiar and unfamiliar islands, and created new maps to help navigation and point out where to drop anchor. In 1935 the Arctic Institute produced a geological map indicating valuable deposits at 228 points, seventy-three of which were coal, to enable local supply of heat and engines, and also peat, graphite, gold (Chukotka), lead and zinc (Vaigach), copper on Novaia Zemlia, oil in the Pechora basin, and so on.[40] As Otto Shmidt explained, "Each year we republish maps of all seas of the Arctic Ocean and supplement them with new surveys . . . If you compare the map of the northern part of the Kara Sea published 6 years ago with the last edition, then it will seem before you are two different seas. The blank spots have been filled with depths and with discovered islands. Instead of a rough approximation of faint lines, you have delineation of the shorelines."[41] He acknowledged that much work remained—for example, ice prediction remained a young science. Eventually specialists created an intergovernmental bureau for ice forecasting that consisted of experienced oceanographers who met in November, March, and May to produce as accurate as possible forecasts of the ice conditions along the entire sea route. They analyzed information on the ice pack, circulation, age, and composition of ice.

As in many other fields of science and technology, the effort to transform knowledge of shipping from something incidental or passed on solely by word of mouth and experience slowly became more experimentally and observationally based.[42] Yet with the onset of the five-year plans, political and economic pressures came to overwhelm science. The plans represented what might be accomplished in a socialist world, but in the absence of sufficient circumspection they led to bottlenecks and shortfalls, to attribution of blame to innocents, and to incongruous orders to the periphery.

Still, a feeling of Bolshevik certainty prevailed in Leningrad and Moscow that, with the proper pressure on the provinces, industry, transportation, agriculture, mining, fishing, and lumber activities could be made to meet ever-increasing plan targets. By insisting that local people—and nature itself—comply with orders from the center, they thereby politicized every aspect of data gathering and analysis. For if local people insisted that various natural uncertainties handicapped their efforts to meet plans, then the solution was to reject those claims. Sniveling lumber mill managers, cautious ship captains, and tentative collective farm brigadiers would have to balance their observations about temperatures, weather, river flow, onset of ice, and the permafrost with the conviction among central planners that the glories of socialist reconstruction of the Arctic required those local officials, managers, and captains to make due.

In Arkhangelsk, to take one example, in the lumber town of Kotlas, Party Committee members requested that a commission discuss the float planned for a local river. In 1932 the men floated 7,000 cubic meters (m^3) of lumber downstream, in 1933 16,000 m^3, and they were cutting for a targeted spring float of 35,000 m^3 in 1934. But the head of the operation, one Gladyshev, an "energetic worker," who had overseen the two harvests, said there was no way the river could handle such a load and refused to try.[43] To offer another example: Study of the flora and fauna of the seas commenced in the 1930s but did not have a firm institutional foundation until the 1950s.[44]

The captain of the *Lenin* explained the difficulty here. He wrote, "Keep in mind also that each year is different. For example, the Iugorskii Strait may not be open and free of ice all year long. In Amderma, sometimes even until

August you cannot stick your nose into the Kara Sea because of ice. We will have more powerful icebreakers, but we don't now and it is necessary to work with the fleet which we have, and therefore we cannot say that we have completely and thoroughly assimilated the Arctic. We cannot say that because most of all we haven't studied fully the regime of ice conditions by years in the Arctic. We do not know why for example now we have a temperature of plus 3 when sometimes it is 1. All these laws are just being studied. Science works on them and when they will be solved, then we will look at our work in another way."[45] Yet the captain faced political and economic pressure to "stick his nose" into unknown or uncertain ice regimes, not to wait for scientists and engineers to present results in a measured and timely fashion.

The centralization of economic, social, and cultural control that was crucial to the Soviet polity contributed to this difficulty, at the same time as the rapid growth of the R and D apparatus after the Russian Revolution enabled some flexibility, although R and D capability remained largely concentrated in Leningrad and Moscow and to a lesser extent in Kiev. The institutes devoted to study of the Arctic tended to be in Leningrad owing to their roots in commissions of the Academy of Sciences, KEPS, in the Russian Geographical Society, the Main Hydrographic Institute, and other organizations. While Stalin moved the presidium of the Academy of Sciences to Moscow in 1934, and while a great number of applied science institutes connected with industrial commissariats were established in Moscow during the 1930s and beyond, Arctic research remained Leningrad-based.

New institutes and societies appeared that reflected the needs of branch industries from transport to forest, from mining to machine tools, and so on. The societies replaced those founded in the 1920s that reflected, in the eyes of party officials, the narrow, technocratic interests of the old, Tsarist intelligentsia, with new ones devoted to revolutionary transformation of the productive forces. For example, the All Union Scientific Engineering-Technical Society of Inland Water Transport (VNITOVT), one of many new societies whose members prided themselves on creating tongue-twisting acronyms, contributed to the mastery of the north. Members declared that all projects and plans in the scientifically socialist USSR would be, tautologically, on a scientific basis. In the mid-1930s members of the society met fairly regularly to hold meetings, conferences, and discussions. VNITOVT attracted institutional members of society whose concerns were closely connected with shipping and freight: the Sea and River Steamship Administration, the Administration of Arkhangelsk Port, the Red Forge Factory, and Glavsevmorput. They addressed a wide range of topics and issues: waste at the foundry of Red Forge, the development of a journal to register the progress of ship repair, new kinds of copper alloys, how to prevent accidents, methods to guarantee maximum efficiency of steam engines, and historical topics. VNITOVT produced brochures on "Stakhanovites . . . in the Red Forge Factory," "Maritime Legal Questions for Ship Captains," and "The Stakhanovite Journey of the Icebreaker 'Sibiriakov.'" Research projects considered assimilation of the small rivers of the northern basin, the shortening of stand-down time of ships in ports, new methods of handling and shipping goods, and raising of the qualifications of workers.[46] In 1939 the society helped organize the publication of scientific-technical columns in the newspapers *Moriak Severa* (Northern Sailor) and *Rechnik Severa* (Northern Riverman).[47]

Arctic conquest relied both on trained specialists and on a workforce to carry out the various production tasks. One of the major problems that faced administrators, captains, and managers was the generally low level of education among many of the people who filled Arctic spaces. Part of this has to do with the large number of peasants and exiles forced northward during the collectivization campaign, many of whom were only partly literate. While I have not found statistics that enable complete national and regional comparisons, the archives indicate that in each sector of the economy—forestry, construction, mining, shipbuilding, shipping—officials were worried about the fact that many of their employees had not finished even four or five years of school. Beginning in the 1930s, but especially in the 1950s, they established a series of on-the-job training programs to improve qualifications. To increase productivity of labor, retain employees, and cut down on accidents, such programs made sense.

Sea-faring individuals from sailors to captains found it a waste of time to improve their qualifications. If they desired to improve qualifications, problems of logistics and time stood in the way. Essentially only Leningrad-based institutes provided training; what captain could afford to leave his ship for months for retraining, and which workers had the opportunity? Glavsevmorput captains suggested organizing Leningrad courses in Arkhangelsk. Other bureaucracies were too miserly with their budgets to be interested in raising qualifications no matter how much officials harped about the situation. N. N. Bredis upbraided the captain of the *Sedov* for not working closely with young navigating officers who lacked experience so that they could learn how to use sextants. Another captain told of a case in which he wished to raise the qualifications of his class one master lathe operator to that of senior machinist and pay him fifteen rubles more per month. But his personnel department said there were strict limits on how he used his salary fund. It appeared he would lose the talented worker to another firm that paid more. (The captain lamented not even being able to replace the ratty flag flying from his icebreaker.)[48] The effort of bureaucrats to make sailing a reliable, scientifically based economic endeavor that followed plans could not succeed in such an environment.

Captain Burko of the *Sadko,* a ship of infamy after its autumn 1937 capture in the Laptev Sea ice, expressed uncomfortable surprise that his officers were poorly trained and had very little interest in expanding their knowledge. They rarely read and had no desire to take the initiative. His navigation officer was one of these people who refused to study. Still, he claimed, through Stakhanovism they had improved the level of technical training with the result that they faced fewer repairs.[49] The difficulty in getting people to improve their qualifications or to take initiative in leadership or safety matters is not hard to explain. First, terrible accidents were a part of daily life. Second, as the Stalinist terror grew around them captains and sailors, bosses and workers, lumberjacks and tracklayers saw their comrades arrested and disappear in the camps. Keeping a low profile and avoiding the risk of failure may have been one likely outcome of Stalinist pressures to overfulfill norms.

Wintering and Icebreakers

On the basis of meteorological and other information and with the impetus of the second International Polar Year and political theater, in 1932 the *Sibiriakov* managed the entire Northern Sea Route in sixty-five days from Arkhangelsk to Yokohama, Japan, even though a broken propeller led the ship to drift for eleven days in the Chukchi Sea; the *Sibiriakov* completed the journey without wintering in part by employing improvised sails.

Throughout the navigation season of 1934 Glavsevmorput caravanned freighters from the Lena River delta to indicate the possibility of tying the economy of Yakutia to that of Moscow. These achievements led Shmidt to predict that by 1937 they would equip the Northern Sea Route so that ships could regularly complete the polar trip without zimovka. But for this they needed more, always more, high-powered lighthouses, revised charts and maps, radio stations, weather forecasts, ports and coal depots, and especially icebreakers.

Powerful new icebreakers were the essential tool. With them, the "industrialization of the north" would follow. On the eve of the revolution, the Russians had some twenty icebreaking and ice-strengthened vessels operating in Arctic waters. Until the 1930s most of the Russian and Soviet icebreakers came from British or other European shipyards. This created a series of problems in maintaining the aging fleet: It was underpowered, it was difficult to get parts, and repairs consumed too much time. To underline the critical importance of the prerevolutionary fleet, keep in mind that in 1918 the *Ermak* participated in the "ice march" of Baltic fleet from Helsinki into Kronstadt and Petrograd through the frozen Gulf of Finland to rescue warships. The *Ermak* continued its service along the Northern Sea Route and according to one source even served Leningrad during the blockade from 1941–43 during World War II, and it was retired only in 1963.

By the late 1920s the shipbuilding industry had recovered sufficiently to embark on modernization of the sector. It built motorboats for rivers, military vessels, lighters, freighters, and icebreakers, although many vessels relied on coal power, which presented serious logistical problems.[50] Yet even though overall some 1,500 new enterprises opened during the First Five-Year Plan, managers, engineers, and party officials struggled with outdated equipment or equipment that was outmoded even before it came on line. This was particularly true in the shipbuilding industry—and especially for the Northern Sea Route—where through the 1950s the number, quantity, size, and power of icebreakers and other vessels was inadequate to the task of Arctic conquest.

Feverish activity in shipbuilding mirrored that in other branches of the economy during rapid, heroic industrialization. Year by year the fleets expanded, the inland fleet, the northern Atlantic and Pacific Fleets, the

Black Sea fleet, and the fleet of Glavsevmorput. For example, the Northern Marine Steamship Authority based in Arkhangelsk, at one time part of Glavsevmorput, operated nineteen icebreakers in its history, twelve of which dated to before the Russian Revolution, were built in England, and included the *Sadko, Lenin, Litke, Malygin, Sibiriakov,* and *Sedov.* Upon being established in 1922 the authority acquired twenty-four ships of various classes including freighters, passenger ships, icebreakers, tugboats, and lighters; in 1927 it sailed twenty-eight ships; in 1941 fifty-six ships; in 1965 113 ships; and in 1990 123 ships.[51] As a reflection of the inflexibility of the Soviet system, however, Glavsevmorput could not requisition sufficient numbers of small vessels for on-the-spot deliveries of fuel and food to the coast. On the other hand, the partial mechanization of Dikson, Kozhevnikov, Tiksi, Ambarchik, and Providence ports facilitated loading and unloading for larger vessels.[52]

Because of growing recognition of the limits of the fleet, in the mid-1930s the Soviet government determined to build new, more powerful icebreakers. Of course, the first ship to be launched was the *Joseph Stalin* in 1938, although two years later than planned. The Ordzhonikidze Shipbuilding Factory in Leningrad launched four ships of the *Stalin* class. The icebreakers roughly shared major specifications: length—about 107 m, breadth—23 m, draft—9.2 m, displacement—11,200 tons, speed—15.3 knots. They could navigate through ice almost 1.0 meter thick.[53] During the war the four new icebreakers earned a reputation for escorting caravans through the Arctic. In the postwar years, the USSR maintained its lead in and expanded on icebreaker technology with larger vessels and eventually with nuclear-powered ships.

Having journeyed into the White, Barents, Laptev, and Kara seas, the captains and sailors frequently found themselves in life-threatened situations—trapped in ice floes for months at a time with supplies of fuel and food falling short. Because this was the USSR of Stalin, where there were no failures, no unforeseen obstacles, no fortresses, as the slogan read, that the Bolsheviks could not conquer, even the dangerous and life-threatening circumstances—months and years of helplessness drifting in ice floes—became spectacles of Stalinist certainty and visionary leadership.

A litany of published reports, celebratory tomes, and radio broadcasts of skyward-steaming indices of tons shipped and kilometers traveled hid a series of accidents and disasters from public view. The Soviet leadership, as is well known, closely controlled media to arrange political, cultural, educational, and other messages. One aspect of this control was to sanitize or fail to report technological failure. Perhaps without thinking this was their goal, propagandists attempted to show that modern technological society was safe and functioned as intended. Train and plane crashes, industrial accidents, and so on did not exist for the Soviet citizen but did occur under capitalism. Yet the archives reveal the significant human costs of living in industrial society, especially in the Stalin years when plan fulfillment took first place over worker safety, pollution, and other costs of modernization. When accidents occurred, whether covered rarely in public or more usually in tense closed investigative commissions, most often the members of the commissions blamed human error and rarely the technology, even as personnel worried about endemic technological lag, shortages of spare parts, and delays in repairs. As a result, scores of pilots, captains, train engineers, and their crews gave their lives in the impatient effort to demonstrate control over the elements.

In one instance, the captain of the *Sibiriak* gave detailed testimony on how he ran aground, took on water, lost power, and lost a man in a terrible accident in late November 1937. The ship left Murmansk for Arkhangelsk with ballast with a stop at Kuzomen to load twenty-two crates of freight to deliver to a Glavsevmorput air force base. *Sibiriak* carried 250 tons of coal, 360 tons of water, six passengers, and thirty-six sailors. Within a few days in near-zero visibility, thick, wet snows and heavy winds began to buffet the ship without pause. The captain sought sanctuary near the southwest shore of Novaia Zemlia. With others joining him on the bridge to endorse his decision—and share his responsibility—the captain followed a map showing safe passage, but he went aground on a shoal. He gave order for full reverse, but after turning 20 percent to the right, the ship stopped. He ordered that ballast be pumped out, but the ship wouldn't budge. Within a few hours, the ship began to take on water and waves overwhelmed the boat.[54] A ship in the area offered assistance. Engineer Molodtsov who came aboard from the other vessel disputed the captain's account. The ship went aground in very shallow water because of the captain's error, Molodtsov testified, and the *Sibiriak* did not so much sit on as run through rocks that flooded the fireman's hall and machine hall, killing a sailor.[55] In any event, all captains and sailors shared great worry that powerful, sudden storms and the absence of accurate maps were a constant risk.

If operator error was a factor in accidents, then conditions on board and with equipment contributed fully. The sailors and their captains were left exhausted by the cold. The ships were engaged in military exercises, shipping, survey work, and scientific expeditions. Their season began in July and continued until November. Ultimately, as part of the process of assimilation of the north, the boats might winter in the Arctic ice to test the reliability of ships and their crews and to chart the ice floes. One boat captain recorded his observations in sailing early July from 72'58" and 66'02" to 77'14" and 80'38" by July 10. He reported ice floes, readings of depths, shorelines, meteorological conditions, and various islands. At the end of August he joined the ships *Vantsetti* and *Belomorkanal* and arrived at the southern edge of the ice. In a journey that lasted until November 18, 1937, he established sixty-five hydrological stations, conducted 3,637 nautical miles of echo-sounding near the Malygin Banks, Sterlegov, Tessema, and elsewhere, undertook topographic work on the Uedinenie, Ruskii, Firnley, and Geiberg Islands, carried out magnetic and astronomical observations, and held a practicum for twenty-one students of the Hydrographic Institute. He accompanied the *Maret, Traident,* and *Pravda* ships from Dikson to Sterlegov and the *Dikson* and *Krestianin* from Sterlegov to Dikson. With the *Vantsetti* and *Belomorkanal,* he sailed to the Laptev Sea. He remarked on the lack of preparedness for the difficult trip: "I don't know how to explain it, but one thing is for sure. We went out to sea without supplies of everything necessary for the Artic." He took responsibility, but also blamed Glavsevmorput. They had little in the way of winter clothing. Outer garb did not exist: fur boots, mittens, gloves, and so on. For the entire crew they had only ten pairs of fur boots. He suggested that emergency supplies of coal be established near Cape Cheliuskin.[56]

For a trip in the summer and fall of 1938, Captain P. P. Karaianov of the freighter *Arkos* reported shocking delays in loading supplies; in fact, many of the supplies were not available. The supplies should have included all sorts of foods, baked goods, salted and fresh fruits, smoked foods, eggs and tea, dogs, cows, pigs, medicines, furniture, gifts for indigenous people, matches, cigarettes, hydrogen, oxygen, wood, oil, kerosene, gasoline, and scientific equipment destined for a polar station. He complained that there should have been seventy-five days of reserves for many of these items, but the freight and passenger journey departed with only ninety tons versus the planned 1,500 tons. Passenger cabins were neither for passengers nor cabins: They had no natural light, ventilation, or ease of entrance and exit. Still, during the sixty-seven-day trip they managed to get by with no accidents or illnesses.[57]

Oftentimes, the ships simply got caught in the ice. The *Dikson* departed port in July 1937 and joined the icebreaker *Lenin* on July 29 not far from ice at 73' 52" and 73' 04" in the Kara Sea east of Novaia Zemlia. The *Dikson* wintered beginning on November 3 with several other ships in a caravan. On the 24th, they turned off the steam boilers and went over to wood from coal heating and kerosene lighting. Only a daily routine broke the monotony of cold and dark: rise, shine, and breakfast at 7:30 a.m., 8:00 a.m. the beginning of the workday to maintain equipment, lunch at noon, dinner at 5:00 p.m., tea at 8:00 p.m., and lights out at midnight. The caravan drifted from 74' 27" and 111' 09" on November 3 to 76' 56" and 120' 4" on August 4, 1938—that is more than nine months and two days.[58] Stalinist icebreaking was clearly sailing by the seats of its frigid, wet pants.

Aviation in the Arctic

By the late 1920s officials and aviators alike had recognized the role that aviation could play in Arctic assimilation. The Soviet aircraft industry was beginning to grow, so the officials sought a way to celebrate technological innovation on stage before the citizens of the world. The aviation wing of Glavsevmorput used aircraft for four main purposes: reconnaissance of ice floes in the Northern Sea Route, freight of supplies for research and rescue missions, exploration, and ideological purposes.

Large-scale technological systems became paradigmatic in the twentieth century as symbols of state power. Achievements in big technology distract attention and budgets from social and political problems at the same time as they engender national pride. Most leaders, engineers, and citizens unquestioningly embrace big technology for economic, military, and other purposes and as icons of national achievement—in the form of modern highways in Germany or the United States, hydroelectric power stations in India and Brazil, rockets and nuclear weapons for North Korea, and the industrial transformation in China.[59] In the USSR, as elsewhere, large-scale technological systems had a purpose that went beyond the technical end of that system. Dneprostroi was more than a hydroelectric power station but a demonstration of Soviet electrical engineering skills and independence. Magnitogorsk represented the verve of Stalinist industrialization through modern steel mills. The aviation feats of pilots proved that the USSR was second to none in a cutting-edge technology and indicated the glory of Stalin's enlightened leadership.[60] This ideological significance of technology might be called the "display value" of technology, that is, its cultural meaning beyond its technical importance.[61]

Precisely in the 1930s, at a time of tremendous social turmoil and uncertainty, Soviet leaders embraced the display value of aviation through a series of feats. If in May 1927 Charles Lindbergh first flew nonstop across the Atlantic Ocean, then on June 18, 1937, pilot Valery Chkalov and two crew members flew from Moscow to Vancouver, Canada, via the North Pole, a flight of more than sixty-two hours and nearly 9,000 kilometers that established a new nonstop, long-distance flight record. A month later Mikhail Gromov established another record in a flight from Moscow to San Jacinto, California, by way of the North Pole (more than 10,000 kilometers in sixty-two hours). Stalin saw these pilots—and Soviet citizens came to see these pilots, Stakhanovites, and other technological heroes—as the quintessential "New Soviet Men," not only for their records of distance, time, altitude, norms of production, and so on, but for their archetypical, utopian, mythological, and superhuman characteristics that would become norms for behavior in the near future. As Berman writes, "Chkalov possessed the qualities of bravery, stamina, cool-headedness, perseverance and technical proficiency—precisely the virtues that enabled him, as a modern Prometheus, to master nature."[62]

Crucial to Arctic exploration, on May 21, 1937, Ice Commissar Shmidt, pilot Mikhail Vodopianov, and several crew members landed on the ice near the North Pole to provide support for Ivan Papinin's drifting North Pole station. A year earlier the crews of two aircraft established a base for the expedition on Rudolf Island with a landing strip and two eight-room houses. Shmidt orchestrated the entire feat, selecting Vodopianov to head the aviation detachment that consisted of five aircraft with forty-three people plus supplies on board. The pilots became national heroes with their "selfless" and courageous personalities, even as some of them later were accused of being enemies of the people and wreckers during the Great Terror.

Polar aviation was important economically, not only for serving ideological functions. Between 1933 and 1936 there was a ten- to twenty-fold increase in indices of freight tonnage, number of flights, and kilometers flown in polar aviation. But a series of technical and organizational problems handicapped further improvements. Flights remained irregular, runways were in poor condition, there was no passenger service to speak of, and accidents seemed to be endemic due to the "problem of training of the pilot-captain who knows how to fly in conditions of the absence of or insufficient ground equipment." The Soviet solution was more "agitation" to combat the problem of the overly self-confident pilot. Adding to the difficulty, Glavsevmorput used nineteen different kinds of planes and eight different motors that made organization of airport safety, supply, and repair extremely difficult.[63]

Landings in Naryan-Mar, Amderma, and other polar cities frequently occurred in winter conditions. Working with common sense and drawing on the experience of individuals connected with northern aviation, specialists at the Institute of Permafrost had developed criteria for how to prepare runways of snow and ice and to use various insulators (e.g., straw) to keep them safe until the early summer months. But not everything could be planned as flight conditions naturally depended upon changeable weather. A late blizzard, an early thaw, early freeze, strong winds, fluctuating temperatures that caused fog—all these things required on-the-spot attention. Because pilots often flew to meet the needs of important, publicly orchestrated expeditions, they might have to fly—and land—precisely when weather conditions in late spring or early summer or early fall were most changeable. This led to orders to factory directors and chairmen of city councils to get "volunteers" out to the runways, to shore them up with snow, to smooth out hillocks, to put out fire pots to illuminate them. Many of these people in Amderma and other towns were in fact gulag prisoners pulled from work in mines or factories into emergency service. Such polar pilots as Mikhail Vodopianov did not notice that those who greeted him were in fact prisoners or tried to ignore that fact. And Vodopianov gives the impression of not caring who had helped out and usually not thanking the work crews.

In 1936, in preparation for the celebratory Stalinist aviation feat—Moscow-Arkhangelsk-Naryan-Mar-Amderma-Franz Josef Land, a distance of 4,000 kilometers to 86 degrees north latitude, using two Polikarpov P-5 planes—Ivan Loiko, a pilot, was sent to Amderma to build an ice and snow runway. The planes were specially equipped, radio stations were set up along the route, and they added instruments for flying blind and reserve gas tanks to permit flight of 1,500 kilometers without refueling. The cabins had extra insulation and were loaded with a 1.5-month store of food in case the pilots had to ditch on the ice and wait for a search party. Loiko arrived in Amderma on March 19, 1936. On March 30, Fabio Farikh arrived in another plane with spare parts and in particular wheels, so that the planes could land at Amderma on the return with skis, replace them with wheels, and return to Arkhangelsk.[64]

Vodopianov landed soon thereafter, while Farikh's fuel cut out, he was forced to ditch onto the ice in the Kara Sea, and all of Amderma organized to find and rescue him. Vodopianov refused to join the search for Farikh because of orders to continue on to the North Pole. He informed Loiko that he would guarantee him "another term [in the gulag]" for interfering in his attempt to conquer the North Pole—which had come at the direct request of Stalin. Loiko, a pilot known for his World War I exploits, who as a young boy lost his parents and brothers, who was sent to Siberia in the Stalinist repression, who lost his wife and daughter and any contact with them, who had looked death in the eyes many times, and who gave his last to the Amderma airport, grabbed his straight edge razor and killed himself. Farikh, who was eventually recovered but lost two of three passengers on the ice, had it out with Vodopianov in earshot of others.[65] Farikh visited the camps later.

Fabio Farikh grew up in a well-to-do family in Petersburg; his father was one of the first automobile enthusiasts in Russia, which may explain Farikh's interest in technological innovation, especially things that moved with great speed. In 1918, a fresh driver's license in hand, Farikh joined the Red Army and suffered through hunger, typhus, and civil war fire fights. After being demobilized he tried his hand at building a bicycle, a motorcycle, and even an automobile. Having "become infatuated" with the heavens, he studied aviation mechanics. He began flights on the Irkutsk-Yakutsk-Bodaibo route.

In 1930 he had graduated from the Moscow School of Civil Aviation. In 1931 he fulfilled the direct order of minister of heavy industry Sergo Ordzhonikidze to fly from Krasnoiarsk to Dudinka, a distance of 4,400 kilometers in thirty-five hours. Farikh succeeded without breakdowns or giving in to inclement weather; Commissar Ordzhonikidze gave him a new Ford. He became a pilot in the well-known "Suicide Squads" for *Pravda* that delivered newspaper drums in any weather from the head editorial office in Moscow to other large cities to be printed by daybreak.[66]

In February 1932 Farikh first flew the Moscow-Arkhangelsk-Ust'-Tsil'ma route without support of airports, maps, or on-the-ground navigation. He was now inserted into the nexus between airplanes and the gulag: his next assignment was to fly the secret police head of the Vaigach mining expedition, Fedor Eikhmans, from Moscow to the island. The flight took over two weeks and put them at the edge of hunger and freezing after fog and blizzards forced them into emergency landings in the tundra. The return trip on May 6 was no easier as Farikh was forced to ditch because of engine failure near the Lai River and collapsed the landing gear. He spent two months on a small island waiting for rescue. In February 1934, Farikh participated briefly in the attempt to save the *Cheliuskin* but ran into personal conflict with one of the military pilots involved who arbitrarily excluded him from the search. In March 1936, as part of the preparation for the Papanin expedition of 1937, Farikh flew from Moscow to Amderma in an orange, two-motor N-120, when his compass failed and he was forced to land in a blizzard in Ust'-Kara about 200 kilometers south of Amderma. On April 2, the unlucky Farikh resumed his flight, but having gone not even twenty-five or thirty kilometers ditched again on ridged ice because his fuel had run out. Three passengers refused to listen to Farikh and set out for the nearest village, and two of them disappeared somewhere near the Kara Sea.[67] Farikh's star finally showed in 1937 when he flew from Moscow to Uele and back—24,000 kilometers, 145 hours, and forty-seven risky landings, all in difficult conditions, in the polar night, often without radio contact, for which he received an Order of Lenin. He followed that up by flying on October 4 to Franz Josef Land to join in a failed search for a lost pilot, Sigizmund Levanevskii.[68]

Glavsevmorput established a base on Franz Josef Land to enable aviation missions to fly from Moscow through Naryan-Mar to the North Pole and back. This way the effort to glorify the Arctic pilots could be orchestrated and celebrated in the Leader's capital. But the missions, that began in March or April while the ice remained firm on rivers and in tundra that doubled as runways, were always at risk. Many pilots crashed the planes while dealing with weather, visibility, engine problems, and fuel shortages, and early thaws, for example, of the Pechora River flowing through Naryan-Mar, and we do not know how many of them died. But the Stalinist system required victorious pilots and sought out guilty parties to be blamed for failure, and in the celebration of Arctic conquest it politicized science and technology.

On one occasion, Vodopianov and Makhotkin damaged their two planes landing at Franz Josef Land. Having repaired one with the parts of the other, Vodopianov flew back to Naryan-Mar, while Makhotkin waited on Franz Josef Land for a steamship during the summer navigation. But owing to capricious weather, Vodopianov flew as far as the Iugorskii Strait only in the second half of May. Ivan Khramov, the gulag director of the Amderma mines, a Chekist of working class background, who himself served a prison term, ordered prisoners to collect snow and ice from the Kara Sea shore to build an ice runway for the big plane.[69]

Vodopianov faced the choice of where to fly further as his plane had skis, and so he could not land on a regular runway. So rather than wait, too, for summer navigation, Vodopianov determined to fly from the Iugorskii Strait to Naryan-Mar, insisting on a Pechora River landing before the ice melted. On May 18, he radioed his request to fly. But having prohibited pedestrian and truck traffic on the Pechora since late April as the warming temps and rising sun melted the ice, regional party officials demurred. Arkadii Evsiugin, a Nenets and loyal Soviet citizen who rose from modest origins to first secretary of the Nenets regional Communist Party committee before being purged, arrested, interrogated, and sent to labor camps in the Far East, and the head of the Nenets Aviation Committee, turned Vodopianov down because they could not guarantee his safety.[70] They were damned if they did and damned if they didn't. Evsiugin would be a wrecker if he interfered with Vodopianov's mission because of its importance to the state. But if the plane crashed through the ice . . . he'd be a wrecker.

The then-first secretary of the Nenets Party organization Ivan Iakovlevich Prourzin, who himself was arrested in 1938 as an "enemy of the people" and served sixteen years in labor camps, called Evsiugin, reminded him of the significance of the polar feat to Party leaders, and ordered him to find a way to build a runway. Evsiugin marched along the Pechora River to conclude the ice was strong enough to accept a plane as winds from the north dropped the temperatures at night to –10C or –15C; the landing had to be without any delay. Everyone—managers, party members, workers, Nentsy, and a special aviation detachment—worked until morning. When the sun rose, the emergency runway was there.[71]

When Vodopianov's plane appeared, they anxiously watched it descend, especially when the pilot flew away from the Pechora and headed toward the city, and not to circle around for a better view, but to land across from the Naryan-Mar docks where barges and tugs were anchored and where the ice was dirty and likely weak. Nothing happened, thankfully. When the men approached Vodopianov, he said, smiling, "From here it's closer to walk to shore. . . ." Vodopianov soon left Naryan-Mar, without a word of thanks for the help he received nor for changing his skis to wheels. Vodopianov learned he could not fly to Arkhangelsk because its airfield was under water. So he decided to fly to the still incomplete Vologda airfield nonstop, a distance of 1,120 kilometers. He made that trip and returned to Moscow in his N-127 on May 21, 1936. On May 26, a week after he landed in Naryan-Mar, the ice begin to move, and two days later the river was a torrent.[72]

In the autumn of 1937 the government ordered the Nenets region party leadership to prepare immediately for the landing of an expedition under G. B. Chkhnovskii that would search for Pilot Sigizmund Levanevskii who flew over the pole from Moscow toward the United States with a six-man crew in a new long-range bomber and disappeared likely due to loss of at least one engine. Levanevskii, who became a Glavsevmorput pilot in 1933, had already completed a number of long-distance flights. In 1934 he flew from a temporary ice runway in the Chukchi Sea as part of the *Cheliuskin* rescue mission and received an award for his participation, although he crashed and did not rescue anyone. In 1936 he flew from Los Angeles to Moscow.[73] (Another pilot who participated in the search for Levanevskii, M. S. Babushkin, stopped in Arkhangelsk on return from the Arctic, the next day took off for Moscow, but never reached altitude and crashed into the Northern Dvina.)

Glavsevmorput sent a specialist to Naryan-Mar to work with local officials in determining a site for a landing strip to support the Levanevskii expedition. They identified the village of Tel'viska, about 5 kilometers from Naryan-Mar. But seeing that it would require about 100 workers and two months to get ready, they chose the village Kuia on the Kuia River, 20 kilometers downstream from Naryan-Mar. This location had trees to the east and the river to the west, and there might be updrafts, but they were given only ten days to ready it. Arriving on a cutter, they realized how much work remained: felling trees, moving earth, leveling the land; B. G. Chukhnovskii successfully landed a detachment of four N-210s from Arkhangelsk.[74] All seemed to go well until October 14 when Evsiugin received a complaint from the head of the Naryan-Mar port, Kiselev, against Chukhnovskii. It seems that Chukhnovskii, like other self-important hero-pilots, decided his requests had the force of law. He commandeered a steam tugboat for travel to Naryan-Mar and back, while he was offered a cutter. Kiselev was absolutely within his rights because tugs were needed to manage the port, especially on the eve of the close of navigation season in late fall. Evsiugin demanded the tug be released; Chukhnovskii threatened to complain all the way to Stalin that the locals had failed to create a satisfactory runway. Evsiugin offered a cutter and faced down Chukhnovskii's anger. On October 19 the pilot flew off to the Arctic, not having told anyone about the departure, in fact not once having said a word thanks. Yet later that evening Evsiugin heard the noise of airplanes. Indeed two planes were overhead, one piloted by Chukhnovskii. They had to send people from Naryan-Mar without delay to illuminate the landing strip. Apparently they returned for weather or some other reason. And they left again on October 21 without warning. These pilots were a courageous, self-involved lot.[75]

Of all the pilots, Mikhail Vodopianov gained the greatest notoriety. Vodopianov, a Stalinist hero of polar aviation, a polar explorer, and in retirement an active writer, was a faithful son of the Stalin era.[76] Imperious and self-serving, he often set out when conditions were unsafe. Vodopianov's autobiography, written in the modest yet self-aggrandizing tone required of new Soviet heroes and ignoring the losses of so many pilots as they pushed into the northern latitudes often with inadequate preparation or support, sheds light on the great challenges that the pioneers of polar aviation faced when opening new vistas, overcoming intransigent nature, and setting distance records in the 1930s.[77] A Soviet hero of good stock—he came from a poor rural family, studied attentively, fought for the Reds in the civil war, became intrigued with automobiles and airplanes as a little boy, and by force of personality, hard work, and the will of the party became a pilot himself after having taken care of horses and served as a driver—Vodopianov joined Dobrolet in the 1920s as a mechanic. Dobrolet, one of many national voluntary organizations to support modern technological culture such as Avtodor for automobiles, promoted civil aviation for passengers, freight, and postage, according to what must be highly exaggerated official statistics saw the creation of 26,000 kilometers of routes in the 1920s along which airplanes flew a total of ten million kilometers, carried 47,000 passengers, delivered tons of freight, and was one of several associations intended to promote a modern technological society.[78] Pilots contributed to agriculture through crop dusting, fisheries and hunting by aerial reconnaissance, and so on.

Early flight was a combination of spit, courage, and guesswork, as well as unrelenting pressure to meet flight plans no matter the weather, for example, as members of the *Pravda* "suicide squad." So rudimentary was the cockpit that radios did not operate or operated poorly. Pilots might have to light matches to look at the compass. They flew by the seat of their pants in fog, encountered unfamiliar icing situations, frequently got lost, struggled to follow railroad tracks to the destination, and had to ditch into fields of grazing farm animals when they could not find a landing strip. Imagine having to land for an emergency repair: the landing strips themselves were hardly safe and certainly not modern, more often than not a mowed meadow or cleared woodland, certainly not an asphalt surface. "Airports" were essentially 1,000-meter-by-1,000-meter fields. Like other pilots, Vodopianov knew that the history of Arctic exploration was the history of tragedy and heroes; Sedov and many others had perished in pursuit of the Pole. But Vodopianov pushed forward almost recklessly to open the Arctic to the new Soviet man even in the polar night that lasted from October to February. He imagined himself and his comrades raising red flags on the Pole.

Capricious weather might mean taking off with wheels but needing skis to land. Or it might mean spending one, two, or six days waiting for weather to change. The pilots hunkered down in their planes—or in well-stocked wintering cabins if nearby—telling jokes, playing chess and dominos. Once Vodopianov played more than 400 domino games while waiting to fly. Icing

Mikhail Vodopianov flying over the North Pole. (Courtesy of the Russian State Museum of the Arctic and Antarctic)

presented special challenges: Should (could?) pilots lose altitude to find warmer temperatures? Could wings and fuselage be heated or covered with a lubricant? And even with all these uncertainties, Vodopianov asked early in his career to transfer to the Far East to open the Khabarovsk-Sakhalin route in spite of legendary gale winds and freezing temperatures that were part of "normal operation." Pilots and their mechanics put flame pots under the engine to warm the block.

Before his polar feats, Vodopianov participated in the effort to rescue the *Cheliuskin,* personally appealing to the government commission supervising the rescue mission for permission to join in, and meeting Valerian Kuibyshev, head of Gosplan, for a personal interview so he might become a hero. The rescue mission required the pilots to jump from landing strip to landing strip across the nation, hopscotching further north, waiting days for weather to break in frustration, then flying again. Vodopianov eventually risked taking off in a blizzard, flying almost 5,000 kilometers before arriving at the

Vyacheslav Molotov, Joseph Stalin, unidentified, and Otto Shmidt welcoming pilots from the Arctic, ca. 1938. (Courtesy of the Russian State Museum of the Arctic and Antarctic)

Cheliuskin camp. He and his colleagues safely removed the *Cheliuskintsy* (the Cheliuskinites) from the camp on April 13, two months after the ship was crushed in the ice and sank.

In March 1937, another aerial assault on the Arctic began. Under direction of the Shmidt, a team of pilots flew from Moscow to Kholmogory near Arkhangelsk, to Naryan-Mar, to Franz Josef Land, waiting for the right weather to continue on to the North Pole. On Rudolf Island, the northernmost of the Franz Josef archipelago, wintering explorers met them expectantly as the polar aviators delivered supplies and LP records on which their families had recorded their "letters." "Hello, Papa!" More than two weeks later, on May 20, the pilots got instructions to leave for the North Pole. Vodopianov was prepared to lead the squadron on three engines when his mechanic informed him of an antifreeze leak, but eventually all four engines were operative. Approaching the Pole, the navigator dropped a "smoke bomb" to indicate better the terrain and winds, and on May 21 at 11:30 the expedition landed. The Glavsevmorput aviation expedition had brought special cargo: I. D. Papanin and three companions to participate in the first polar "drift station"; it drifted into the Greenland Sea some 274 days later. (On the history of these drift stations, see Chapter 4.) On June 6 the four planes took off for Moscow, after Shmidt and others delivered "moving speeches," Soviet flags were proudly anchored in the ice, and the gathered crews sang the Internationale.

In large part because of Soviet achievements, polar aviation became an international military and commercial reality. The increasing frequency and success of Arctic aviation during World War II led a number of commentators to see a revolution in commercial aviation through "the linking of the continents by a network of air routes traversing the Arctic basin" that would substantially cut distances between people living in northern countries. One writer spoke sanguinely about the challenges to be faced, pointing instead to the great savings in time, distance, and fuel. Temperatures at 30,000 feet and higher were no problem because they were relatively the same throughout the stratosphere, and in fact in polar regions there was less diurnal change and less temperature variation. Icing dangers could be overcome through efforts to prevent or substantially reduce formation of ice covering on wings and fuselage. As for oil lubrication, the problem was starting an engine, not running it, and special coverings on the hoods preheated the engine. The greatest problem was poor visibility because of low-lying clouds and fog, a common occurrence in polar regions where warmer air from lower attitudes flows over pack ice, especially in the summer months. Yet "fog in the Arctic is less dense and lower lying than it is elsewhere . . . [and] seldom rises to a height of over 3,000 feet" so that planes could skirt it. The real challenge was regular and emergency landing. Water landings were difficult because water was often filled with small lumps of ice, on top of which were the dangers after landing of sinking or getting hemmed in and crushed. Ninety percent of pack ice was too rough to be landed successfully, and pilots could not discern hazards due to the lack of visible relief even after several passes. Soviet polar aviators encountered all of these problems and then some.[79] Yet Vodopianov, Farikh, and the other polar pilots opened the possibility of aviation in support of Glavsevmorput research and development, commerce, and reconnaissance.

Stalinist Zimovka: Even When Planned, Unplanned

The exploits of Arctic explorers captured the imagination of citizens and leading officials alike. Newspapers, magazines, and radio covered the travels in great detail. Leaders—Lazar Kaganovich, Vyachslav Molotov, Joseph Stalin himself—sent congratulatory telegrams (in some cases "radiograms") to the scientists and sailors who sailed and drifted through the icy waters. The telegrams praised their Bolshevik heroism, selflessness, and determination. The campaign to create heroes operated both from above and from below as Arctic exploration, polar aviation, and other feats captured the public imagination.[80] The heroes pushed further north, charting weather, depths, and mineral wealth as they were dragged by Arctic currents, with Arctic ice floes, and under coal, diesel, and gas power. Many of the exploits resulted not because of the all-powerful Stalinist plan, but because of accidents, uncharted waters, and other failings. The explorers found themselves in these difficult straits that were not their responsibility. One reason was technological lag. In spite of the insistence that Soviet technology would enable the glorious future, in fact in all ways and in all sectors of the economy it lagged significantly behind that in the United States, Germany, England, and elsewhere. From tractors for agriculture, to machine tools in the metalworking industry and smelters in metallurgy, to tugs, freighters, and lighters in the shipping industry, and from the grain fields in Ukraine along the Dniepr River to the iron and steel furnaces of Magnitogorsk, and to the icebreakers that plied the Arctic Ocean, the lag persisted.

Still, in the 1930s explorers, scientists, and sailors pushed the limits of their ships further into the Arctic. As they had celebrated aviation records, the nation's citizens celebrated their great achievements of charting depths, producing more complete maps, tabulating natural resources, observing flora and fauna, and conducting themselves in a truly Bolshevik fashion. With regard to the latter point, this meant fearlessly engaging the elements to show what the new Soviet sailor could accomplish—that his Russian and European capitalist predecessors were unable to achieve. Yet this heroic effort often led the ships and their crews to disaster. The experiences of the *Sibiriak, Cheliuskin,* and *Sedov* reveal in startling relief all of these features of Stalinist exploration.

The only certainty was that exploration was an uncertain endeavor. It seemed that never did things work out as planned. In 1926, a team of Soviet explorers, equipped with three years of supplies, landed on Wrangel Island, now a large nature preserve at 7,600 square kilometers in the Arctic Ocean between the East Siberian and Chukchi seas. Clear waters that facilitated the 1926 landing were followed by years of continuous heavy ice surrounding the island. Attempts to reach the island by sea failed, and it was feared that the team would not survive its fourth winter. In 1929, the icebreaker *Fedor Litke* sailed more than five weeks from Sevastopol to Vladivostok and then to the Bering Strait, where it was caught in ice for nearly two weeks. Then it reached Wrangel, evacuated all settlers, and returned to Vladivostok in September, for which the *Litke* won the "Red Banner of Labor" award.[81]

Party officials and leaders of Glavsevmorput intended the *Cheliuskin* expedition to show the *Litke*'s achievement was not an anomaly and to demonstrate the power of the Soviet people under Stalin to conquer the Northern Sea Route. But when the ship, sailing from Murmansk to Vladivostok, was trapped in the ice in December—and sank in February 1934—the expedition instead underlined the extensive challenges to create a normally operating trade route and to supply Soviet outposts along the Arctic Circle with needed supplies. Hence propagandists instead used the event to reveal how the Party cared for the people, all people, even sailors, women, and children who were trapped in the Arctic ice, and that Stalin personally followed each day with concern. The crew members and the pilots who rescued them eventually were hailed as heroes, unlike those who wintered and drifted on the *Sedov, Malygin,* and *Sadko* in 1937–39; many of those who returned were accused of wrecking and were sent to labor camps, while several leading personnel were executed.

The *Cheliuskin* left Leningrad on the night of July 16, 1933, and headed for Murmansk under the direction of Captain Voronin and Glavsevmorput director Shmidt. Sergei Semenov, the secretary of the expedition, acknowledged that they were a quickly assembled bunch, not in any way a "collective." They came from different professions, backgrounds, and educational levels, all of whom would be needed for any polar expedition. Many had not even been introduced before departure. In Murmansk they added a group of construction workers who were headed for temporary work on Wrangel Island. In all, 111 people sailed for the Kara Sea and beyond, including ten women and one girl, taking 3,500 tons of coal and food supplies for up to eighteen months. The complete list ran from the illiterate carpenter and housewife to the professor of mathematics and philosopher. Soviet writers claimed that before the disaster they had become precisely a "collective" and that all had come to understand that whether "the heavy work of the stoker or the uninterrupted work of the janitor, all labor was a matter of honor, good name and heroism"—and that they loved their Soviet motherland.[82]

The *Cheliuskin* was not an icebreaker and in fact had been specially fitted with extra plates on the hull to strengthen its hull. Likely it should not have been forced deeper into the autumn ice, but these were heady times of Stalinist achievement of superhuman plans. With great difficulty under Shmidt and Captain V. I. Voronin, the expedition made it almost to the Bering Strait when it was caught in ice. The nearby *Litke* was unable to reach it before weather forced the latter to retreat to safe harbor. The crew drifted for several weeks toward the Pacific Ocean, when the boat was pushed back to the northwest into the Chukchi Sea. In February the ship's hull began to fail. On February 13 the ship began to sink; it had been stuck in ice for nearly five months when the ice pressure crushed the hull. The members of the expedition set up an ice camp—"Camp Shmidt"—in the Chukchi Sea where they wintered. In the spring they fought a constant battle against changing conditions and drift to set up an ice runway; they were saved by polar pilots as a result of which the pilots earned the recently established "Hero of the USSR" medals.[83] As the future director of AARI wrote in 1945, "From a formal point of view the travel of the 'Cheliuskin' ended unsuccessfully. However, the entire world saw of what the Soviet nation and the Soviet people where capable in the example of the 'Cheliuskin' epic."[84]

When Shmidt gave the order to abandon ship, the sailors, scientists, women, and children removed what they could—coal, kerosene, food, tools, wood, and, luckily, tents. They threw everything overboard onto the ice. The holds filled rapidly with water, and the ship sank into the sea, taking one man down with it. For a few moments a chaotic boiling water of the wreckage, logs, planks, and pulverized ice covered the surface.[85] From boards, logs, casks, detritus, and blocks of ice they built a headquarters barracks and a kitchen, both of which they located within 100 meters of the sinking to limit the distance they had to carry supplies and materials across the ice. Six carpenters and two masons joined in the effort. Six-by-eight beams gave the structures some stability. They used empty jars to fashion windows. They finished the task in three days.[86] The *Cheliuskintsy* were forced to live in tents filled to capacity and anchored to ice floes in Camp Shmidt. As the camp drifted it was constantly threatened by fissures in the ice; on March 6 the camp broke apart. A crevasse shot through the center of the barracks, although everything turned out OK: Fortunately not one timber or plank fell into the water, the barracks broke neatly into two pieces, and by noon the crew had put Camp Shmidt back in order.[87]

The *Cheliuskintsy* were face to face with the power of the polar winter, and there was little hope for quick rescue. They had enough food for about three months and batteries for the radio to maintain contact with the outside world had about a one-month charge. According to radio operator Ernst Krenkel, who participated in the first North Pole Drift Station (see Chapter 4), "People didn't feel the 30 degree cold, nor the force seven wind." One-third of the expedition members were Communists or members of the Komsomol. As Soviet propagandists duly noted, "This of course was a huge advantage that not one of the past bourgeois expeditions knew about."[88]

They encountered great problems putting up tents because of the solid ice. For the first few days several tents had sixteen to nineteen residents and people "slept literally on top of each other." They broke out new warm clothes that consisted of parkas used by local inhabitants and learned how to don them properly.[89]

The first step of course was to reestablish radio contact with the "motherland" and proceed, in heroic fashion according to official sources, to secure a properly Soviet existence on the ice. They had great difficulties with the antennae. It was nearly impossible to drive the towers into the ice, which was as hard as cement, then the wind blew them over, and in the dark it became harder and harder to find the axes, knives, rope, and other tools; if the sailors put them down, they lost them in the snow. Once they got the towers standing, they had the problem that the radio apparatuses were in Fakidov's tent where the women, children, and other individuals suffering from the cold had gathered because the other tents weren't up yet. The radio men discovered a problem in reaching Wellen (Uellen in Russian, Ugelen on USCGS charts), a small village with a population of several hundred located near Cape Dezhnev where the Bering Sea meets the Chukchi Sea; the antennae could produce only 300-meter waves and they required 450 meters. On

February 14 they succeeded finally in increasing the amplitude and from that day forward sent and received regular radiograms at Camp Shmidt.[90] From Moscow, officials monitored their condition and position and planned a rescue airlift.

The *Cheliuskintsy* saved the following food items: sixty boxes of tinned meats, five boxes of tinned fish, fifty boxes of crackers with eighty packages each, sixty rounds of cheese, nineteen boxes of butter at twenty kilograms each, vegetable oil, bottled tomatoes, twelve bags of dried potatoes, three of rice, three of bread crusts, three of candy, a half bag of cocoa, some chocolate, dried peas, ten boxes of dried milk with forty jars each, chair, flour, kasha, sugar, chicken, and pork. They calculated a modest daily ration, but as people were evacuated bit by bit by airplane, people could eat "whatever they wanted." They had very little in the way of cookery: one cauldron, a large number of forks, but very few spoons, no pots, and little else. They rescued fifty mattresses, sixty-seven pillows, and seventy blankets. Last, they secured twelve boxes of bullets, gunpowder, five rifles, and seven revolvers to use against polar bears.[91]

Breakfast at 7:30 a.m. was likely the hardest meal to prepare. For water, the cooks used older ice rather than snow because it had a lower salt content that they melted on top of the cook stoves; each morning they awoke earlier than the rest of the camp to defrost the day's water. Because no one had any idea how long they'd be there, it was a challenge to fix daily rations accurately. They saved cocoa as a treat to raise the general mood after a stretch of particularly bad weather or to celebrate on another occasion when the weather was fantastic and the next convoy of airplanes had arrived to carry the next group of *Cheliuskintsy* away. Daily labor consisted of splitting wood, shoveling snow, and garbage detail. Lunch at 12 was a simple event of soup, rice or kasha, and sometimes potato puree. At 3 p.m. they got 150 grams of crackers and two pancakes. At 16:30 the leaders gathered in Krenkel's tent to get the news from TASS. Camp leaders organized evening meetings on politics and ideology, for example a circle on dialectical materialism in Shmidt's tent that resulted, we are told, in hours of lively discussion. The *Cheliuskintsy* played cards and chess and held literary evenings . . . but they could count only four books in the entire camp.[92]

Once safe and secure, having taken an inventory of supplies, and having established radio contact with the nearest base, Shmidt ordered the construction of a rudimentary airstrip to facilitate evacuation of the survivors. The men worked in three shifts to transform a seemingly stable ice floe into airstrip some 600 meters long with a reserve field in case the ice broke up. They identified a site about five kilometers from camp. They faced the constant challenge of recognizing fissures and cracks, of determining which they could repair with ice, water, and snow and which they could not. More than once they had to abandon work when huge cracks appeared and formed crevasses and the ice broke up. Otherwise, they focused on smoothing wind-formed and rock-hard ice drifts that formed after each blizzard. Hillocks and moguls created their own set of challenges. On one occasion they nearly lost two sailors who had set out to the landing strip—Valavin and Gurevich—who returned disoriented and barely alive in the dark.[93]

Polar aviators began constant flyovers to study the possibility for a rescue attempt that generated international concern; the United States allowed Soviet pilots to fly search missions from Alaska. In early March the airlift began. Over the next month a caravan of six pilots removed all of the *Cheliuskintsy* safely from the ice. What had been an Arctic disaster that resulted from political pressures to challenge the ice in a ship totally inappropriate to the task turned into a momentous, victorious spectacle that melded feelings of national belonging to the technological feats of Stalinism covered extensively in the Soviet press.

Each year under Glavsevmorput the number and riskiness of expeditions grew substantially, until the critical year of 1937 when bad weather and poor planning combined to result in a terrifying event. Equally the result of political pressures and Arctic weather was the twenty-six-month drift of the *Georgii Sedov*. The focus of a series of books, movies, and other media events, the plight of the *Georgii Sedov* filled a 1940 Glavsevmorput publication. Drawing on newspaper and wire reports, telegrams, and reminiscences and on the bulletin board newspapers of the *Sedov*, the volume revealed the bleak loneliness of the scientists, passengers, and crew as they drifted helplessly through the Kara Sea. Yet these unfortunate men truly reveled in their record-setting drift and indeed learned a great deal about man–nature interaction, while sharing this knowledge with the nation. After a risky aerial removal from the ice of the majority of passengers and crew, the icebreaker *Sedov* drifted for twenty-six months from the Novosibirsk Islands to

Spitsbergen with a skeleton crew of fifteen sailors. The drift "entered the history of our people and Soviet sea-faring as an example of the self-control and courage, courage and will, unlimited love for and allegiance to the socialist motherland." They lived through nights of blizzards and violent storms. But no matter the challenges and dangers, the sailors, Soviet journalists assured readers, calmly worked "for the glory of the motherland."[94]

The unanticipated disaster that the authorities transformed into a heroic passion play of Soviet determination against the elements began on October 23, 1937, in the Laptev Sea somewhat northern of the 75th parallel when the *Sedov,* along with the steamships *Sadko* and *Malygin,* suddenly were trapped in the ice. There were 217 people on board who wintered until February 1938 when the weather cooperated sufficiently for the authorities to send an air rescue mission that brought in supplies and evacuated all but the remaining expeditionary force of scientists and sailors. Although the ships were 1,100 kilometers from the shoreline, the pilots managed to land three times to evacuate 184 individuals who had wintered thus far. In summer of 1938 Glavsevmorput sent the *Ermak*—the oldest icebreaker of the fleet—to the northern ice. While losing two of three propellers, the *Ermak* managed to free the *Sadko* and *Malygin,* but the *Sedov* remained iced in in part because its steering mechanism had been fully damaged. Later on, the "flagman" of the Glavsevmorput fleet, the icebreaker *Stalin,* attempted to reach the *Sedov.* This was the first Arctic voyage of a new generation of ships built at the Ordzhnikidze Factory in Leningrad. Setting out in the late summer and encountering extremely hazardous conditions, the *Stalin* sailed within fifty nautical miles of the *Sedov.* Any further effort would have risked the ship and its sailors, too, so it turned back, and from August 29, 1938, the *Sedov* drifted alone. During the drift, the sailors slowly repaired the steering mechanism, and they tried to keep the engines, which had been fully mothballed to protect them from deep freeze, in condition to restart when they were freed.

On the first and second anniversary of the drift, Stalin and Molotov personally greeted the *Sedovtsy* as they came to be called. The "warm words . . . energized the Sedovtsy." The sailors claimed that Stalin's words "were that guiding star that guided us in the fog, and in the gloominess of the polar night, and in time of blizzards." Children and teachers and ordinary citizens sent greetings and poems and drawings to the Sedovtsy care of the main office of Glavsevmorput. Propagandists claimed that while Scott and others who had wintered in the north experienced the misery of loneliness and wrote parting letters to their families, the Soviet explorers and researchers did not know these feelings "of isolation . . . of detachment from society, from their government. . . ."[95]

According to the official narrative, the remaining crew of the *Sedov* accumulated scientific achievements befitting heroes. Because of the fact that many of the sailors lacked scientific training, all men were put to work carrying out temperature, ice, depth, and other measurements. They took a series of oceanographic observations, and they measured distribution of water masses and other factors that had an impact on the amount and distribution of ice, ocean currents, and weather. They were able to hypothesize about the entire ice cycle. When the ice set in around the Novosibirsk Islands in the Laptev Sea, the ice was about thirty centimeters thick; when the *Sedov* reached the very same ice near Spitsbergen, it was deeply iced in. Meteorological observations on Bolshaia Zemlia gave the Meteorological Service several keys to developing forecasts for the entire northern hemisphere and confirmed the work of the drifting "North Pole" station of Ivan Papinin that cyclones frequently develop in the central Arctic. The crew carried out gravitational observations every 15 miles. In several cases the drift of the *Sedov* carried further than that of the *Fram*, with which it intersected.

The technological feats of the Stalin era were turned on their heads, not to reveal the dangers of living in a spasmodic industrial society, the risks of battle against the Arctic, and certainly not the recklessness of the leaders, but the solidarity between Stalin, polar explorers, and the masses and the victory of the USSR over the ice. As an official publication noted, "With unwavering attention the entire Soviet people followed the drift of the . . . 'Georgii Sedov' on board which . . . the sailors carried out their Stalinist watch." Notwithstanding the great difficulties, the crushing ice, hurricane force winds, and deadening deep freezes, they managed regularly and according to plan to carry out a broad scientific program. In the summer, the winter, in the polar night, and in the foggy days of the Arctic spring. They "always carried to the end that task they had begun . . . the Soviet people, who fulfill the task of the motherland, of Stalin, will not tolerate any difficulties." The publication continued, "Heroism, dedication, courage in struggle with all dangers—all of these are the highest moral qualities which the Party teaches in the Soviet people."[96] And from the beginning Stalin took a direct interest in the drifting ship and the sailors, their lives and their scientific works. He "encouraged and inspired" them from afar. In summer 1938 he was directly involved in the decision to send his namesake icebreaker to free the *Sedov* from the ice, although allowed it to return to safe harbor rather than risk the *Stalin* to become icebound on her maiden rescue voyage. When the drift sped up, Comrade Stalin ordered a final rescue mission. The *Stalin* reached the *Sedov* after 750 days in the Greenland Sea at the 80th parallel and accompanied the ship to Murmansk. The unplanned zimovka led to charges of "wrecking" against top officials and scientists and a purge of Arctic explorers in which leading personnel perished, but which Shmidt managed to avoid.

On the eve of World War II polar explorers from the Soviet Union had established a heroic imprint on the northern latitudes. Expeditions had become almost common, in the way that, for many people, space exploration was a mundane endeavor by the late 1970s. Drifting past the North Pole and stepping on the moon were regular scientific activities. Glavsevmorput, no longer a bureaucracy but an empire of icebreakers and airplanes, mining and lumbering enterprises, food production and delivery services, reindeer farms, and natural and social scientific research institutes, conducted activities in an increasingly public atmosphere. Its successes played out in *Pravda, Izvestiia,* and *Pravda Severa,* on bulletin boards, and on the radio. Its pilots flew over the North Pole, its captains pushed the limits of the ships along the frozen Northern Sea Route, and when disaster struck—the sinking of the *Cheliuskin* or the 812-day drift of the *Sedov*—these too became achievements of the Soviet people only one generation after the glorious October socialist revolution. Stakhanovites and Party members had been victorious over the elements and enemies and "wreckers" who had apparently also sought to occupy the Arctic.

A parallel Arctic and empire existed in the polar night—and in the midnight sun of the summer—where bureaucrats, Party members, researchers, pilots and captains, workers, and reindeer herders worked toward conquest of the elements and those enemies. The gulag system of prison labor camps filled northern latitudes at the same time as Otto Shmidt expanded

Glavsevmorput activities in pursuit of the economic, scientific, and military assimilation of polar regions.

As for Glavsevmorput, in 1953 the empire ceased to exist by the stroke of a pen; it was first subsumed directly under the Council of Ministers as part of the Soviet Navy, but with significantly fewer rights and responsibilities, then entirely incorporated into the navy. In 1963, after yet another administrative change, AARI found itself under the Main Administration of the Hydrometeorological Service. The directors of AARI wrote in 1991 that "the decision of the government about the liquidation of Glavsevmorput was, in our opinion, mistaken. Its impact told negatively on the development of the Soviet far north."[97] Yet Glavsevmorput had grown too large, and it had too many responsibilities ranging from transit to mining to shepherding indigenous people to ship repair. On top of this, Arctic research fell on hard times after World War II as budgets were focused on rebuilding the industries and collective farms of the European parts of the nation.

3

The Role of the Gulag in Arctic Conquest

Very dear leader, glorious Stalin, my light,
He fitted swift steam icebreakers,
He sent them to the *Papanintsy,*
To save each and every one.
From Murmansk, from the city,
From that harbor, from the ship harbor.
The icebreakers headed out
The icebreakers, our Soviet falcons.

—MARFA SEMENOVNA KRIUKOVA (1939)

The vast tundra and taiga of the north were home to fisherman and trappers, peasants and Old Believers who had escaped persecution of the Tsar and the Russian Orthodox Church, and Nenets, Saami, and Komi people who also trapped, fished, and raised reindeer. The region was sparsely populated, with perhaps one or two people on average per every square kilometer in Arkhangelsk and one person per every four square kilometer in the Nenets Autonomous Okrug. Soviet rule added two kinds of settlers: exiles and gulag slave laborers. Many of the scientists, explorers, pilots, and captains who explored the Arctic fell into the Stalinist labor camps—the infamous gulag system—arriving as prisoners or later accused of crimes that led to their incarceration.

Perversely, at the same time as Soviet officials praised the polar explorers, party officials, workers, and sailors for subjecting Arctic climate, geography and geology, latitude, rivers and lakes, the ocean and seas to human control, they destroyed the lives of many of the same individuals who had devoted careers to the study of polar regions in slave labor camps, depriving the state of experienced and dedicated individuals with scientific knowledge of the region. Simple workers, polar pilots and Arctic explorers, selfless party members and criminals worked in an ever-growing network of camps that were founded to subjugate Arctic resources to state programs. The prisoners established new norms of digging, lifting, boring, dredging, felling, laying, pouring, planting, harvesting, draining, raising, and lowering as they completed the tasks of road building, rail laying, canal construction, mine excavation, and factory construction.

To a great extent the raw human material of the gulag was intended to make up for shortages of technology.[1] The camps were, in fact, forms of technology: human machines with limited operational capacity and a short, if lifetime, guarantee. By the early 1930s the secret police had determined to use hundreds of thousands of prisoners in such hero projects as the White Sea–Baltic Canal and also in forestry, agriculture, and such other prison camps as Sevdvinlag, Kargopol'lag, Mekhren'lag, and Iagrinlag (the Northern Dvina, Kargopol, Mekhren, and Iagry Island camps). The camps served virtually all operations—including those of the Glavsevmorput empire, for example mines on the Arctic coast at Amderma that were tunneled deep into the permafrost and produced spar for the cement industry, fluorite, copper and nickel for smelters on the Kola Peninsula, and rare metals in smaller operations across the northern latitudes. Like managers in the lumber, fisheries, and other industries, the Amderma mine's managers lamented the rudimentary technology available to them and frequently requested supplementary budgets for road construction, dump trucks, and machines for loading and unloading, but resolutely turned to more human labor to build, dig, pour, load, and dump.[2]

Employees of Glavsevmorput, the Commissariats of Water Transport, of Ways of Communication, of the Forest Industry; the civil aviation administration and Gidrometsluzhba (the Weather Bureau); specialists and graduate students in geology, topography, geodesy, and cartography; employees and managers in Arctic industrial enterprises, shipbuilding, and steamship authorities; and indigenous peoples engaged in hunting, fishing, and reindeer herding activities themselves were arrested, in some cases released, then arrested again, interrogated and shot, or sent to serve years of hard labor in the gulag.[3] They were accused of counterrevolutionary activity, terrorism, spying-terrorism, diversionary terrorism, counterrevolutionary Trotskyism, wrecking, and anti-Soviet agitation. Generally, the terror acted against the upper-ups rather than the lower workers, although in the search for mortal enemies of socialism the secret police tracked down mechanics, tractor operators, and lumberjacks. When a death sentence was handed down, especially in the period of the Great Terror between 1936 and 1938, and including three public show trials of leading Old Bolshevisks, a "troika," a martial committee of three men, deliberated for fifteen or twenty minutes, without the presence of the accused, cared little whether he had pleaded guilty under duress, and sentenced him to death. Usually the sentence was carried out within moments. Even with the archives opened after the collapse of the USSR in 1991, it has been difficult to determine the fate of many victims of the repression, let alone where they were buried to inform their families, for the current Russian leadership seems determined to take offense over the criticism that the nation has yet fully to come to grips with the legacy of Stalinism—although the Russian Federation is a different country.

One cannot rationally explain the reasons for the repression of so many individuals irrespective of their innocence, class background, nationality, employment position, and age. Why would the authorities, and their secret police accomplices, arrest and destroy so many innocent people? Among the many explanations explored in dozens of books are the paranoia of the Stalinist regime; the fear of enemies within and without the nation in view of the USSR's international isolation; fear of popular uprisings in burgeoning cities that the authorities could not feed or house; the specter of fascism in the 1930s; industrial backwardness; shortages of capital but willingness to use coercive labor; the momentum that grew within the NKVD the further one got from Moscow as local and regional agents of Stalinism sought to show their vigilance through more and more arrests by uncovering of more involved and absurdly convoluted plots; and the use of torture and denunciation to identify still other counterrevolutionaries, Trotskyists, spies, wreckers, and enemies of the people.

Many individuals were arrested because they were of the wrong social class: members of the bourgeoisie, somewhat wealthier peasants called kulaks, or from families of priests. Individuals who had belonged to political parties other than the Bolsheviks and even "Old Bolsheviks" became enemies of the people. Minority nationalities were accused of crimes against Stalin—entire groups of them—Crimean Tatars and Volga Germans—and forced into exile. So millions of these people—kulaks, minorities, old Bolsheviks, Mensheviks, individuals who had belonged to other political parties, members of the bourgeoisie, people with foreign contacts, foreign language skills, experience, or relatives abroad, and many others because of denunciations—were interned in the gulag system.

An economic explanation exists, too: The authorities determined to fill such seemingly empty spaces as the tundra and taiga with cheap labor dedicated to securing the landscape, identifying resources, developing them, and shipping the booty in one form or another to the privileged people who lived in Moscow and Leningrad. In addition, there were great uncertainties of relying on volunteers, settlers, and exiles to secure the Arctic until the Khrushchev and Brezhnev periods, when the state relied more on incentives to individuals—subsidies in terms of housing and salaries for workers, officials, and specialists who chose to move there.

Granted, political reasons were crucial on several counts. First, at least initially, "reforging" or remaking and reeducating prisoners had a central place in the establishment of the camps. Second, even in the late Stalin period the authorities did more than give lip service to reeducation, but hoped to see it come to fruition and believed the camps might be "redemptive" in some ways. For many a prisoner, his crime was important for it determined the length of sentence and type of service. Third, securing Arctic resources supported the leadership's legitimacy and the USSR's military strength. But by the 1930s the political leaders and camp authorities had focused on economic issues, and many, if not most, of the people who survived the camps were released not as reeducated, but required to live far from their homes. The camps helped to educate the illiterate, but reforging ultimately was a secondary or tertiary aspect of camp life, and the prisoners knew that getting more labor out of them was the real goal, and they must have realized that their sentences and crimes were arbitrarily determined.[4] The camps quickly became military construction, transportation, forestry, mining, canal, and other water works; they were hydroelectric, irrigation, agricultural, and other technologies based on human fodder to meet the ever-growing demand of the state to industrialize rapidly.

The White Sea–Baltic Canal: Stalin's First Arctic Gulag Project

The gulag system originated on an archipelago of the beautiful Solovetski islands in the middle of the White Sea, the site of a monastery dating to the fifteenth century and where a botanical research station was opened in the late nineteenth century. The islands resemble those of coastal Maine in their gentle relief, rocky shores, and rich and vibrant flora that include cherry trees, dog rose, cedar, maple, guilder rose, rowan, and hazel. The Solovetskii Camp of Special Significance (Solovetskii Lager' Osobogo Naznacheniia, or SLON) was founded in October 1923. Thousands of prisoners, including writers, biologists, poets, playrights, priests, philosophers, and others died here. Several vicious camp leaders transformed SLON into a construction operation and exported thousands of prisoners to the mainland, to the village of Kem on the White Sea coast, to build a canal to the Baltic Sea.

Canals and hydroelectric plants were the first modern technologies embraced by the Bolsheviks for their utopian promise of creating the electricity and transportation networks needed to build communism and for bringing workers together in collectives dedicated to creating an industrial society of plenty. Those canals and dams in the Arctic did not really move forward until the creation of the gulag as a source of labor. Yet even during the civil war of 1918–20, engineers and the political authorities planned enthusiastically to electrify the nation through hydroelectric stations augmented with boilers fed with low-grade peat and coal and to build irrigation systems and majestic transportation routes. Many of these projects had prerevolutionary roots. At the beginning of 1919 in Leningrad at a meeting of the Council of People's Commissars, Bolshevik officials approved the formation of an office to oversee the design of a White Sea–Baltic Canal that cut across Karelia just at the Arctic Circle that I shall refer to here in shorthand as Belomor (White Sea). V. M. Nikolskii, who in 1915 put forward a variant of the project, was assigned project design responsibilities, which he completed by 1922. He and his colleagues set forth two variants, one with sixteen locks and twelve dams and the other with thirteen locks. But without financing and an economy that continued in a free fall until the mid-1920s, neither project moved forward.

When Stalin came to power, many large-scale nature transformation projects that had languished due to political and economic uncertainties, several of which were in Arctic regions, found political will, financial support, and the labor to proceed. Large-scale projects, which as Hughes pointed out involved increasing complexity and included institutional, financial, governmental, and other components,[5] became emblematic of Stalinism.

Stalin and his followers believed that those projects were the most efficient way to transform the economy, nature, and the people within it, and especially to transform Arctic regions. They facilitated the gathering of a huge labor force in one construction site for political indoctrination and technical training. They allowed for priority marshaling of machinery and equipment. With the establishment of the system of the gulag camp system, they enabled the harnessing of a coercive labor supply system to hero projects far and wide, and especially nature transformation projects, forestry operations, and road and railroad networks.

On February 18, 1931, the Council of Labor and Defense approved a decision to build Belomor. The canal would address one of the significant military concerns of the USSR—gaining access to Baltic ports from the White Sea and giving the USSR another "warm water" port in addition to those on the Black Sea. (Murmansk harbor would become such a strategic port only after World War II.) But given the low population density in Karelia—one-half that in the USSR on average—and very poorly developed industrial and transportation infrastructure, the planning for the canal created immediate difficulties. Planners took advantage of the burgeoning slave labor camp at SLON system to feed Belomor—which then fed an ever-growing number of other projects and camps throughout Arctic and subarctic regions through the 1930s, 1940s, and 1950s. A motley and vicious group of former Cheka officers—and former prisoners—directed the operations.

Iakov Davidovich Rapoport oversaw the construction of a number of hero projects of the gulag from the 1930s through 1950s and was central to the murderous effort to build a canal through Karelian forest and swamp. He was born in a white-collar family in Riga, Latvia, in 1898. His family fled to Voronezh in 1914 so he could to avoid being forced into service in the German army. In 1917 Rapoport graduated from a high school, and in the fall he entered medical school at Derpt University, which had been evacuated to Voronezh, and completed three years before being drawn, like many young men, into full-time work as an investigator of the Cheka. He became deputy director of the Voronezh Cheka, served as an agent in the Main Police Administration (GPU) of the Ministry of Internal Affairs (NKVD) of the Russian Federation, and in 1922 moved into the NKVD bureaucracy. (In 1934 the NKVD assumed secret police functions for the entire USSR.) Rapoport then returned to the economic division of the OGPU. From July

1930 he held a series of leadership posts in the camp administration, in November 1932 he was designated deputy director of the slave organization to build Belomor, Belbaltkombinat (BBK), and from August 1933 to September 1935 was its head. On the basis of this experience, and his ruthless ability to organize prisoners to achieve targets in inhuman conditions with inadequate machinery and equipment, Rapoport was next assigned to head construction of the Rybinsk and Uglichsk hydroelectric facilities on the upper Volga from 1935–40 and simultaneously in 1940 was head of the Volzhsk labor camp. Rapoport was the logical choice to be the head of the newly created Main Administration of Hydroelectric Construction of the NKVD. This meant that parallel with construction at Rybinsk and Uglichsk and the Volga–Baltic Canal, Rapoport also supervised the construction of the Northern Dvina water route, reservoirs, harbors, and other facilities connected with the Moscow–Volga Canal—and river engineering as far as Nikolaevsk-na-Amure and the Nakhodka port.[6]

Nature transformation projects, and especially geoengineering of rivers, captured Stalin's imagination. At that time the United States was engaged to bring technology and democracy through the Tennessee Valley Authority and the Bonneville Power Authority in Washington state through new hydroelectric power stations and irrigation systems. Stalin competed with the United States to build the largest stations in the world as symbols of socialist superiority, even in the Arctic region.

The modest Rybinsk and Uglichsk stations moved ahead very slowly and reached full operation parameters only in the 1950s, in part owing to World War II. The head of the NKVD, Laventry Beria, issued a proclamation on June 28, 1941, to cease work on a large variety of projects owing to the war. Arctic and subarctic projects in the Russian northwest included the Kandalaksha and Kola Aluminum Factories, the smelter Severonikel', the Belomor port, the Segezhskii Pulp and Paper Mill, and various Karelian facilities conquered from the Finns in the Winter War. In all the NKVD constructed dozens of hydroelectric power stations, fertilizer factories, the Konosha-Kovzha, Lakhi-Khizavarskaia, many other railroad lines, and nineteen different road and bridge projects that filled the northern latitudes. The gulag remained central to Stalinist Arctic projects until the late 1950s, while Rapoport's career continued from the Arctic to the Ural Mountains and into the nuclear enterprise.[7]

In several of these projects, Rapoport joined another of Stalin's water workers, Naftali Aronovich Frenkel (1883–1960), a Ukrainian Jew with a career as a bandit and huckster during the New Economic Policy (a "NEPman" who became a millionaire). He was arrested for his efforts, then demonstrated the ability to organize construction when he himself was a prisoner in SLON and was promoted to supervise Belomor—and later other gulag projects as an NKVD colonel—with inhuman detachment. Before World War I Frenkel worked in a series of business ventures and was sent to a technical school in Germany to study transport.[8] He returned from Germany to occupy a position in the Odessa port. During the civil war he worked for all sides, joined a criminal group, also served the Cheka, and was arrested. The police took him to Moscow, pronounced a death sentence, but commuted it to a term at SLON where eventually Frenkel developed his system of hard labor.

As the camp filled with alleged enemies of the state—intellectuals, priests, scientists, writers, and crafty political opponents—sanitary conditions deteriorated. Frenkel told the camp directors that he would supervise the construction of a public bath, or banya, within twenty-four hours if he could pick his crew and if the men got a hot meal with meat and vodka; they got the coveted meat and vodka and succeeded in building the bathhouse. (Even the gulag camps required bath and disinfestation facilities, for epidemics of respiratory and gastrointestinal viruses and diseases and plagues of insects lowered worker efficiency.) Working up from prisoner to someone in charge of construction, he converted his status to that of a boss. He had also demonstrated that the camp should demand labor from the prisoners to contribute not only to the camp, but to the national economy. SLON, which had used various cultural forums to reeducate the political prisoners, shut its "theaters" and newspapers, ended evening strolls, and set up various commercial enterprises. In 1927 Frenkel was transferred to the small port of Kem on the White Sea where he ran Karelian forestry enterprises and a section of the Murmansk railroad. On July 19, 1930, Frenkel was named senior consultant in the production-economic department administration of all gulag camps in the country under Rapoport.[9]

SLON was thus no longer an experiment to reeducate prisoners, but a cruel source of cheap labor. Its operations had convinced members of the Politburo to use slaves in a series of correctional camps to build infrastructure in the seemingly empty taiga and tundra—regions of low population densities but with important resources. By June 1, 1930, seven camps with roughly 170,000 prisoners fell within the Administration for Corrective Labor Camps, while in the forestry, construction, and reclamation departments of SLON already more than 62,000 prisoners toiled. From these immodest beginnings in the Russian northwest, the gulag spread across the northern latitudes wherever geologists found deposits of valuable ores or chemicals, foresters determined that nearby fir, spruce, and birch would suffice for construction lumber, fisheries specialists estimated that rivers, lakes, and seas had adequate catch, and party and labor camp officials could figure out how to supply labor to build roads and railroads.

Even as a prisoner Frenkel reorganized the prison administration not to fulfill the goal of "political education," but as a business venture to operate according to "self-financing" principles. The prisoners began to fell forest, lay rails, raise buildings, and raze political prisoners. Hard labor suddenly became attractive to the Moscow bosses who realized what to do with prisoners and just as suddenly came to believe they were catastrophically short of them. Requests flew into the Lubianka Prison in Moscow, the central processing facility in the nation, where initial interrogations and tortures took place. In 1930 Frenkel took over direction of a construction camp in the Komi republic, renowned for its timber resources, and by 1931 he was back on the White Sea as construction manager of "Belbaltkombinat" and a year later deputy director of the Belbaltlag (the labor camp itself). Economic, not political or ideological, concerns had become the major motivation.

According to one reminiscence, Frenkel was single-minded about the canal to the point that, besides kasha and bullion and tea, he didn't eat anything, never took time off, nor went to the theater. He knew only how to force the pace of construction. He insisted to begin excavation at various sites in the spring when a foreman knew it was too early, when the ice stubbornly blunted shovels or the sodden ground swallowed men to mid-calf. He might stand for an hour at the edge of some foundation pit, then call the engineer over and with two or three well-chosen, poisonous phrases destroy the self-confidence of that individual. He remained a distant, unknown quantity, and no one can say with certainty what he had read or enjoyed doing.[10]

For all his work in November 1932 Frenkel won an Order of Lenin and, after the construction of the canal, transferred to a new trans-Siberian railroad, the Baikal-Amur Mainline (BAM), being built by BAMlag. This camp also controlled vast quantities of manpower, money, and other resources. BAM was designated an "all-union construction project," so that agitators descended to recruit people especially in Amur province. But they did not provide anything in the way of creature comforts in the way of housing and dining halls for the first volunteers, and there were deep freezes, on top of which the project was poorly financed, so they left as fast as they came. The solution was to send Frenkel. Frenkel ordered prisoners to work summer from 8 a.m. to 12 midnight and in the winter from 9 to 9. The laborers could hardly escape because there was only one road in, and the worksites were generally surrounded by swamp and lakes. Those who tried were shot on the spot by guards, and a legend spread that if they nabbed someone, they brought back his ears as proof. Frenkel served as head of BAMLag until 1937 or 1938 when he was sent to Dalstroi—apparently as a prisoner again, but where the NKVD drew on his experience to build railways. (Dalstroi—the Far North Construction Trust—was dedicated to gold mining and construction in the Kolyma Region. One of the largest gulags, it was closely connected with Glavsevmorput, but because it was far from the Russian northwest it does not fall within this Arctic history.) All work on BAM ceased in 1942 when rails were ripped up and transferred to build a supply route into Stalingrad for the crucial World War II battle.

During the Great Terror, of course, Frenkel was arrested, and his remaining living enemies certainly rejoiced to hear that he likely had been shot. But NKVD chief Laventy Beria freed him in 1940, and he received the title Corps Engineer and a second Order of Lenin—and in 1943 a third order. But Lenin's medal did not provide enough security. In April 1947, sensing fast-approaching purges, he retired for reasons of personal health. With people like Frenkel and Rapoport running the show, the Soviet vision of industrialized nature, manpower, and a coercive political system were in place to build canals, roads, railroads, mines, and factories in the Arctic.

In Glavsevmorput and Belomor, Stalin and his politburo colleagues had abandoned any measured approach to develop Arctic resources. The coercive effort spun out of control in spite of central planning. There was no time to listen to reason or to pause to catch one's breath. A blizzard of resolutions, orders, and findings enveloped the tundra, accompanied by armies of prisoners and volunteers sent dutifully northward and eastward with minimal supplies, yet with maximum certainty that conquest was inevitable, even if the extent of resources, the cycles of the seasons, and the fluctuations of rivers, winds, and temperatures remained unclear.

By the summer of 1932 some 125,000 prisoners, but virtually no machinery or equipment, had been assembled, most of whom were de-kulakized peasants or political prisoners accused of participation in counterrevolutionary organizations. Within twenty-one months, on August 2, 1933, the "Stalin" White Sea–Baltic Canal, at 227 kilometers from Lake Onega to the White Sea, had been finished; tens of thousands of prisoners who worked with shovels and wheelbarrows fashioned on the spot perished, although according to several estimates as many as 300,000 prisoners died. An estimated 5,600 men died largely from malaria on the Panama Canal during the U.S. phase of construction from 1904–14; more than 100 men died in constructing the Hoover Dam and another forty-two from pneumonia over a roughly five-year period beginning in 1931; five perished building the Empire State Building in the mid-1930s. Worker safety was a significantly greater concern under capitalism than under socialism. Two submarines, two destroyers, and two escort vessels made the first trip through the locks indicating the military significance of the canal for Stalin. What had the prisoners cobbled together with their hands and souls? The canal included nineteen locks, sixty-four dams, twelve spillways and water discharge outlets, and reservoirs with a storage capacity of 7.1 billion cubic meters of water that inundated forty-two villages and approximately 80,000 hectares of forest. They fixed the cost of the canal at roughly 100 million rubles, which convinced them to rely on slave labor for other Arctic nature transformation projects.[11]

During construction, to keep costs down, in addition to using slave labor, they attempted to employ lumber where possible instead of steel, for example in the doors at the locks. The major requirement was that the lumber had to be either fully submerged to slow rotting or accessible for quick and easy replacement. Frenkel observed that the lumber used in ships lasted twenty-five years. The point was that in twenty-five years the USSR's metallurgical industry would be outproducing America's, and so it was reasonable to save metals and use wood in the canal.[12] Ultimately, according to patriotic

Stalinists and patriotic post-Soviets alike, the BBK was built inexpensively, but "with high quality" and on schedule through miraculous labor- and equipment-saving innovations.[13]

Upon the canal's completion, nearly 13,000 prisoners were freed, but 60,000 "canal-armyists" were sent to toil on the Moscow–Volga Canal, BAM, and forestry operations and other projects. In fact, few of the canal's construction brigades were released, but continued under Rapoport to assimilate the lands contiguous with the canal. In 1937 laborers for the Severonikel Metallurgical Factory in Monchegorsk were hived off the BBK, in 1938 others were sent to Segezhlag to erect the Central Pulp and Paper Mill in Segezha, and in 1940 prisoners were amalgamated in a series of other construction, metallurgical and mining, port and shipbuilding, and other gulag camps. So unwilling were the bosses to part with their slave laborers that even those who had finished their terms were booked again, found guilty again, and given additional terms; in 1939 86,567 prisoners remained in the canal's gulag, of whom 29,706 had been found guilty of counterrevolutionary activity, and also roughly 28,000 individuals in 8,505 families of dekulakized peasants. If in 1925 there were roughly 5,000 workers in Karelia, then in 1932 their number had grown to 23,000, and at the Belomor canal the number oscillated around 50,000 persons, while on the eve of World War II BBK had 81,446 "employees" of whom 70,811 were slaves, including 4,180 women and 4,097 guards.[14]

Arctic paper mills, an aluminum smelter, a metallurgical factory, hydropower stations, and a canal required more than armies of poorly fed prisoners with hand tools. They required poorly fed engineers with pencils. The secret police had already begun gathering engineers in Moscow in Special Design Bureaus, many of whom had been swept up in the Industrial Party, Shakhty, and other "affairs" orchestrated to indicate to society and specialists that they, too, were under suspicion for potential and imagined crimes against the state. These design bureaus fed special engineering labor camps connected to gulag operations called a *sharashka.* In this way the gulag had access to coerced scientists and engineers for projects ranging over time and space from canals and smelters to submarines in the shipbuilding city Molotovsk (see Chapter 4) to jet airplanes, rockets, and nuclear weapons.

The water to supply the BBK was sufficient at all times, even in drought years, and this was the result both of natural circumstances (the prevailing landscape of swamp, bogs, and lakes) and of the damming of several lakes to supply the canal, in particular Khizhozero.[15] In the fall of 1929 the NKVD had uncovered a "counterrevolutionary organization engaged in wrecking" in the central and local administrations of water routes of the Commissariat of Ways of Communication. They arrested and convicted approximately fifty engineers, hydrologists, and transportation specialists including Natalia Kobylina. The uncovering of the plot served several purposes. One was to put all Soviet engineers with independent minds on notice to subordinate their technical concerns to others' political dreams. A second was to secure engineers to design and build the canal. Kobylina received a ten-year sentence. Kobylina was born in 1889, the daughter of a naval officer, in Sevastopol and finished gymnasium. She entered higher women's engineering courses at St. Petersburg Polytechnical Institute, married a student from the Institute of Ways of Communication, Kobylin, and changed her family name. They joined not only in matrimony, but in lock and canal construction together. In 1918 when the government formed the Main Committee of State Construction in the Supreme Economic Council, she moved there as engineer, then to Central Administration of Water Transport. She divorced and in 1921 married an employee of the forerunner of Gosplan, Georgii Chernilov. She lived relatively peacefully in Moscow until October 1929 when the OGPU arrested her. Four months later, the investigator determined she had engaged in wrecking and sentenced her to ten years in prison.[16]

When the circle of counterrevolutionaries was enlarged to fifty people through interrogation and torture, the police had their canal specialists. The secret police sent them to Karelia to a remote site with no housing or roads, where they made food deliveries only in the summer because of the absence of connections to outside world. The engineering task was to exploit roughly 1,000 prisoners at Khizhozero to build an earthen dam and flood gates. On December 15, 1932, a nearly completed dam ruptured, threatening construction elsewhere. One writer characterizes the effort to stop the breach as one of "the most clear and heroic pages of the selfless struggle of the shock workers and canal solders for the on-schedule completion" of the canal. The rupture not only destroyed the floodgates but the watershed canal and other components. An icy wind blew off the lake, and they could see very little of the churning water and remains of the earthen dam in the dark. Out of courage or perhaps fear they realized they could not wait for the torrent to subside, but waded into the water carrying bags of sand and gravel. One canal soldier simply stripped to his underwear and jumped in. This had the effect of encouraging others to follow the courageous fool. They worked in the icy water and never asked for a spell. The next day shock workers from another camp were ordered into the frigid breach, and two days later they had staunched the flow. Kobylina described the accident and the successful battle with the rupture in a report to the authorities, and surprisingly none of the engineers was punished. Kobylina later was transferred to the Moscow–Volga Canal and was rehabilitated fully only in 1965.[17]

And so, when the White Sea–Baltic Canal had been completed, the renamed Belbaltstroi Construction Organization proposed another series of Arctic canals, further to the north, to be built through Kola granite, not peat, while other workers and prisoners were shipped to Siberia to join new prisoners and workers at BAM in 1934–42 (not completed and restarted again without prisoners but as a Komsomol site in the 1960s) and at the Moscow–Volga Canal (1932–37 and completed under Matvei Berman, who flowed along the White Sea–Baltic Canal to become head of the Gulag system in the mid-1930s and under whom some three-quarters of a million men toiled in fifteen different projects).[18]

No sooner had BBK begun to admit ships and Communist Party dignitaries than the Council of Ministers in August 1933 discussed a plan to build a second series of locks and deepen and widen the existing canal because of the realization that Belomor did not meet expectations nor technical specifications and concluded their meeting with an astounding resolution to investigate the possibility of a new canal across the Kola Peninsula from Kandalaksha to Murmansk some 250 kilometers in length. Subsequent documents referred to the "Big Belomor-Baltic and Kola Canals" and "Deep Water Gulf of Finland–Barents Sea Route." As with many things, the canal across the Kola Peninsula had prerevolutionary roots.[19] The Leningrad Design Bureau of Belomorstroi produced a 153-page proposal in September 1934. One author has pointed out how absurdly unrealistic the canals were given the terrain, geology, and climate. On top of this at the time 370,000 people lived in Karelia and in Murmansk 150,000 people—in an area of 273,000 km^2, or two people every square kilometer.[20] Where would demand for use of the canal arise? Who would operate the canals? And what would they ship along them since fertilizer, nickel, copper, and other mines and enrichment facilities would not be fully operational for years?

Planners were not to be deterred. They counted on both the insatiable appetite of the secret police for more prisons and on the tremendous mineral wealth of the region that they claimed justified any such costly development projects such as canals. They pointed to the two million kW of potential electrical energy of the Karelo-Murmansk region that would more than meet future demands of industry, especially with only 68,000 kW currently being generated from hydroelectricity on the Niva, Kondopoga, and Tuloma rivers and 44,000 kW being installed on other rivers. Finally, the planners had no doubt that the Monche tundra would soon become the Soviet "Sudberry (Canada)"—with production of platinum group metals, nickel, and lead of national significance. To build the canals and jumpstart industry, they projected the creation of a "production army of one-quarter million people" by the end of the Third Five-Year Plan. Yet in 1936 in thirty-three branches of the slave labor combine (BBK NKVD) there were but 30,000 "free" men, by which they apparently meant kulaks exiled from central provinces of the USSR. The engineers planned to set up seventy-five more villages in the taiga primarily along rivers, and by 1942 they anticipated a doubling to 60,000 laborers.[21] Would the other 200,000 workers "volunteer" to join the heroic effort in Stalinist fashion—through trumped-up charges of arrest that resulted in incarceration?

A Short Shelf Life

Gulag camps spread rapidly across the Russian northwest. The camps had both names to specify a region and a number, like a post office box: Viatlag (K-231, a forestry camp in Kirovsk province), Dubrovlag (DKh-385, in Mordvinia), Izvdel'lag and Sevurallag (N-240 and AB-239 respectively, both in Sverdlovsk province), Kizellag and Nyroblag (VV-201 and Sh-320 respectively, both in Perm Province), Koslanlag and Pechorlag (KL-400 and PL-350 respectively, both in the taiga of the Komi ASSR), Sevkuzbaslag and Siblag in Kemerovsk region, Indigirskii Camp in Yakutia, and scores of others filled with millions of emaciated prisoners who were clothed in rags and armed with hammers, shovels, axes, and wheelbarrows. If SLON was established to reeducate political prisoners, then there was no question about the role of the other camps whose battalions of prisoners were dispersed across Arkhangelsk, Karelia, and Murmansk to build canals, roads, and railroads. In Arkhangelsk province dozens of camps—at Sevdvinlag, Kargopol'lag, Mekhren'lag, Iagrinlag—used prisons in reclamation, forestry, shipbuilding, and other economic endeavors. The camps filled and refilled owing to high mortality rates or as new construction projects were indicated across the northern latitudes; in these cases prisoners from one camp were often forced to march to a new site and many perished en route.

Archival materials enable us to create a more complete picture of the central importance of the gulag to Arctic construction and emphasize again the economic importance of the camps. One of the strategic concerns of Russian military leaders since Peter the Great at the turn of the eighteenth century has been the establishment of a world-class navy and warm water ports. In 1936, as part of the burgeoning NKVD empire, the authorities established a major shipbuilding facility on the White Sea, Molotovsk, not far from Arkhangelsk, to supplement shipyards in Leningrad. Later, major shipbuilding yards would be added to Murmansk and Vladivostok. Molotovsk, renamed Severodvinsk in the late 1950s, was built by gulag prisoners, tens of thousands of gulag prisoners from the Iagrinlag camp, and it is jealously guarded by the inheritors of the KGB—the FSB—in the twenty-first century.

The Molotovsk prisoners lived on Iagry Island. Iagrinlag prisoners were employed to construct shipyards and ships for the inevitable war against the capitalist powers. For Iagrinlag they dredged channels and rivers, dug foundation pits and ditches, opened trenches dozens of kilometers long, poured concrete, laid bricks, built docks and dry docks, welded steel plates, and manufactured shells and bombs—all of this in the worst winter conditions.[22] Most prisoners did brute force construction work. The erection of "Factory 203" (the Shipbuilding Factory) demanded a huge number of workers. Echelon after echelon of prisoners, most of whom were "enemies of the people," marched to work each day from Iagry Island over a wooden bridge to Molotovsk. They came from all over the empire—from Ukraine and Belarus, from the Baltic states and the southern Urals, and many from Moscow, Leningrad, and Nikolaev. The location and number of prisoners changed over time from 1938 through 1961, as did the qualitative makeup.

In 1938–39 there were probably 40,000 prisoners, and in the 1940s perhaps up to 80,000, before settling at roughly 50,000 individuals. Few among them were members of the intelligentsia; the intelligentsia numbered perhaps in the hundreds according to one source. During the war such prisoners as deserters, shtrafniki (entire battalions of criminals forced into the army as punishment), and even German prisoners of war were impressed in Iagrinlag. After Factory 203, prisoners built Factory 893 and Factory 402 (today Zvezdochka and Sevmash respectively) that became centrally involved in the nuclear submarine building effort.

The gulag also made use of engineers and others with special expertise. From 1939, Special Bureau 4, a sharashka of engineers, worked on ship design, and from 1942 in Factory 402 they organized a special ship repair division that employed engineer-convicts and in its illustrious history launched 139 Soviet and foreign military and transport ships and later submarines.[23] The roots of the engineers' labor camp was recognition among Iagrinlag commandants—and importantly their counterparts in the Butyrka Prison through which much of the meat of the Stalinist grinder passed—that the Soviet state might take special advantage of expertise by arresting engineers and harnessing them to state programs. Such special camps, captured in fiction by Alexander Solzhenitsyn in *First Circle,* existed for rocketry, jet engines, shipbuilding, biological weapons, and other research programs. Iagrinlag's sharashka began operation with seven specialists already engaged in Iagrinlag and was augmented by six others harvested from Butyrka, some of whom had worked on rockets. The sharashka grew to thirty engineers.[24]

One of the prisoner-engineers recalled how they arrived on an early August night at a dreary train depot near Arkhangelsk. A. P. Eppel', former deputy head of a foundry at the Leningrad Shipbuilding Factory, looked out the window and remarked with irony, "Snow? Well, brothers! You won't be buying tomatoes here." They spent six years in the cold, but unlike ordinary slaves got special treatment. A truck carried them to and from Iagrinlag, so other prisoners rarely saw them. They lived in sanitized barracks that smelled of carbolic acid. They were able to forgo the usually strict controls over their conversations, communications, or letters. The head of the sharashka was an NKVD Major Andoshuk, a man with some technical training. At first Andoshuk kept them separate from other prisoners because of the top-secret engineering projects. But of course they had to speak with people working in the factory to solve technical problems and discuss blueprints. So eventually they were allowed to wear typical factory work clothes and were accompanied by guards who wore civilian outfits. Everyone knew they were special prisoners. When the engineers got to know each other better, they remarked on the difference between engineers trained in the Tsarist period and those of Soviet education. The older engineers usually knew several languages, appreciated art, loved music, enjoyed games, and knew chess and checkers; they played "blind" chess on the way to and from work;[25] younger engineers were narrower in training, interests, and appreciation of European culture.

The pressures of Stalinism to meet heroic production targets meant redoubled emphasis on engineering, applied science, and technology. The deans and rectors of universities and institutes worked with authorities from the central educational ministry to turn out engineers who met that demand and hence gained such narrow specializations as ball bearings, pipe manufacture, lumbering, heat engineering of oil boilers, and so on. These new young engineers—and their bosses—could not be troubled with foreign languages. Indeed, contacts with foreigners, however indirect, even knowledge of a foreign language, might lead to suspicions that someone had foreign interests or could be a spy. Whatever the reasons for their arrests, many of those people dragged into the sharashki understood that knowledge of the profession required the ability to read a foreign language.

The Molotovsk engineers included P. A. Al'bov, the main engineer of the Kharkiv Tractor Factory; A. S. Tochinskii, the main engineer of *Iugostal* in the southern Urals region; and the main builder of the *Kirov* cruiser, V. L. Brodskii. At Iagrinlag they designed a torpedo boat destroyer, *Project 30*, and prepared blueprints for the *Bol'shoi Okhotnik* submarine ("Project 122"). They studied British and American ships that the USSR received through lend-lease programs and tried to reverse engineer. They also had mundane tasks, for example, establishing a production line of mess tins for the Red Army. According to one source, they worked without material stimuli, without hopes of cleansing of the lies and slanders from their interrogation records, but simply for the joys of creative work and intellectual satisfaction.[26]

Iagrinlag had nine camps, each of which had dozens of barracks and each of which had a barely functional steam bath and a club. Camp No. 1 housed roughly 5,000 to 6,000 prisoners. There were men's and women's divisions, the latter being wives and daughters of "enemies of the people." They were engaged in construction of Factories 402 (Sevmash) and 893 (Zvezdochka) and also utilities and housing on Iagry. In 1957 this division was closed and the prisoners were distributed among other camps. Camp 2 was connected with Sevmash with 10,000 to 12,000 prisoners involved in building the factory and the port. It had a special sharashka in which there were about 100 highly qualified engineers including professors and academicians. They carried out cutting-edge technical projects. Camp number 3 had about 5,000 women involved in plastering, painting, and other interior finish work for Sevmash and housing in the city. One of the barracks was for children where 50 to 180 babies born out of wedlock lived until they were two years old, at which time they were shipped to relatives or to an orphanage; the authorities created orphans with their policies of arresting fathers and then mothers and then punished the orphans for being related to "enemies of the people."[27] The fourth camp was involved in construction of Sevdormash, which made ordnance and other orders for the front. The fifth camp in Glinniki (near Tsiglomen) had 1,000 prisoners who baked bricks. Camp 6 had about 2,000 prisoners until 1946 involved in road and other excavation work in the city. From 1946 more than 1,000 prisoners were involved in woodworking of various sorts, including furniture, and also in clothing repair. Many of these prisoners were elderly—from forty-five to seventy-five years old—Estonians and to a lesser extent Lithuanians and Latvians rounded up after Stalin's Red Army conquered the nations.[28] More than 7,500 Estonians occupied labor camps in Arkhangelsk region in ten different military brigades and such forestry operations as Kotlasles.[29]

Camps 7, 8, and 9 had about 10,000 prisoners each. The prisoners were involved in construction and utilities work on a cogeneration boiler, roads, and housing. There were also workshops that repaired clothing and shoes—mostly staffed by enfeebled elderly prisoners. At a faraway camp near Sheleksa station in Plesetsk region 500 women worked in the forest to fell raw lumber for construction back in Molotovsk, and at the Lomovoe station roughly 2,000 men quarried aggregate for the Molotovsk concrete factory. The Plesetsk camp served as the foundation for the Plesetsk Cosmodrome from which the USSR launched military satellites into polar orbit. A series of other forestry, railroad, and woodworking facilities in small villages in the area, for example Solve and Rikasika, now the site of a massive chicken broiler factory, contributed supplies and labor to the military effort.[30]

Iagrinlag hence was not one camp on the island of Iagra, but a series of camps each holding thousands of prisoners. The prisoners slept on mattresses filled with sawdust and under two light summer blankets. Hundreds of prisoners filled these beds, and they "died like flies, each day they carried out more and more dead," Valish Khabibulin, a prisoner from a poor peasant background recalled. "In the camp all people were used up to the very limit, skin and bones, and almost everyone had scurvy."[31] There was one nurse Khabibulin implored to help him: "Nurse, I have nothing wrong with me. Help me get out of here while I am alive." Khabibulin had worked as a tractor driver. The nurse informed the administration who called him to work in the agricultural division as a tractor driver, which saved him from trudging through deep snow to fell trees and from nearly certain death. The camp commandant acknowledged the need for a tractor driver and wondered how the emaciated man could handle heavy equipment, but took him on anyway. They put Khabibulin to an American tractor, no doubt acquired through U.S. lend-lease programs. Khabibulin was surprised he could be trusted with such a thing, given that he had been convicted as a counterrevolutionary. He was told it wasn't his business, to read the manual, and to master the machine. He discovered it was much easier to operate than a Soviet one.

Khabibulin survived his camp experience out of luck and common sense. He did not make the mistake of other prisoners of eating too much too quickly when given the chance. He happened to see the nurse months later in the dining hall. She could not believe her eyes and said to him, "There are so many young guys like you who are dying here." Khabibulin got ill again, nearly lost his sight, and ended up in an infirmary. After a month he recovered and was sent back to work. One day, resting in a copse of small pine trees on his way back to camp from making birch branch brushes, he noticed carvings in the bark of a family, first name, patronymic, and number. He realized that this was a burial ground for prisoners.[32] This was Arctic assimilation: inhuman conditions, underfed, inefficient slave labor, one American tractor, vast taiga and tundra filled here and there by strategic tasks of the Stalinist economy, and an occasional cemetery.

At the end of the summer season—in October 1943—the agricultural workers were sent back to Iagrinlag Camp 2. For Khabibulin this was an impossible adjustment; the massive beehive of prisoners with the diseases, injuries, and odors denied him any solace. The guards escorted the prisoners to and from worksites every day in every kind of weather, under rain and snow, walking, waiting, working, with long lines in the dining hall. Their task was waste, sewage, and heating systems for the shipbuilding factory that manufactured cutter-attack ships, hunters, and submarines. The men built trenches four to five meters deep with shovels, from morning to night, without a day off, and yet the guards fed them only twice daily. They got 250 grams of bread and wish-wash for breakfast; in the evening, he who had worked got kasha, wish-wash, and bread. Khabibulin remained in camp number 2 until December 1943, when he was transferred to camp number 7 and a slightly better life: They fed them better, some of his old buddies from the agricultural division were there, and women prisoners tended a few animals.[33]

Slave labor was poorly organized, and there were few incentives for slaves or guards to do more than the bare minimum. The guards naturally resorted to cruelty. At Iagrinlag at times roughly 6,000 prisoners were standing around waiting for something to do. When these men entered the work zone they addressed such tasks as seemed appropriate to them because no one supervised them closely. On top of this, the camp commandant, one Masloboev, fraternized with prisoners, perhaps drank with them, and took women *zeks* (prisoners) home at night, for which he was eventually relieved of duty.[34]

Absurdly, many of the prisoners had two roles. During the night the guards treated them as incarcerated prisoners, but during the day many of these individuals worked as heads of departments and construction projects because of their civilian skills. This gave them the right to move about freely with various passes that they themselves sometimes filled out—and not only at projects sites but also at planning meetings in which they had authority by virtue of their technical skills. Even more absurd was the fact they had to produce written accounts of their workdays and thereby were in the position to evaluate their own performance positively and thus to accelerate the pace of serving out their terms. One official complained about the fact that bookkeepers who kept track of such things as plan fulfillment were themselves prisoners. During the evening shift they just got drunk and made a mess of things. Given that this was not just a camp, but one of "government importance"—military shipbuilding—this clearly was not the best way to do things.[35]

Of course, a better way to ensure higher productivity of labor would be to feed and clothe prisoners adequately and provide them with proper medical care. In June and July 1939 the new Iagrinlag commandant, Kronov, issued an order to improve the medical dispensary that included measures to fight scurvy, antibacterial efforts to prevent dysentery, and mosquito eradication to prevent malaria. But Iagrinlag had only eighteen doctors (four of whom were prisoners) who resided and worked in Molotovsk, and the poor pay and conditions made it hard to attract others. Beginning in July they sent out seven experienced doctors to Glinniki and other distant camps. By acquiring more beds they isolated some of the TB patients in the central hospital, which led to more beds available at the other urban centers. Throughout 1939 the health of the 22,520 prisoners improved slightly judging by a survey rating their ability to work. But the number of deaths remained high: officially it ranged from forty-six in June and forty-three in September to twenty-nine in July and thirty-one in October. In the absence of a dental clinic you can imagine the cavernous spaces in what used to be mouths of thousands of men with gingivitis. An X-ray machine was a fantasy. Another problem—many prisoners had begun to mutilate themselves, chopping off fingers, taking poison, burning themselves to avoid work. Epidemics of flu often felled 35 percent of a labor contingent in some forestry and other operations.[36]

Arctic commandants also had a short shelf life. After the war a new commandant took over, a Captain L'vov, who instituted a much more harsh regime, with punishments, beatings, the equivalent of chain gangs, and isolation. It was rare than anyone returned from that camp, and if they did they were flesh and bones; they looked like "skeletons." In May 1948 Khabibulin reported to the deportation department, learned he was being freed, was told to keep his mouth shut and to find a spot far from people in the tundra or steppe, or he'd end up back again in the camps, and he made his way to Abaul, his home. He never forgot the real enemy of the Soviet people: "the Stalinist regime."[37]

Sevdvinlag for Railways and Forestry

Like the company towns they were meant to serve, the Arctic camps were often geared to specific civil engineering tasks. Sevdvinlag (Northern Dvina Labor Camp, also Camp 283) was located near Velsk on the Arkhangelsk-Vologda highway and open from 1940 through 1946. The prisoners were involved mainly in the construction of the Konosha–Kotlas and the Konosha–Kozhva railways to connect resources and laborers of the Komi Republic with Arkhangelsk region. Sevdvinlag prisoners undertook small agricultural operations along the rails to dent their hunger. The camp grew from 15,365 prisoners on January 1, 1941, to 42,117 on July 1 and declined over the next four war years to 34,000, 21,000, 10,000, and 7,000 prisoners before being closed.[38]

In addition to its agricultural and forestry operations, Sevdvinlag was involved in "general civil engineering" of such structures as railroads, bridges, apartment buildings, and railway stations. According to NKVD reports, the stations were filthy, the waiting rooms filled with piles of garbage and overcrowded. In postwar Arkhangelsk region, efforts were under way to improve passenger travel with better stations and platforms. But the Obozerskaia station was in a miserable state even though it was crucial as a junction for a large number of passengers, and repair of platforms at other stations had not even begun. Often, because of shortfalls of such simple materials as cement and gasoline, wagons, hand tools, and so on, work stopped for months at a time.[39]

The most important activities of Sevdvinlag involved roads, lumbering, and civil engineering projects divided according to administrative districts of the province: Velskii, Kuloiskii, and so on. In the Kuloiskii Construction Brigade, the commandants gladly reported that productivity of labor was up with 1,317 men in general work and 1,152 in forestry operations out of a general fund of 5,790 people. But they lamented that the supply and care of horses in the camp was problematic.[40] If other camps are any indication, this means that transport and excavation relied on manpower more than horses, and more on horses than tractors, and the trains moved slowly between camps on poorly anchored rails.

By February 1946 Sevdvinlag prisoners were involved in excavation of

13,400 m^3 of trenches and foundations, raising 1,860 m^3 of living space, and splitting 9,180 m^3 (approximately 500 cords) of firewood. Such capital-intensive projects as bridges were rarely tackled; capital shortages persisted across Arctic and subarctic regions. While the Kotlas Bridge Factory had manufactured the supports and trusses, a 250-meter bridge over the Vaga River would have to wait until other supplies became available.[41] Dozens of labor camps engaged in similar, fruitless operations, sometimes filled with Poles and Estonians enslaved during the war.

Like those at Iagrinlag and other camps, the mortality rate among prisoners of Sevdvinlag was high, although how high is difficult to determine. The statistics regarding mortality rates among prisoners may have been doctored to demonstrate the relatively "healthful" environment in which prisoners toiled, even if they lacked sufficient, let alone healthful, food, appropriate clothing, and rudimentary medical care. Camp commandants must have worried about mortality rates because they relied on prisoners to meet construction targets, although they could count on more prisoners to replace fallen ones and so were quite callous about the slaves. In 1946, Major Makarov, the head of Sevdvinlag, reported that the health status of prisoners was stable or improving. Some 10 percent of the prisoners were sick and unable to work at a given time, while "only" 0.21 percent of the labor force died. Yet this is one prisoner per 500, a tragically large number, and a month later the rate had quadrupled to one in 115.[42]

Throughout the gulag network officials were worried not only about forging infrastructure through the taiga and tundra, but giving at least lip service to political education of prisoners and proper political upbringing among guards. That is, the political functions of the gulag persisted although, judging by low attendance if the sincere effort to indicate that political education was robust, it appears to have been an afterthought to commandants, guards, political officers, and prisoners alike. In February 1946 one Grishina, deputy director of the political department of Sevdvinlag and a lieutenant in the NKVD, gave lectures on "Lenin-Stalin as Organizers of Victory of the Red Army in the Period of Foreign Military Intervention" and "Soviet and Bourgeois Democracy and Their Differences." I cannot judge if his speeches were sufficiently acute to keep communists dedicated to the causes of Arctic assimilation. On January 1, 1946, in Sevdvinlag there were 295 communist guards, soldiers, and other employees, 145 of whom were engaged in independent study on Marxist-Leninist theory, but 122 of whom, according to the archives, were poorly prepared and so were engaged in study circles on the short course of party history, the charter of the party, and the 1936 Stalin constitution. They had organized seven agitation collectives with 139 agitators and in a stunning quantitative achievement carried out 597 discussions that involved 8,588 people and held another 2,013 newspaper readings that likely thrilled 23,242 people. Most agitators were "serious," reflecting the fact that they had studied the material deeply, consulted with other specialists, and even illustrated the material with examples! The camp newspaper, *Za Magistral'*, put out thirteen issues to promote political education.[43] Soviet political correctness accompanied each kilometer of rail and each tree felled in the taiga.

Meetings of party "aktivs" frequently led to discussions of difficult issues that were not orchestrated beforehand, even if the office of the party secretary produced an agenda. At a July 2, 1946, meeting of a party aktiv of Sevdvinlag, just before the camp was closed and its prisoners transferred to camps in the Komi autonomous republic and Siberia, some five dozen loyal party members discussed two major issues: improvement of political education of military guards and improvement of the material and cultural condition of the workers. In the Soviet times, there was always room for improvement in political work, but two aspects of the issue stood out at Sevdvinlag. One was that the guards were so overworked, sometimes going eight days and longer without a day off, that they were indignant that, on top of work, they had to attend political instruction meetings when their days were done. As a result, apparently, they had sunk to the level of prisoners, engaging in black market sales of prohibited items to prisoners and resorting to booze to fill the time (one drunken guard was seriously injured when he fell under a train), and there had even been several shootings and murders.[44]

Regarding "material and cultural conditions" of the workers, discussion revealed that the families of guards lived hardly much better than prisoners. Food was of poor quality; they had no firewood for heat or cooking (where did those cords of wood go?) and they had to scavenge it from the forest. Except for a few sergeants, none had boots or uniforms. They received vegetable oil, but did not have pots, pans, or dishes. The aktiv resolved to work on these conditions—and to organize meetings at the propaganda center—

the "red corner"—of the organization,[45] although how this would provide more pots and pans was unclear.

Party organizations of gulag camps also considered the sacred: membership in the Communist Party. A guard, Andrei Antonovich Smolinsky, was seriously wounded and lost consciousness in a battle near Kotel'na of Poltava region on September 6, 1943. He was transported to the hospital without documents including his party membership card, which were lost or destroyed during battle. Smolinsky had written the political department of Sevdvinlag about obtaining a new card; party officials agreed to ask the Arkhangelsk Party Committee if this wounded and decorated veteran of the war might be reissued his card.[46] After all, the loss of a card, in any but the most serious of circumstances, was a mortal insult to Stalin and the party and the subject of a 1936 film, "Party Membership Card (Partbilet)"; the party leadership did not like the film even if the point was well taken and instructed Mosfilm to fire the director Ivan Pyr'ev. As for Smolinsky, I assume the wounded soldier acquired a new card.

The gulag prison camps made mistakes in controlling Communist Party membership, although one might think that an organization charged with the identification, punishment, and reeducation of enemies of the state would have been more vigilant. As did all organizations, party officials of Sevdvinlag worried whom among their guards and other staff to accept and exclude from Communist Party membership. Offenses against the party included systematically avoiding meetings, insufficient struggle with enemies of socialism, as well as the eternal struggle against drunkenness. A candidate's personnel folder preserved every scrap of paper concerning his career, when he entered the party, his nationality and educational level, social background, any bad behavior or other marks against his record, and administrative admonishments. One Mikhail Nadrezov (I have changed the name of this and other individuals so as not to run into the ever-changing rules of the FSB), a Ukrainian with a third-grade education who had been a candidate for party membership since September 1943, worked in Sevdvinlag as an armed guard. In 1933 like many other innocents he ran afoul of the system, was found guilty of section 59–3 of the criminal code ("violence and banditry against the administrative order"), and got a four-year sentence. In 1937 he was released after completing his term and joined that very organization that had incarcerated him, the NKVD, but did not indicate his prior conviction on his personnel form, and in his autobiographical statement indicated that he had been freed early and the judgment against him removed. In addition to hiding his crime, he had a brother and a stepbrother who were "traitors" against the Motherland, although Nadrezov said he had had no contact with either since 1928. Further, at the Konosha Labor Camp he "systematically gets drunk and engages in hooliganism," for example by shooting his pistol into the air. His application for full membership was declined.[47]

Arctic Islands in the Archipelago of the Gulag

Each forest, every wind-blown plain, each landmass and island had to be explored, secured, and exploited. As Alexander Solzhenitsyn noted these were islands of prison camps; on occasion—Solovki and Vaigach—they were literally islands. The desolate Vaigach Island labor camp existed from July 1930 to October 1936, with a main camp at Varnek Bay. Roughly 1,500 prisoners were engaged in mining and exploration for polymetallic ore. By 1934 they had found sufficient lead-zinc deposits for the industrial development of zinc deposits—if by "industrial" we mean the use of ill-equipped prisoners. The prisoners also found a vein of copper ore along Dyrovataia Bay at the northwest end of the island.[48]

Vaigach Island, largely unexplored by Europeans until the late nineteenth century, sits across from the Eurasian landmass on the Iugorskii Strait. At 3,400 square kilometers (1,300 square miles) the island has a large number of rivers, swamps, and small lakes during the summer thaw. A place of Nenets shrines, many of which have now been desecrated, in 2007 it gained a 243,000-hectare nature reserve to protect such threatened Arctic species as polar bears, Atlantic walrus, and the white-beaked loon. The preserve obscures the violent nature of the island's history.

The geologist N. A. Kulikov led an expedition to Vaigach in 1921 that first discovered polymetallic ores. In 1925 an Academy of Sciences expedition returned to Vaigach for more samples. In 1930 under the former deputy chief of SLON, Fedor Ivanovich Eikhmans, the OGPU set up administration of the camp in Vaigach Bay. They brought 125 prisoners including several geologists with seven guards and administrators on the ships *Blowing Snow* and *Gleb Bokii,* which wintered over the 1930–31 season. They did not think about the need to test ore samples on Vaigach itself in some rudimentary laboratory; they evacuated the ore, but not prisoners, in summer by boat to Arkhangelsk and then by plane to Moscow. The samples apparently were promising enough for the OGPU to bring another 209 condemned men in September 1931, this time including geologist P. V. Vittenberg who had been convicted of anti-Soviet slander. Over the next four years under Vittenberg's direction the prisoners discovered another fifty-eight potential deposits.

Arctic researcher Pavel Vladimirovich Vittenberg was born in Vladivostok, one of many children of a political exile punished for participation in the Polish uprising of 1862–63. International by blood, upbringing, and worldview, Vittenburg was admitted to the Riga Politechnical Institute in 1905, one of five such newly founded institutes in the Tsarist empire under the leadership of finance minister Sergei Witte who hoped to jumpstart Russian industrialization through railroads and engineers. Vittenberg could not attend the institute because of student strikes. Instead he studied natural history at Tuebingen University in Germany. He undertook research in Ussuriiskii district of the Far East of Russia in 1908, no doubt working with the ethnographer and traveler Vladimir Arseniev, best known for *Dersu Usala*. Vittenburg received a doctorate in 1909 and returned to St. Petersburg. He immediately began active field research for the Geological Committee and the Academy of Sciences, especially in the Far East. He turned to the Arctic in 1913 when he joined Rudolf Samoilovich in an expedition to western Spitsbergen (Samoilovich is covered later in this chapter). In 1918–20 Vittenberg explored the northern Kola Peninsula; in 1921 he joined Samoilovich on another expedition to Novaia Zemlia. He wrote a large number of articles and monographs and engaged in extensive correspondence with scholars at home and abroad including Nikolai Knipovich, Alexander Fersman, and Fridtjof Nansen. At the same time, he actively participated in the organization of science and as a pedagogue, taking part, for example, in the creation of the Standing Polar Commission of the Academy of Sciences and several museums. In 1925 he moved to Leningrad State University where he headed the division of polar countries in the Geology Department and taught courses on hydrometeorology and in the same year became academic secretary of the Commission for Study of Yakutia.

They arrested Vittenberg in 1930 as part of an "Affair of Historians of the Academy of Sciences," in which militant communists accused leading specialists of hostile views toward the USSR and counterrevolutionary activity. Perhaps the real reason for the historians' affair, in addition to the tenor of the time, was the fact that the Academy refused to admit new members who were Marxist and party members; in 1929 of 1,158 employees of the Academy there were only sixteen Bolsheviks.[49] Vittenberg was also suspect as a "bourgeois specialist"; a member of the intelligentsia, he surrounded himself with family and such friends as the children's writer Kornei Chukovskii and the architect Albert Benua (who emigrated to Paris in 1924).

At first Vittenberg was forced to work on the White Sea–Baltic Canal and then was drafted into the Vaigach Expedition as head of its geological division. While freed from his term in 1935, he did not hurry home, but instead continued research in the Arctic and joined the Main Geological Administration of Glavsevmorput. He hoped to explore mineral deposits on Northern Zemlia and set sail on the *Sibiriakov*, but heavy ice sent the expedition east to the Taimyr Peninsula. His experience carried him to Amderma where his researches led to the publication of a monograph on Vaigach deposits. In 1941 he published *The Geology and Mineral Resources of the Northwest Part of Taimyr Peninsula*. During World War II he carried out research in Arkhangelsk, Syktivkar, Ukhta, Vorkuta, and the Northern Urals.[50] Vittenberg worked in the Karelia, teaching at Petrozavodsk University, and was fired in 1950 during another wave of repression. Freed again, from 1954 he worked at the All-Union Geological Institute of the Ministry of Geology.

The sadistic Vaigach commandant, Fedor Eikhmans (1897–1938), began his career in the Cheka during the civil war in Turkestan. He seemed to take glee in abusing prisoners. He rose to become director of the second division of the eastern department of the OGPU in 1922. From 1923 he was head of SLON. According to witnesses he actively participated in execution of prisoners. He also apparently enjoyed parading prisoners in the nude. His next posting was as first deputy of the Gulag Administration from which position he supervised all corrective camps of the OGPU. In June 1930 he was removed from this position and put in charge of the Vaigach operation. He was arrested in July 1937 and shot as a spy and participant in counterrevolutionary organizations—perhaps because the Vaigach mines were so poorly built that they flooded all the time.[51]

Conditions for the prisoners on Vaigach Island improved slightly during the second zimovka. The authorities sent two manufactured log houses to house the head of the secret police, a laboratory, diesel generator, and materials for six drafty barracks for the prisoners. The supply of food got better with a diet consisting of potatoes, onions, carrots, and cranberry extract as a hedge against scurvy, and a small canteen sold alcohol and clothing. Yet this was an Arctic labor camp and many prisoners perished; if they died in winter, their bodies were stored on ice at the bay for later burial. Opposite the camp on the other side of the bay sat the mine; prisoners took motorboats (in the summer) or walked along the shore and in the winter followed hand ropes anchored in the snow on stakes for to be lost in a blizzard might lead to death. Another mine existed on Ugra Peninsula.[52]

The geologists determined early on that further deposits, say of gold, silver, or platinum, were insufficient to justify further efforts. And yet on October 30, 1933, the *Gleb Bokii* brought another contingent of prisoners; their total now was 1,100. These prisoners built a runway at the village Varnek, where an iron ore mine opened. In winter of 1932–33 they also built a House of Culture run under prisoner Vatslav Ianovich Dvorzhetsii. Of Polish gentry descent, Dvorzhetskii worked in the Kiev Polish Dramatic Theater when arrested as a Polish-Ukrainian nationalist and sentenced to ten years in prison. He survived Vaigach, then used his literary talents to build a railway across the tundra and fell trees in the Komi Republic, and served another prison term during World War II.

At the height of mining activities, 1,498 prisoners toiled on Vaigach. Yet work collapsed in 1934 because of flooding of the mine—as happened again in Amderma mines, although Vittenberg had warned against the design chosen for operations. The design emphasized speed, not permanence. Even so, officials pondered a second Vaigach expedition, but when the war broke out, these plans were abandoned.[53] With the mine closed, the OGPU sent the laborers on foot to Uchpechlag to lay the Vorkuta-Iugorskii Shar railroad line, plus a narrow-gauge rail 70 kilometers long to Vorkuta-Usa and 274 kilometers from Ust'-Vym' to Chib'iu. This project, too, collapsed, with several Russian visionaries hoping to restart the project in the 2010s. Other prisoners were left to dismantle equipment and move it to Amderma, and still others worked on the Vorkuta-Ugra Ball and the Isakogorka-Molotovsk railways. Large numbers of men died on the railway projects.

Coal and Oil from the Gulag

Glavsevmorput and burgeoning industrial facilities had a growing hunger for coal, and the hope was to avoid shipping coal from Ukraine or Siberia to the Arctic but to find deposits close to demand. Another gulag tundra mining operation with an insatiable appetite for slave laborers, the Ukhtinskaia Expedition, opened in 1931 and closed in 1938. Its prisoners were harnessed to the development of oil and coal deposits in the Ukhta-Pechora region, essentially southern areas of the Nenets Autonomous Region centered on the Pechora River basin, which flows north into the Barents Sea, as well as coal in Inta and Vorkuta to the north and east in the Komi Republic and also radium, oil, gas, and asphalt. Fossil fuel deposits in the region have acquired growing significance in the twenty-first century and will require significant efforts and investments to attract, pay, and house workers.

Ukhpechlag grew out of the Ust'-Vym' unloading point established in 1929 in southwest Komi region for the hundreds of and thousands of exiles and prisoners, likely peasants and kulaks, forced into the Stalinist migration by the collectivization campaign. By summer 1931 there were nearly 2,000 prisoners housed temporarily at the point in barracks and tents in this subarctic village among lakes and swamp and forest. This was less than one-quarter of the shelter necessary, but established the pattern for labor camps of miserably inadequate housing. (In October 1936 the political prisoners began a mass hunger strike that lasted 132 days. They demanded protection from hardened criminals, better food, and an eight-hour working day; I have been unable to learn what happened in the strike, but assume that OGPU guards treated survivors ruthlessly.) In spring 1931 the Supreme Economic Council awarded 2.6 million rubles to develop the fuel industry of the Russian northwest. The prisoners built three or four exploratory shafts in the regions of Vorkuta and Adz'va with the goals of producing no less than 7,000 tons of coal, expanding oil exploration, and solving the problem of transport of the fuel. The rails in the Arctic rarely kept up with the freight for, as we have seen, construction moved slowly and targeted capacities for projects were seldom met. The industrial assimilation of the entire north was a massive, chaotic endeavor that consumed people and resources irrationally. Ukhpechlag was dedicated to the "industrial assimilation" of Pechora region. But this was an impossible task given the vast size of the area at 1.5

million km^2 from the Urals to the Pechora and Vychegda rivers. Ukhpechlag consisted of three mining and prospecting divisions according to geography and one dedicated to railroad construction. The "expedition" held 9,012 prisoners in 1932, 23,482 in 1936, 31,035 in 1937, and 54,792 in 1938.[54]

A number of specialists assisted in the endeavor either as volunteers or prisoners. They included a professor of geology who specialized in oil, Ivan Nikolaevich Strizhov (1872–1953), who suggested exploring for gas in 1931 by using shafts first drilled in 1907. An expert on Ural region sites, before the Russian Revolution he worked in the business of the Nobel brothers, moved to the Grozneft Trust that operated in the oil fields near Groznyi, Chechnya (the rich oil reserves explain substantially the Russian determination to hold onto the multinational tinderbox of Chechnya), and became director of Baku oil operations. Arrested as a "wrecker" of the oil industry in 1929, likely as part of the Industrial Party show trial, he was sent to the village of Chib'iu on the Chib'iu River—the left tributary of the Ukhta River, a godforsaken stream—to work in the geological-topographic sector of Ukhpechlag to find gas for "Gazstroi" during a ten-year sentence.[55]

Another specialist, Andrei Iakovlevich Krems, came from Baku, apparently with oil in his blood, having overcome losses of his parents as a sixteen year old and of his hearing to typhus to become a geologist. He earned a degree in evening school from the Azerbaidzhan Oil Institute. He was chief geologist of Azneft and joined the Main Administration of the Oil Industry in Narkomtiazhprom (the Commissariat of Heavy Industry), working with the leading Soviet oil specialist, Ivan Gubkin. He was sent with a group of specialists in the oil industry to the United States to study. At the end of the 1930s, with two five-year plans underfulfilled and with an easy target to blame—a group of individuals who had studied with capitalist devils in the United States—the entire group had been executed except for Krems, who was sentenced to eight years in the gulag for "participation in a Trotskyist organization." He, too, found himself in Chib'iu and after drilling a successful well was released to work as a "volunteer" in Ukhta to become chief geologist of a regional prospecting organization; the good words of Fersman in testimony before the NKVD helped gain Krems's release.[56]

Even with horrible climatic conditions, miserable equipment, and underfed laborers, Ukhpechlag businesses grew rapidly. In 1931 and 1932 the mines produced 9,800 tons of coal, and in 1933 27,870 tons. Oil and gas prospecting expanded to some 10,000 to 15,000 tons in 1933, with kerosene capture also growing. Transport also grew along both rails and rivers. Narkomvod (the Commissariat of Waterways) provided twenty tugboats to move the oil downstream through Naryan-Mar to the White Sea and then to Arkhangelsk and was ordered to build a port on the Pechora near Shchel'iaiur to facilitate more deliveries.[57] The deputy chairman of the Supreme Economic Council, Valerian Kuibyshev, declared at the seventeenth congress of the Communist Party on February 3, 1934, "One of the greatest problems of the northern province, the solution of which we will achieve in the Second Five-Year Plan, is the development of coal and oil reserves in the Pechora River basin which will guarantee efficient (high calorie) fuel for the northern navy, the industry of Murmansk region, and the northern province."[58]

Camps on Rock and Ice

The camps pushed northeast into Vologda and Arkhangelsk provinces, into the Komi Republic, and north into Murmansk region where exiles and gulag prisoners were central in settling the desolately beautiful Kola Peninsula. During the First Five-Year plan, the Kola Peninsula became a central focus of Bolshevik efforts to build heavy industry. Over the next twenty years, the Soviets founded Apatit Trust, Severonikel, Kandalaksha Aluminum, Olenegorsk Mining-Enrichment Combine, Pechenganickel, the Lovozerskii and Kovdorskii Mining-Enrichment facilities, and a series of "picturesque" modern cities connected with them: Khibinogorsk (Kirovsk), Apatity, Monchegorsk, Olenegorsk, Kovdor, Zapoliarnyi, and others.[59] Prisoners were critical to each endeavor.

Of significantly low population densities, cities in the Murmansk region required settlers and prisoners to be subjugated to the Soviet economy. At least 8,000 families totaling 45,000 people—21,325 in Khibinogorsk, Monchegorsk, Murmansk, and another 45 towns and villages—"made a decisive contribution" to Apatity, road building, and hydroelectricity. The resettled workers were the backbone of new gulag construction organizations essential to the erection of the industrial artifacts of Stalinism. While the number of prisoners deported to the Kola Peninsula was relatively small compared to the millions exiled and imprisoned elsewhere in the USSR, their numbers grew rapidly. Murmansk territory was 145,000 km^2, but had only 13,000 inhabitants in 1920 and 27,000 in 1928. Then the prisoners, kulaks, and other deportees came. On November 10, 1932, the regional party committee passed a resolution to accelerate colonization with another 20,000 to 25,000 workers. These workers were imported from SLON and other camps. In 1932–33 prisoners were sent to Apatity, Kandalaksha, Nivastroi, Khibinigorsk, and Murmansk. A major component of the prison camp echelons was minority nationalities: Finns, Poles, Germans, Estonians, and others forced from their homelands into the Gulag. The population of the province grew from 55,000 in 1931 to 192,000 in 1935 and 318,000 in 1940, a sixfold increase in eight years.[60]

De-kulakization forced the migration of eight million people of whom four million died in 1932–33 of famine. In 1930–31 alone more than two million people were transported in more than 1,700 freight trains in each of which the guards had stuffed 1,500 to 1,800 people, usually without any heat or water.[61] People were sent to special settlements, normally consisting of no more than a surveyor's mark, which they had to build from scratch. These people found themselves in Magnitogorsk, Novokuznets, Vorkuta, and Karaganda and in the north to Khibinogorsk, Monchegorsk, Apatity, Belomorsk, Vaigach Island, Amderma, and so on.[62] In addition to industry, the planners pushed even agriculture above the Arctic Circle. So-called special settlers on the Kola Peninsula turned thousands of hectares of thin and poor soil in the effort to establish fruit and animal husbandry.

One of the incongruities connected with the utilization of slave labor was the creation of educational and cultural institutions to support and transform the human material just as the authorities hoped to rework nature itself. Quantitatively the effort was impressive; qualitatively, judging by anecdotal evidence, it was a cruel attempt to provide cover for propaganda efforts to proclaim the humanity of the camps. Kola Peninsula authorities built twenty-one schools, ten preschools, twenty-five clubs and reading corners, two houses of culture, six hospitals, and twenty-five rural ambulatory units, as well as cultural centers, libraries and clubs, and orphanages.[63]

Except as samizdat (literally "self"—or underground—publishing), until the appearance of Alexander Solzhenitsyn's *One Day in the Life of Ivan Denisovich* (1962) Soviet authors never addressed the lives of these prisoners and exiles except in a roundabout way. They might write a story in which the hero exposes and defeats the dangerous elements in society—the kulak, wrecker, or counterrevolutionary. Socialist realism became state policy in 1934 when the Union of Soviet Writers voted to set standards for literary works that limited or prohibited erotic, religious, abstract, and other themes. These standards were repeated in art and music. Literature was to be accessible to the masses, reflect the spirit of socialist construction and the party, and avoid the mistakes of decadent bourgeois forms and styles. In a children's literary work, "Severnaia Zvezdochka Gaidara" (The Northern Star of Gaidar), Evgenii Kokovin wrote, in good Stalinist fashion, about the kulaks as exploiters who steadfastly refused to join the collective farms. In "Shumit Mud'iuga," published in *Pravda Severa* in February 1930, Gaidar had described the fishing farm on Mud'iug Island in the Northern Dvina delta where during the civil war the Whites had a prisoner-of-war camp for communists and Soviet sympathizers. From a young boy, Gaidar learns of a kulak who exploits other fishermen, is engaged in "wrecking," and like many others refuses to join the collective farm.[64] Children's stories thus left no doubt about good and evil and why wreckers merited the camps.

In fact, collectivization and de-kulakization did not have significant play in the Murmansk region because fish, dairy, reindeer, and other farms were not extensive. But Murmansk region officials hoped to expand each activity to feed burgeoning towns and cities and their expanding industrial infrastructure. They planned to increase the number of head of reindeer from 67,500 to 101,000 from 1931 to 1934, but fell short in numbers and did not achieve full collectivization.[65] Still, de-kulakization among indigenous people—identification of somewhat wealthier reindeer herders—began in spring 1931 and was essentially complete in September, although the total numbers of nomads forced into collective farms remains uncertain.[66]

Heroes Become Enemies of the People

A special category of individuals who served in the labor camps were former Soviet heroes, individuals who had been awarded special titles, awards, medals, or all three, who then found themselves accused of wrecking, counterrevolutionary activity, or espionage. As for those people who moved from anonymity into the meat grinder of the NKVD, those who moved from notoriety into the gulag were also innocent of charges. For the latter, another step was required: The authorities had to fall silent about the awards, for how could they admit having mistakenly permitted a "wrecker" to penetrate such a citadel of Soviet power as the Red Army, a research institute, an important factory, or an Arctic expedition? For Arctic explorers the chance of leaving the ice as a hero and ending up in the camps as an enemy of the people must have seemed even more remote than the ice floes they inhabited, given the fact that they willingly took risks to conquer the Arctic. They had to sail late into the navigation season, often with inadequate supplies of food and fuel, with icebreakers held together by solder, Bolshevik faith, and old parts. They willingly took these risks in the name of science and exploration, but might pay with prison sentences for their efforts.

In May 1928, an Italian dirigible under the director of General Umberto Nobile sailed to the North Pole. Another goal of the trip was to identify an island to name "Mussolini Land"—an Italian "Novaia Zemlia" or "Franz Josef Land." Nobile had successfully built rigid dirigibles for flight in extreme conditions. But on the way back toward Spitsbergen, the dirigible crashed, killing a number of the crew who were never found and stranding nine others with one dead among them; they would be rescued by an international effort seven weeks later. A ham operator in Russia, Nikolai Shmidt, picked up the SOS, and soon the leaders of the USSR had ordered the icebreaker *Krasin*, under the direction of explorer and professor Rudol'f Lazarevich Samoilovich, to rescue the Italians. After furious preparations, on June 12 the *Krasin* sailed out of Arkhangelsk toward Spitsbergen with 136 people on board, twenty of whom were leading officials.[67] Both Samoilovich and Nikolai Shmidt would be arrested in the purges, both for spying. Their fate reveals the profound perversity of Stalinist propaganda. During the rescue mission journalists on board the *Krasin* filed daily reports about the heroic effort. On July 12 the *Krasin* rescued the survivors of the *Italy*, with General Nobile himself having been pulled from the ice by the Swedish pilot Einar Lundborg. Larkov reports that, according to *Izvestiia*, upon returning to Leningrad on August 5, 1928, some 250,000 people met the great icebreaker.[68] No one saw Samoilovich off to the camps.

Other participants in the glorious *Krasin* expedition were arrested. Fedor Gavrilovich Gavrilov, thirty-eight-year-old father of four children, was arrested in January 1931, got five years in the tundra prison camp Svir'lag, and was released in October 1934, although nothing more is known about his further career. Several others on the ship—mechanics, stokers, journalists, navigators, people of different class backgrounds and nationalities—were also swept up for counterrevolutionary activity, wrecking, and espionage, including a number who had earned medals and awards. Ten participants of the legendary *Krasin* rescue besides Samoilovich were shot in 1937–38. Even the ham operator who picked up the SOS of Nobile, Shmidt, was arrested in Taskkent in 1941 for anti-Soviet activity and shot six months later. The head of the NKVD in Uzbekistan, who signed his order of arrest and his death sentence, A. Z. Kobulov, was himself arrested and shot with Beria in 1953.[69]

Surely, heads would roll for the zimovka of the *Malygin, Sadko,* and *Sedov,* the latter from October 23, 1937, through January 13, 1940, a total 812 days of deprivations and challenges that became a passion play of the good of Soviet socialism over the dangerous Arctic weather. To keep the spirits of the men, women, and children up, the captains organized study circles on Marxism-Leninism-Stalinism, they had English and mathematics lessons, they played soccer and volleyball on the ice with balls that hardly bounced, they held checker, chess, and dominoes evenings, and they even completed degree programs. In the deep polar nights the *G. Sedov* hosted eleven fourth-year and eleven fifth-year students who had come along on the expedition to continue their upper-level course work in the Hydrographical Institute. The Committee on Higher School Affairs in Moscow approved opening a "branch" of the institute on the ship. V. I. Vorob'ev was the floating rector of the "drifting university." It was not hard to put together a faculty from among the many highly qualified individuals trapped in the ice of the Laptev Sea. I. D. Zhongolovich taught astronomy and gravimetry, S. A. Ianchenko also astronomy, B. A. Morzhov hydrology, and so on. Any school on solid ground in the tundra would envy this faculty. The students lived on the *G. Sedov* and each morning escorted the teachers from other ships to class under armed convey in case of encounters with polar bears. They undertook laboratory exercises by kerosene lamp, in some cases using the reverse side of can labels as paper, but they tried not to take notes in any event because the ink froze. They studied oceanography, hydrography, astronomy, and navigational instruments. At the same time, the expedition directors organized classes for sailors, and a number successfully passed examinations for promotion to navigator or mechanic.[70] Finally, in April polar pilots managed to land on ice runways and evacuate 184 people to the mainland: all the women and children and many members of the crew.

When—during the terrible travail—Glavsevmorput director Otto Shmidt signed order 605 on October 15, 1937, accusing expedition leaders of not having taken the "necessary measures for the liquidation of wrecking, and interference with the apparatus," and ordered personnel from their posts, the heads of the expedition fell into stunned (radio) silence. They were shocked that Shmidt had fallen under the spell of Stalinist paranoia, convinced that Moscow simply did not comprehend their situation, determined not to follow Shmidt's order, and planned to clarify things upon return to Moscow.[71] On more than one occasion, the leaders of the expedition were forced to ignore or countermand orders from the mainland to ensure the safety of the people trapped on the capricious ice, an extremely bold and dangerous practice in the Stalin era. Ultimately, they were forced to acknowledge that "wrecking" must have played a role in their zimovka and the hardships in the "Camp of Three Ships." Several of the crew named names. They asserted that "wreckers" included the very head of the expedition, Samoilovich, who heroically saved Nobile, a number of his coworkers and students, and M. M. Ermolaev who had been at the wheel of the *Sadko*; Ermolaev, a leading polar explorer, had wintered on Novaia Zemlia during the second International Polar Year.[72]

In the arbitrary Stalinist legal system, one individual might be found guilty and executed, the next might be interned in a labor camp, later to return to society. The cases of Samoilovich and Ermolaev, two leading AARI researchers, reveal these two paths. Samoilovich, deputy director of the Arctic Institute, was rescued from the "Camp of Three Ships" by airplane, was taken to the mainland, celebrated his rescue with local officials, and traveled to Kislovodsk, an inland resort in Stavropol' Region, to recuperate. Within a month, he was arrested and taken to Lubianka Prison. The police searched his apartment, fortunately after his colleagues in Leningrad had already hidden his most valuable scientific papers. As a mining specialist—and a socialist—Samoilovich was valuable to Stalinist industrialization. He graduated from the Mining Academy in Freiburg, Germany, in 1904. Already a member of the Russian Social Democratic Labor Party, he was exiled to Arkhangelsk province for political activity. In 1912–13 Samoilovich conducted research on the coal deposits in Spitzbergen. After the revolution he pushed the national effort to develop northern resources through Sevekspeditsiia, participating in more than twenty Arctic expeditions. Yet a six-month investigation found him guilty of treason, wrecking, and counterrevolutionary organizational activity. An NKVD tribunal perfunctorily sentenced him to death by firing squad as a German and French spy.[73]

Mikhail Mikhailovich Ermolaev came from a gentry family of a military engineer. A student of Samoilovich, an organizer of the Arctic Institute, he participated in twenty-one expeditions to the Far North, including that of the *Krasin* in 1928 to save Umberto Nobile. He joined Sevekspeditsiia, where he met Samoilovich. By the time he was thirty he was considered a specialist in geography, geology, oceanography, geochemistry, and permafrost studies. For an expedition in 1932–33 to help starving Novaia Zemlia hunters and a marooned German scholar during the second International Polar Year, he received the Order of the Red Banner of Labor. But he was arrested June 5, 1938, and the NKVD investigators spitefully destroyed his nearly finished doctoral dissertation on Novaia Zemlia. He was freed, quickly rearrested, found guilty of the ludicrous charges of spying and wrecking, the latter in connection with "fruitless study of the ocean depths," and sentenced to ten years in a labor camp. Still a patriot, or perhaps out of desperation, he developed a novel approach to lay track on permafrost along the Vorkuta railroad. He was freed one month after the death of Stalin with an amnesty and fully rehabilitated in 1955. He returned to full-time teaching at the geography department at Leningrad University and finished his teaching career at Kaliningrad University.[74]

Ermolaev recounted his camp experience in 1999. On July 6, 1938, he and his wife Maria, their sixteen-year-old son, Alesha, and his seventeen-month-old boy, Misha, were asleep when the doorbell rang at 2 a.m. They lived in a huge apartment building that took up an entire block, largely inhabited by members of the Leningrad executive committee and sprinkled with intelligentsia, in particular AARI polar explorers, including Samoilovich who was married to Ermolaev's sister. It seemed that almost each night NKVD agents hauled away individuals and entire families.[75] Ermolaev opened the door, and three men entered, presented him with a warrant for his arrest, packed up what they could not comprehend, for example Ermolaev's doctoral dissertation, and whisked him away. One of the officers said he did not recall such a "pleasant" arrest.[76]

Miserable conditions and torture were the path to Arctic camps—or execution. Ermolaev was kept in the Shpalernaia Jail in a cell built for ten prisoners in the Tsarist era but in which there were sixty-five men and to which they added at least thirty-five more. Each of the prisoners was convinced *he* was there because of some mistake.[77] Torture led most of them to confess. One such technique was the "conveyor" in which a series of investigators took turns interrogating a prisoner over several days of sleep deprivation. The conveyor did its job. Ermolaev admitted to being a French spy. They rewarded him by taking him out of isolation. He began to think his sentence might not be that bad, or perhaps that they had forgotten about him, when, in July 1939, they dragged him off to interrogation again and insisted he confirm that a massive anti-Soviet affair had embroiled the Arctic Institute and that Samoilovich had orchestrated the activities of wreckers across the Arctic. They hardly needed to remind Ermolaev that Samoilovich was his brother-in-law to keep him on tenterhooks. Ermolaev confessed without naming co-conspirators and was sentenced to twelve years in prison. Ten days later he was working as an engineer-calculator in a sharashka called "Krest." This was a kind of freedom, for he lived in a dormitory with three to four men to a room and was allowed to speak freely.[78]

About three months later in February 1940 the "Krest" commandant informed Ermolaev that the military collegium of the Supreme Court of the USSR had freed him for lack of evidence. He was more in shock than joy; if he kept a stoic appearance, he felt sick for Samoilovich, his mentor, who was forever gone. He returned home, took up teaching geography, and began to write anew. On August 24, 1940, they arrested him again. He met this process with a sense of fatalism. He asked his wife to make tea for the road; the arresting officers also had tea and thanked her for her hospitality. This time they interrogated him in a business-like fashion without any wasted formality in asserting he was a German spy for his brother-in-law, Samoilovich. He confessed to wrecking in geology along with other leading specialists as part of what he assumed was some "Geology Affair." This time they gave him eight years.[79] He was sent to Arkhangelsk region, then transferred to an outpost of Soviet power in the empty Komi region forest where he felled trees for three years.

Ermolaev's tour of duty extended to Pechorlag, where he and thousands of other prisoners and engineers were involved in building railroads with their hands, picks, and shovels, then to Sevzheldorlag. Sevzheldorlag opened in 1938 with 25,200 prisoners, grew to 85,000 in January 1941, and was closed with about 28,000 prisoners in 1950 when it was joined to the Northern

Pechora Camp to lay railroads across the Komi Republic from Vorkuta to Kotlas, to Konosha and Kozhva. Ermolaev survived this to be released in January 1944. He was proud to have contributed to the war effort by building railroads, but he also knew the railroad was built on human bones; he calculated that 40,000 prisoners must have died from hunger, disease, and anguish. At the end of the war he was allowed to move to Syktivkar, the capital of the Komi Republic, where his wife and sons joined him. He was amnestied in April 1953 and given a full pardon on January 5, 1955. One of the lucky ones, he and his family returned to Leningrad—seventeen years after his ordeal began.[80]

The arrests began as soon as the ersatz heroes of the *Sedov* were pulled from the ice. Orlovskii was yanked off the platform on his arrival at the Leningrad train station, and N. I. Evgenov was picked up a week later. Petr Vladimirovich Orlovskii was crucially involved in the first post-revolution attempts to establish a Northern Sea Route. Before becoming head of the hydrographic administration of Glavsevmorput, Orlovskii worked in the North Siberian State Stock Society of Industry and Trade (Komseveroput'), first organized in 1919 under Kolchak to carry out trade expeditions from European Russia to West Siberia, and contributed to the "Enesei Expedition" under Vilkitskii. In 1934 Orlovskii headed the Kara and Second Lena Transport expeditions. Orlovskii led the hydrographic expedition on the *G. Sedov* in 1936 that discovered six new islands in the Nordensheld Archipelago. But when the *G. Sedov, Sadko,* and *Malygin* were caught in ice and had to winter in the northern Laptev Sea, they fired him for "not having taken the necessary measures for liquidation of the consequences of wrecking, obstruction of the apparatus, and lack of understanding of how properly to handle the cadres."[81] When he returned from the Perm forestry labor camps fifteen years later he was a broken man. He lived in Leningrad when friends helped him to return to the sea, but this time to the warm Sea of Azov where he directed fisheries being built by the Lisichansk Metallurgical Combine (today "Azovstal'" and one of the filthiest and heaviest polluting of all Soviet-era factories). Orlovskii died shortly thereafter in Mariupol of a heart attack.[82] As token recognition of his innocence, his name adorns an island in the Kara Sea and a cape on Franz Josef Island.

Even before they got home from Arctic expeditions, the NKVD snatched E. S. Gernet, S. V. Nikolaev, and K. K. Petrov—each one of them a prominent polar explorer. Every day the authorities pushed to intensify class war to root out "enemies of the people" and "harmful elements." Spies and "diversionary counterrevolutionaries" were arrested and shot: the former old Bolshevik and former head of the Commissariat of Marine Transport, deputy director of Glavsevmorput, N. M. Ianson; first secretary of the Far Eastern Region (Krai) from 1931–33 Sergei Adamovich Bergavinov; two heroes of the *Cheliuskin* expedition and deputies of Shmidt, A. N. Bobrov and I. L. Baevskii; and P. K. Khmyznikov, a hydrologist and hydrographer who planned the 1937 expedition to the North Pole.[83] Innokentii Mikhailovich Suslov, active in the study of the Tunguskaia, a massive explosion believed to have been caused by a comet or meteoroid,[84] who from his student days had been drawn to Bolshevik causes and contributed significantly to the work of the Committee of the North with indigenous people, also disappeared.[85]

The careers of many prisoners took unexpected turns from freedom to prison and back as they moved into and across the Arctic. N. A. Shtern, head of the sector of northern geological prospecting of the Geophysical Institute, whose work took him to Arkhangelsk province, was arrested in 1922, then released, and arrested again in 1929 for espionage. V. K. Kotul'skii, who explored the Norilsk region in 1920s, was arrested in 1930 in Leningrad, released, studied sulfide deposits on the Kola Peninsula, was arrested again in January 1931 in the "affair of wrecking and spying activity of counter-revolutionary groups in geological prospecting industry," and from 1934 worked in a labor camp in the study of apatite in the Khibiny tundra—side by side with "free workers," scientists, and volunteers of Apatitrust and the Khibiny Science Center. Freed again, he then moved in 1941 to Norilsk and became a main consultant of Severonikel, won an Order of Lenin, and in May 1949 was again arrested in the "Affair of Geologists"; in the increasingly paranoid life of the elderly Stalin, there were many such affairs. Kotul'skii was sentenced to twenty-five years and in 1951 died in Noril'lag.[86]

The purge of one person offered space for another to move up the hierarchy, although the beneficiaries never wanted it this way. One student, Viktor Kharlampevich Buinitskii, stayed behind to support the 812-day drift of the *Sedov,* as a result of which the directors of his institute, thinking he had dropped his studies, posted an announcement on the AARI bulletin board that he had been expelled. But when he returned from the drift as a "Hero of the USSR," they fixed this mistake and he got his diploma as an engineer-hydrographer. In any event, many of the "graduates" of the drift university became polar explorers: Irinarkh Fedorovich Tytkin, Vladimir Iosifovich Vilchinskii, and Valentin Petrovich Kozukov.[87]

Viktor Buinitskii was born in 1911 in Chita, beyond Lake Baikal, into a white-collar family. From 1932 he lived in Leningrad where he entered the Mining Institute, and in 1936 he transferred to the newly opened Hydrographical Institute. The following summer with other students he took part in research in the Laptev Sea that led to the disastrous capture of the *Sadko, Malygin,* and *G. Sedov* in the ice. The students continued their studies, participated in important research, and of course were involved in daily maintenance. Zimovka challenged all expedition members: fuel was short, as was warm clothing. They turned off generators and the steam boilers, drained the latter, and put them on mothballs. They turned empty barrels into "fireplaces" for cabins—if only to heat them a bit above freezing. From April 1938 until January 1940, the *Sedov* continued its drift under the command of K. S. Badigin with fourteen men including Buinitskii. For his heroism and steadfastness, Buinitskii was awarded the title Hero of the USSR, an Order of Lenin, a "Gold Star" medal, and a 25,000-ruble prize.

Buinitskii continued his upward trajectory. He graduated from the Hydrographic Institute in 1940, joined AARI as a senior scientific worker, and in July 1941 became its director. During the war, he served in the Red Banner, Baltic, and then Northern fleets, was a navigator on the destroyer *Uritskii,* and was a flagman navigator of a special group of the northern fleet headquarters involved with lend-lease acquisition of allied war materiel.[88] In 1942, in the face of increasing German Arctic activities—the Germans penetrated the Barents and Kara seas with airplanes, submarines, and destroyers—specialists at AARI were evacuated to Krasnoiarsk for the war where they studied how better to ensure the safety of the convoys.

After the war, AARI set out to conduct more extensive research at high latitudes to transform the Northern Sea Route into a normally functioning system. Buinitskii led the first postwar high latitude aviation mission to survey the ice regime. At this time Buinitskii joined the Communist Party and was sent back to AARI as director. In 1946 he defended his candidate dissertation, in 1948 his doctoral dissertation, in 1949 participated in the high latitude expedition "Sever-3" near the North Pole, and from 1950 served as professor and chair of the department of oceanology at Leningrad State University.[89]

If the repression of workers of Glavsevmorput through 1937 hit at lower and middle ranks, then a "bacchanalia" of denunciations and expulsions from the party began during the Great Terror. At an aktiv of Glavsevmorput, Director Shmidt, in accordance with demands of NKVD chief Ezhov, read out a list dozens of names of spies and enemies uncovered in the central and regional apparatus of the organization, many of whom had already been arrested, but also several still employed who might expect the secret police to haul them away at any moment. The terror accelerated in May and June 1937 from Murmansk to Yakutia. The police uncovered a "wrecking anti-Soviet organization" in Murmansk and meted out death sentences for M. N. Maksimov, deputy director of the Murmansk administrative division; V. I. Ivanov, director of employment; and L. K. Kuz'min, head of gold prospecting group; and sent scores of others to prison camps. The terror left land for research vessels (e.g., the research vessel *N. M. Knipovich*), freighters, and icebreakers and headed along rivers and out to sea.[90] Soon the Institute of Peoples of the North was attacked after the NKVD uncovered a counter-revolutionary spy ring—working for Germany—and shot the specialists Ia. R. Al'kor, who had studied colonization of Kamchatka and Chukotka in earlier years and was a leading expert on the Enisei River basin; ethnographer E. R. Shneider; and historian and anthropologist A. S. Forshtein.[91] Ethnographers, cartographers, and hydrographers disappeared into an endless Stalinist zimovka.[92]

An aktiv (mass party meeting) of political officers connected with Glavsevmorput freighters and icebreakers held in June 1939 after the Great Terror had abated gives a sense of the tense environment of vigilance, self-criticism, and superficial overconfidence that existed among party members who had so far avoided the camps. Two hundred party and Komsomol members of the Arctic Fleet, including workers of the political departments of Murmansk and Leningrad, participated in a meeting to discuss the purges and their continuing efforts to eradicate enemies and other "vermin" in Glavsevmorput. The throng opened the aktiv with greetings to Stalin and Molotov for their enlightened leadership, and after a speech by the head of the political administration of Glavsevmorput, Comrade Belakhov, on the party-political work of Arctic navigation in 1939, they voted unanimously to approve an order from the Council of Ministers in Moscow of May 15, 1939, on "Preparation for and Carrying out Navigation in 1939." The order read, "The 17th Congress of Communist Party (Bolsheviks) placed before Glavsevmorput the honorable task in the third five-year plan to transform the Northern Sea Route into a normally operating sea route that guarantees the systematic connection with the Far East."[93] The order noted that a "shameful interruption" of navigation occurred in 1937 when twenty-six ships were frozen at sea and the plan fell short. The reasons for failure were not only "the wrecking activity of enemies of the people, but the clumsy organization and carrying out of navigational operations, the obstruction of cadres by hostile . . . and adventurous elements, political carelessness, complacency and conceit which existed in the leadership of GSMP [Glavsevmorput]." After the staff and fleet were purged of worthless and hostile elements, Glavsevmorput liquidated all problems, freed ships from the ice, and fulfilled the 1938 plan successfully. The embrace of Stakhanovism and socialist competitions on the ships, on shore, and in steamship administration had improved production, while political departments and party organizations of ships, ports, and other marine operations ensured the political reliability of staff and crews.[94] The aktiv's participants did not mention either the role of weather or of being pushed by Moscow to pursue navigation into the treacherous autumn weather as causes of disastrous Arctic shipping adventures.

According to the central authorities, unplanned zimovka was not the only the problem, if the most visible one. Coastal enterprises and organizations failed to meet targets for loading, unloading, sailing, and docking from Arkhangelsk to Vladivostok. This story was a constant feature of the Northern Sea Route in all years, but acquired greater urgency in the late 1930s when the authorities insisted on ratcheting up of plans for freight along the route as if weather had been subjugated to Soviet power. Officials noted 160 idle ship days in 1937. In 1938 the ports of Tiksi, Dikson, and Providenie all handled loading and unloading poorly. And 1939 promised, if anything, greater challenges because Glavsevmorput had significantly increased the plan for passenger and freight transport as part of the assimilation of the Arctic. They would not be able to meet targets given tardiness in repair of icebreakers, freighters, and tugs; late sailing; challenges guaranteeing safe passage of ships; safety of cargos; and continued questions in training and posting new commanding ship personnel, not to mention insufficient leadership to carry out socialist competitions and advance Stakhanovism throughout the Arctic.[95]

Not ones to accept blame for their reckless push into the Arctic, even with a growing, more powerful icebreaking fleet connected with the launching—with two-year delay—of the *Stalin* class ships, political operatives pushed targets, ignoring circumspect consideration of "objective" meteorological conditions and a fleet that remained underpowered. Instead, they identified wreckers, counterrevolutionaries, and Trotskyites, removed them, and declared that in 1939 they could now and must fulfill and overfulfill plan targets. "Until this time," a Glavsevmorput official declared, "sailing in the Arctic carried an expeditionary character. In practice this led each normal trip to acquire a special kind of heroism with exaggerated details that did not always correspond to the truth." He pointed out that past standards to judge success were mechanical: "The main criterion in the evaluation of the work of any ship was considered sailing along the Northern Sea Route to a defined point and its return. Now such work cannot satisfy our government. [We are] at the given stage of development of the sea route where it is necessary to advance the problem of economic exploitation of the route."[96] Heroism had given way to regularity.

Given the multiple messages that the Soviet citizen received, the shoals of self-criticism he had to navigate, and the ice floes of wrecking and counterrevolution that surrounded his daily life, it is still surprising that so many polar specialists were repressed. Roughly fifty individuals in the central apparatus of Glavsevmorput were repressed from 1934 through 1954, while in the Leningrad and other regional administrations of Glavsevmorput 461 staff members gained "zimovka" in the gulag, with the most murderous activity during the Great Terror from September 1936 through October 1938 when 230 of the individuals were shot, and in the Krasnoiarsk and Iakutsk administrations the largest number of individuals were shot or imprisoned—177 and 75 people respectively.[97] In terms of personnel in scientific and higher educational institutions connected with the study of the Soviet Arctic, from 1920 through 1954 233 individuals were repressed with twenty in Moscow, in Leningrad sixty-four, and in Murmansk region fifty-two, most of them in the period September 1936 through October 1938 when seventy of them were shot. Finally, regarding those who participated in polar

Table 3.1 Leading personnel repressed, "Enemies of the People" in the Soviet Arctic, 1920–54

	Total arrested	Of which number executed
Glavsevmorput central apparatus	50	39
Glavsevmorput regional apparatus	461	230
Scientific and higher educational institutions	233	70

Source: Sergei Lar'kov, Fedor Romanenko, *"Vragi Naroda" za Poliarnym Krugom* (Moscow: Paulsen, 2010), pp. 131–132.

explorations or were employed in shipping, thirty-three from the Murmansk steamship authority, twenty-two from the Arkhangelsk authority, and 133 from the Far East authority were repressed, of whom forty-eight were shot in the period 1936 through 1938 (see Table 3.1).[98]

The terror across the Arctic extended into the skies. A number of polar aviators, heroes all, suddenly became anti-heroes. In 1985 at a birthday celebration of Mark Ivanovich Shevelev, the first director of the division of polar aviation, a general and "Hero of the USSR," General I. P. Mazuruk declared with a shaky voice, "In polar aviation at the end of the 1930s we had several hundred pilots, navigators, and mechanics, and Mark Ivanovich did not let one person fall into the paws of the cult, not one! Let's face the truth, we had disciplinary violations, accidents and even catastrophes occurred, and it was easy to turn the guilty into 'the public enemy.' But Mark Ivanovich did not condemn one person! You yourselves understand how much he risked then although he was a Hero and a Deputy—as if we few know heroes and deputies, and leaders of the party and government, who disappeared into Stalinist torture chambers!"[99]

Polar pilots fell to the Stalinist "cult of personality." On the basis of trumped-up charges, gossip, and slander—and no doubt a little envy—they arrested the pilots F. B. Farikh and V. M. Makhotkin and the mechanic N. L. Kekushev, who had won medals in the civil war and an Order of Lenin for a polar expedition in 1937 and who, during World War II flew nearly five dozen missions into the Leningrad blockade and evacuated staff of the Arctic Institute to save its scientific legacy. But for all this Kekushev was sent to a gulag camp in central Kazakhstan to mine copper and build railroads. With tongue in cheek, Kekushev later told his friends about his activities as a brigadier (leader) of a gulag construction brigade: "My brigade was distinguished. Imagine, one was a former Kombrig [commandant], a second—a brigadier Fuhrer of the SS, a third—from the Spanish civil war brigades, and the others—were no worse!"[100]

The hazards of polar aviation in the 1930s, when pilots literally flew by the seats of their pants in airplanes roughly modified to handle difficult conditions, with inadequate if not sporadic radio communications, with hurriedly built runways and ever-changing snow and ice conditions, poor supply of fuel and food, and a cold cockpit, were mercilessly compounded by propaganda bosses who pushed them to fly further, longer, colder, higher, and faster and to achieve aviation records before their capitalist competition. This meant there would be accidents, lots of them, and yet it was symptomatic of Stalinism to assert that any misfortune on the ice resulted from enemies. If the pilots managed to avoid the repression during the war, then in the postwar years dangers worse than ice and lack of fuel abounded. It did not matter if pilots were decorated heroes, for the secret police easily found enemies everywhere, including in the cockpit. Iu. K. Khlebnikov, with an Order of Lenin, who also won an Order of Nakhimov, was arrested in 1948 and forced into hard labor at the Vorkuta mines, the "Stalinist resort," as he came to call it.[101]

Return to the Motherland?

Stalin's death left a series of questions for labor camps: How would they be operated? Would officials relax hard labor in any way? How many of the camps would remain open? How crucial were the camps to the processes of assimilation of the Northern Sea Route and the exploitation of rich ore and other resources? Stalin's death triggered revolts among prisoners in a number of camps that had to be put down by NKVD troops. In Vorkuta after the execution of Lavrentii Beria, prisoners went on strike in July and August 1953. Eventually, the commandant ordered guards to fire on the prisoners; some forty or fifty or sixty men were killed. In Arkhangelsk two groups of prisoners managed to escape, some of whom were dangerous criminals and not recaptured, and seven other escape attempts followed. In the heat of the moment, one guard used his rifle and killed a lumberman. So-called banditry and hooliganism became widespread in the camps as more and more prisoners refused to work.[102]

Nearly a year and a half after Stalin's death, after a June 1954 decree of the Central Committee of the Communist Party, many camps began to carry out reforms to adopt a more humane prison regime and to examine cases of individual prisoners to determine if any errors had been made in prosecution. The authorities dreaded to admit that innocent individuals filled the camps. The reforms spread slowly northward into Sevdvinlag, Kargopol'lag, Mekhren'lag, and Uchpechlag and later to such camps as Iagrinlag. In theory, prisoners had received new rights that offered a more accurate accounting of their labor—if fairness could be an aspect of an innocent prisoner's camp life.[103]

Even as they began to carry out review of sentences, the authorities still insisted upon political education of the prisoners, worried about its low level—perhaps because they realized the camps would likely be emptied and would lose their prevailing economic justification—and hence refocused on ideological retraining as their true task—as originally intended. They lamented the fact that few camps organized regular propaganda lectures and that discussion groups failed to meet; in some camps the administration had organized no such political work. While prisoners and camps might have had subscriptions to newspapers, they did not read them out loud to the large contingent of illiterate prisoners;[104] this was a blessing because, after all, they stood in the cold all day. In Arkhangelsk province in 1954 fourteen camp directors were fired for reasons not entirely clear. The official reason was that apparently they and their underlings had not paid adequate attention to politico-cultural education and few guards or other staff attended cultural councils on a regular basis. In many camps there were far too many prisoners per inspector and guard, so it was difficult to require political reeducation in any event.

In Arkhangelsk region, many commandants reacted slowly to Stalin's death and the reforms it triggered because so much uncertainty about the extent and outcome of reforms remained. Perhaps because of this, "gross violations and illegal activity of the administration in relation to prisoners" continued.[105] The guards continued to put "politicals" and petty criminals together with recidivists and violent criminals. That "not only makes work on reeducation . . . and their labor usage more difficult, but permits the spread of harmful influences among the prisoners." Bands of criminals preyed on the defenseless; women were at special risk if not kept isolated; and youthful offenders seemed the most vulnerable. One day, three prisoners were murdered by criminals in Mekhrenlag in Plesetsk region. On top of this, the guards assaulted prisoners wantonly and threw them into isolation without reason. One Sergeev at "Kovzha" camp of Kargopol'lag was in isolation for 125 days, during which time they did not permit him to wash. Isolation for six weeks was a common occurrence.[106] Corruption remained widespread. One inspector reported that at Camp 7 "Shkarba" the guards sold vodka to prisoners.[107]

While, according to a mid-1954 report, the living, food and sanitary conditions in the camps corresponded to needs—with each prisoner having his own bunk and bed clothes and three meals daily—these prisons never had a laundry facility. Prisoners had to put on their filthy work garb after the weekly steam bath. They had no matches, paper, or pencils. In fact, food was miserable: they had absolutely no canned vegetables, meat, or fish.[108] On top of this, of course, medical care was spotty and perfunctory. The camps had few medical workers, medicines, or medical equipment. Prisoners waited a long time for medical help, often getting treatment, especially emergency treatment, too late, and TB and other infectious diseases were rampant in the general prison population. Dysentery frequently broke out: at one camp an epidemic felled 123 prisoners. Mekhrenlag had open positions for thirty-five doctors, twenty-four feldshers (rural paramedics), thirty-nine nurses, and seventeen public health inspectors that they could not fill.[109]

Not surprisingly, workers stopped fulfilling norms, refused to work, or were not even brought to worksites because of absence of guards. In August 1954 17 percent of the prisoners of Kargopol'lag failed to fulfill norms, and in the first fifteen days of September 4,680 man-days were lost. In the first eight months of 1954 at Mekhrenlag 21,870 man-days were lost due to refusal to work, and 34,740 man-days were lost because there were not enough guards to escort the prisoners.[110] According to another source roughly 10 percent of labor was lost daily to prisoners unable—or unwilling—to work.[111] Was this an Arctic thaw that followed the accelerating de-Stalinization thaw under Nikita Khrushchev?

If the daily life of prisoners in Arctic (and other) camps improved only slowly, then the review of cases moved forward at a glacially slow pace, and the quasi-military tribunals set up to consider them were reluctant to find anyone innocent. Officials hesitated to reconsider convictions judiciously, perhaps for fear that a true criminal might be wrongfully released or because they feared that wholesale release of gulag prisoners would lead to social dislocations and unrest. One Arkhangelsk provincial commission made up of eighteen individuals from the district attorney's office, the secret police, and ministries of internal affairs and transportation considered the cases of 241 individuals over the second half of 1954. The commissioners found ninety-one sentences to be correct, pardoned only two individuals completely, but released 108 with amnesty or with sentence reduced to time served.[112] The commissioners expressed confidence in their conclusions and worked hard at the endeavor, meeting twice weekly to consider forty-five to fifty different cases. On July 28, 1954, seventy-six lists of names with a total of 818 individuals arrived at the commission. Lists continued to arrive, and the commissioners anticipated having to consider 1,500 to 2,000 names in short order.[113] On August 26, 1954, the commission produced a report based on analysis of 131 lists with 1,316 prisoners' names. The commission had examined the personnel files of 698 of the individuals, yet fully rehabilitated only four, shortened terms for 234 individuals, amnestied another fifty persons, and refused to reexamine the rest. In typically grotesque Soviet fashion, the work was held up by absence of typist.[114]

Elsewhere review lagged as well. In Kargopolag, a tribunal determined that the government order to review cases applied to only 704 of 824 men, and only sixty-five of them were released "conditionally early." Virtually all of 1,244 invalids (handicapped) remained incarcerated. In Mekhrenlag as of September 1954 a commission had still not been formed to hear cases. The head of Camp 7 of Mekhrenlag had received 100 declarations of innocence from prisoners, but examined only thirteen of them, and in Camp 2 the commandant had not answered 241 of 822 declarations.[115] Prison labor continued to play an important role in local and regional economies for decades after the 1950s, particularly in those places where camps had been prevalent under Stalin—in the Arctic, in the Urals, in Siberia, in the Far East.[116]

Stalin Leaves the Arctic Circle

One great polar researcher, A. P. Babich, served years on floating laboratories, in zimovka, to understand ice floes and currents and cold and wind. Sentenced to death for crimes against the state, his sentence was commuted to suffering instead in a mining camp. In the last letter to his loved ones that arrived after his death, he wrote, "It seems to me that I am not in a camp, but in the farthest zimovka, and can in no way return from it to the Mainland. But if only at sometime this zimovka would end."[117]

Stalin died on March 5, 1953, of a cerebral hemorrhage, his death was announced to the public on March 6, millions mourned, and tens of thousands of people waited in lines to see his body lie in state. His body was put in the mausoleum that had been hurriedly re-anointed "Lenin-Stalin." Most people realized that Stalin was a vicious dictator whose policies had led to the deaths of millions and they knew that the labor camps continued to hold innocent relatives and friends, 2.5 million of them. Yet most Soviet citizens were shocked that the tyrant who had led the country for twenty-five years was dead and worried what might come next in the national and international arenas. How could the nation, now engaged in a cold war and still rebuilding from the Nazi invasion, move forward?

Arctic research and development remained an open question after Stalin's death. Because of the costs of the nationwide reconstruction effort after World War II and of engaging in the burgeoning and expensive cold war weapons programs, budgets everywhere—especially outside of Moscow—were tight. The postwar five-year plans focused on rebuilding heavy industry and infrastructure. Richly promising Arctic resources were difficult to extract and required the allocation of greater manpower, capital, and investment from the center, even as the local authorities in Murmansk, Karelia, and Arkhangelsk were pushed to resurrect smelters, mills, hydroelectric power stations, and factories, reopen mines, and expand production, and they called for significantly higher financing. The efforts of scientists in Glavsevmorput, AARI, and other institutions lagged because of shortfalls of machinery, equipment, instruments, and financing, even as specialists wandered back from the tundra to resume full time research. But Arctic exploration and exploitation activities would expand rapidly from the mid-1950s as part of the de-Stalinization reforms in which the national scientific enterprise also saw tremendous growth in numbers of researchers and institutions and as specialists regained autonomy in the formulation of their research programs.

Three of the men sparring to lead the Communist Party—Nikita Khrushchev, Viacheslav Molotov, and Georgii Malenkov—realized the danger of the fourth pretender to Stalin's mantel, secret police chief Lavrenty

Beria. In June 1953, the three men ordered Beria arrested in the Kremlin. He and a number of his associates were executed six months later. For two to three years a succession struggle played out in which Khrushchev rose to the top. Khrushchev then triggered what became known as the "Thaw," after a novel of the same name by Ilya Ehrenburg. Khrushchev attacked Stalin for his murderous policies and his "cult of personality," although he did not attack the Communist Party itself, nor as a leading official in Moscow and later in Ukraine as party secretary had Khrushchev rejected those policies.

In the domestic arena, Khrushchev reoriented the economy more toward the benefit of the consumer while modernizing industry; sought to improve agricultural performance—and thereby the diet of the citizen with more meat, dairy, fruits, and vegetables; applied science and technology to the production process under the banners of "mechanization" and "automation" to free workers from manual labor; and promoted greater openness in society through criticism of Stalin's abandonment of so-called Leninist norms of party democracy. In housing, millions of people lacked their own living space with modern utilities. Khrushchev encouraged a crash construction program where speed often was more important than quality; but the authorities built millions of apartments. In this environment, AARI resumed its presence on Arctic drift stations—without fear that specialists might return to face arrest for wrecking.

In February 1956, in a special session on the last day of the Twentieth Communist Party Congress, Khrushchev read out a litany of details of Stalin's crimes against the nation before stunned party members in his so-called Secret Speech. He condemned the excesses of Stalinism, its "cult of personality," the suffering of millions of citizens, and so on. He revealed that Stalin personally had ordered party favorite Sergei Kirov to be murdered in his office at the Smolnyi, the Leningrad region party headquarters and the site of Lenin's first government, and then used the murder as a pretext to purge the Leningrad party organization; Kirov had personally pursued assimilation of Arctic resources in concert with leading polar explorers.

Khrushchev's efforts to reform society created great uncertainty with regard to the gulag and those prisoners who would be pardoned and returned to society. As Miriam Dobson explores, party leaders recognized a need to return to the rule of law, but not to punish the perpetrators of the terror. They determined to empty the camps and to recognize and reintegrate the victims of Stalinism into society. The reasons for emptying the camps included economic liberalization and the need to promote economic growth in competition with the west, as the costs of running gulag labor—across inhospitable bands of unsettled northern latitudes—were very high. In addition, the camps had become restive after Stalin's death with uprisings and protests. If, under Stalin, the masses were informed every day in the media and at meetings in their places of work that they had built socialism, then many of them could not reconcile their constant vigilance against enemies internal and external to the USSR with this socialism. Technological feats in the Arctic regions and in aviation and icebreakers only temporarily diverted attention from their low standard of living. The new leadership therefore sought to recreate the message of an impending glorious future, a future that had essentially disappeared in the Stalin era. This also meant repudiating terror, although no one fully answered such questions as: Who are the victims? Were only political prisoners victims? And who was to blame for the mass terror? The release of millions of prisoners led to a difficult dialogue about identify, politics, belief, honor, honesty, and community, with community now to serve the role of reeducating offenders of the Soviet morality.[118]

And if the pursuit of reforms during the thaw meant rediscovery of constructivist visions of socialist society, then it also raised disorienting and politically unsettling concerns. When Molotov, Malenkov, and Kaganovich were dismissed from power in June 1957 as an "anti-party opposition," this act led many party members to resist further reforms, for now they were required to question the honor of an entire generation of leaders. When Stalin was removed from the mausoleum in 1961, it further offended more conservative members of society. Khrushchev attempted to maintain control of the process of de-Stalinization at the 22nd Party Congress in 1961 when he pushed the unanimous adoption of the Third Program of the Communist Party. In the program Khrushchev reiterated the Soviet truth that the Bolsheviks had eliminated the exploiting classes and built socialism. Now he promised to create a communist garden of plenty within twenty years.[119] And yet society had not come to grips fully with Stalinism, with the labor camps and the returnees, many of whom had no place to go, no family or home. On top of this, the camps began to grow again. In the Brezhnev era, a period of relative calm on the domestic front and steady, if uninspiring, economic growth, leaders rejected de-Stalinization. They "remilitarized" society. The authorities determined to deal with dissent harshly, with trumped-up charges, long prison terms, psychiatric prisons, and punishment through drugs and torture. In the international arena, determined to defend the socialist world from reforms, the Brezhnev leadership ordered the invasion of Czechoslovakia in 1968.

For Arctic researchers, the political changes under both Khrushchev and Brezhnev ushered in a golden age of well-funded, long-term, cutting-edge research. Arctic research and development in the natural and social sciences grew massively from the 1920s into the Brezhnev era, alongside efforts to tame the Arctic regions through gulag camps. Arctic researchers surely welcomed the changes in policies after Stalin's death. Khrushchev's reforms meant that some researchers who managed to survive the gulag might return to work. While Vittenberg was permitted to resume his studies at Petrozavodsk University in Karelia, for example, such talented explorers as Samoilovich had been murdered or disappeared in the camps, lost forever to the frantic process of Arctic conquest. Yet Khrushchev's determination to raise the standard of living of the Soviet citizen meant not only better housing and food, but increased funding for science and technology as a path toward a better standard of living. Arctic programs benefited from this support in ever-more ambitious research, nature transformation projects, and acceleration of efforts to urbanize Arctic regions. This led directly and indirectly to a bigger budget for Arctic research in AARI, Gidrometsluzhba, and other institutes and universities and a much larger research program generally as demonstrated in a growing number of expeditions and drift stations that filled the northern latitudes.

4

The Arctic Sciences of Places and People

Soviet leaders and scientists reached an accommodation in the 1920s that permitted rapid expansion of the scientific enterprise. Most senior researchers and professors and many of the students were representatives of the middle or upper classes and so remained mistrustful of the Bolsheviks and their public commitment to cultural and political revolution. Yet they welcomed financial and administrative support from the new leaders that contrasted with the inattention of the Tsarist regime. Soviet leaders in a variety of commissariats and bureaucracies supported the growth of the modern scientific enterprise, intervening directly in many cases to support representatives of the physical, chemical, and life sciences. Under Stalin, science and technology policy was centralized more vigorously, as the makeup of the student body changed to reflect greater class consciousness, and the scientific enterprise continued to grow rapidly in numbers of institutes, researchers, project, journal publications, and so on.

In the developing fields of polar research, government support was crucial to enable explorers, hydrographers, zoologists, geographers, and others to push northward and eastward along the Northern Sea Route. Through a series of ad hoc and formal institutions, they began much more systematically to study the circumpolar USSR. By the late 1920s and Stalin's "revolution from above," polar research and development expanded even more rapidly in a series of settings: the Institute for the Study of the North, the Arctic Research Institute, the Geographical Society, and others. With the creation of Glavsevmorput in 1932, Arctic research became regular, planned, and systematic as with other fields of science. From this time into the twenty-first century, Russian/Soviet polar researchers have had a constant presence in the Arctic on military ships to secure the borders; on research vessels to evaluate resources, chart currents, and study water chemistry; and on so-called drifting stations in the central Arctic basin to establish firm understandings of meteorological trends, ice regime, oceanography, and other subjects necessary to operate the Northern Sea Route.

Most important for Arctic assimilation were the physical sciences. With government and Communist Party support, specialists rapidly expanded the number of researchers, programs, and independent institutes devoted to the study of the physical nature of circumpolar regions, weather and climate, and mineral resources. More and more confident of their understandings of those regions, and of the support they could count upon in terms of financial resources, machinery and equipment, and food and supplies, leading polar explorers trained a large cadre of specialists who filled the Arctic, carried out extensive studies, and returned to branch and central research institutes with extensive data that enabled exploitation of mineral, fossil fuel, fish, forest, and other resources. By the 1960s standardized prefabricated housing units, tractors, and bulldozers enabled some comfort for two- and three-shift research crews that wintered in the central Arctic basin for months and even years at a time.

The emphasis on geophysics meant less support for the "human sciences," here broadly defined to include the medical, veterinary, and social sciences. If, for example, the government provided significant resources for the modernization and expansion of medical care throughout the USSR, and the percentage of the annual nation budget reached roughly 6 percent in the 1950s, by the 1970s it had dropped to 1.5 percent as military and other expenditures consumed investment, and even in the 1930s and 1950s the emphasis was on medical research, not delivery of services even in major urban centers. People in newly developing Arctic regions lacked access to the same resources available to Muscovites and Leningraders.

Soviet specialists also brought the social sciences to bear on Arctic regions. Ethnography, anthropology, linguistics, and sociology were critical to understanding the lifestyles of local and indigenous people. They facilitated "modernization" once Communist Party officials had determined to transform northern peoples into self-aware Soviet citizens. While there may have had great successes in applying the physical sciences to the study of the Arctic, specialists in the social sciences encountered much greater difficulty in bringing modern Soviet understandings to the tundra. This chapter examines the application of the polar sciences, broadly defined, to Arctic development, including the social sciences, and their role in subjugating resources and people in the Arctic to central Soviet economic and political desiderata.

Arctic science and technology developed within the framework of Soviet science generally, although with several nuances.[1] First, the tension to conduct applied research with immediate applications made it difficult at times for specialists to conduct basic research. Given the fact that all Arctic research seemed applied in its many connections to the Northern Sea Route, trade, and industry, the scientists managed to secure sufficient support, especially in the Khrushchev and Brezhnev eras. Second, the highly centralized scientific enterprise created special challenges. Leading specialists preferred to remain in Leningrad and Moscow institutes from which they ventured northward for Arctic exploration. It was harder to staff branches of institutes and create new ones in Arctic latitudes. Generally, scholars who settled in the Arctic latitudes in many fields had lower qualifications than those in Leningrad and Moscow, and employee turnover was greater, at least until the 1960s. Third, ideological interference played a significant role in shaping many fields of science from physics with interference in quantum mechanics, to biology and Lysenkoism with its rejection of genetics, to the social sciences. In polar research there was less direct manipulation, perhaps because the geophysics of polar regions—water, atmosphere, ore—was less amenable to the heated ideological battles that consumed much of science elsewhere, although many innocent explorers and researchers fell in the purges. Last, because of the Khrushchev reforms scientists gained greater autonomy. And with the largesse of Cold War funding, the scientists were able to establish new institutes and centers; the number of scientists grew rapidly in all fields. The sciences, including the Arctic sciences, benefitted in this new environment.

Glavsevmorput's Arctic Research Institute

The most important institute for research and development of polar regions was the Arctic and Antarctic Research Institute (AARI). It was subordinated to Glavsevmorput until the 1950s and then moved to Gidrometsluzhba (the State Meteorological Service); its scientists maintained a great degree of autonomy in determining research programs no matter its bureaucratic affiliation. In the first twenty-five years of its activity and that of its predecessors, AARI personnel organized roughly 300 different expeditions—a third of which were geological. It acquired a fleet that, over time, included more than fifty ships from motor-sail boats to icebreakers and research vessels. Its researchers discovered dozens of islands: Bolshevik, October Revolution, Komsomolets, Pioneer, Vize, Isachenko, Dlinnyi, Domashnii, Shmidt, Sedov, "Izvestii TsIK," Ushakov, Arctic Institute, Voroshilov, Kirov, Proliv, Shokal'skii, Iungshturm, and Red Army, as well as bays and fjords (Rusanov, Neupokoev, Mikoian, Akhmatov, and Matusevich). Even during the war specialists noted "significant achievements"—research in meteorology, geophysics, and climate studies generally continued.[2] Its scientists established a comprehensive network of polar stations and observatories along the Arctic coast and on islands. If before the revolution scientists had established only six such stations, then by 1937 they operated fifty-five of them, and in 1947 already seventy-five, which carried out not only meteorological research, but hydrological, visual (ice), aerial, geomagnetic, and actinometric (solar radiation) observations.[3] Many of the results were published in *Trudy Arktichestogo Instituta,* the institute's journal that continues to be published into the twenty-first century. Scientists also issued dozens of monographs.

The structure of the institute evolved in keeping with technological developments in shipbuilding, aviation, and drift stations; analysis of a growing database; and more accurate measurements. It its early years AARI consisted of four departments: Ice (later Ice Regime and Forecasting), Marine Hydrology (later Oceanography), Meteorology, and Geophysics.[4] Specialists created a sector of riverine hydrology (later the Division of Hydrology of Rivers and Water Resources) whose members won a State Prize for their research in 1981. In the second half of the 1930s work began on formation of a system for hydrometeorological observations and forecasting. AARI added departments for marine operations in Dikson and Pevek in 1939 and later in Tiksi to support the sea route.[5]

World War II interrupted regular Arctic missions as personnel were siphoned off to the front, the institute was evacuated to Krasnoiarsk, and ships were used for military ends. Yet by 1946 the institute's research program began to return to normal. An eastern high-latitude expedition in 1946 under Captain K. K. Vyzov resulted in a 519-page tome with seventy-six graphs and seventy-three photos. Staff included biologists, chemists, and other scientists who carried out hydrological, ice, geological, hydrobiological, meteorological, aerological, geophysical, and actinometric research, in all thirty-five participants plus eight graduate students; Vladimir Frolenko, a filmmaker; Igor Ruban, a prolific artist, author, and participant in Arctic and Antarctic expeditions; and several journalists.[6] The growing interest in Arctic transport and polar aviation led in 1945 to creation of Department of Long-term Meteorological Forecasts to produce accurate long-term and short-term forecasts.[7]

Organizational developments at AARI also reflected changing and growing economic interests that enveloped the burgeoning Northern Sea Route. Study of the hydrological regime of the Arctic was crucial to tie the economy of the far north and its "great steamship rivers"—the Enisei, Ob, Lena, Indigirka, Kolyma, and others—to central markets through the Northern Sea Route. From 1936 researchers undertook systematic publication of materials on climatology of polar regions and in 1944 finished a large monograph on the subject. Yet often industry responded to scientific results with great delay, a phenomenon that repeatedly frustrated party officials and economic managers. They believed that the highly centralized organization of planning and production in the Soviet system could force the pace of innovation. In the 1930s efforts to design ice-worthy vessels led to the formation of the department of ice worthiness of ships, but the shipbuilding industry embraced theoretical advances slowly. From 1961 marine ice science developed into a new department. The institute added a department of innovation and research for hydrometeorological stations and outposts in the 1940s in which—during the 1970s—the engineer Iu. K. Alekseev invented a large number of instruments, for example, a hydrometric recording current meter for measuring marine currents, the "North Pole" winch, and other devices.[8]

In the 1950s a kind of "normal" research unfolded that included a series of long-term drift stations to study the central Arctic basin, analysis of the regimes of Siberian rivers, significant expansion of research activities to Antarctica, and active participation in the International Geophysical Year (IPY, 1957–58). V. V. Frolov, meteorologist and polar researcher, became the institute director. In spring 1950 M. M. Somov launched the second Arctic drift station ("Severnyi Polos" [North-Pole] or SP-2), and in 1954 almost simultaneously A. F. Treshnikov (SP-3) and E. I. Tostikov (SP-4) launched stations, guaranteeing essentially an uninterrupted Soviet presence in the

Table 4.1 Arctic and Antarctic Research Institute scientific fleet ships placed into operation, 1967–87

Name of ship	Date put into operation	Displacement
Professor Vize	November 1967	7,000
Professor Zubov	November 1968	7,000
Mikhail Somov	May 1975	14,000
Akademic Shuleikin	July 1982	2,100
Prof. Mul'tanovskii	July 1983	2,100
Akademician Fedorov	September 1987	16,500

Source: B. A. Krutskikh et al., "70 Let Ordena Lenina Arkticheskomu i Antarkticheskomu Nauchno-issledovatel'skomu Institut," *Problemy Arktiki i Antarktiki*, vol. 66 (Leningrad: Gidrometeoizdat, 1991), p. 23.

Arctic basin until 1993. In 1953 the institute organized Arctic observatories in Dikson, Tiksi, and Pevek, building on smaller facilities in place since 1939, to ensure support of shipping and complex hydrometeorological study of the region, as well as to improve weather and ice forecasts. Work on deltas of Siberian rivers proceeded in four directions: study of hydrological regime of deltas and coastal regions; clarification of laws of mutual influences of rivers and the seas; the search for regularities in data to improve long-term and short-term forecasts; and work with local organizations of transport to improve shipping. At the same time institute efforts to understand the ice worthiness of ships expanded. On September 5, 1955, the institute opened an ice experimental pool to carry out modeling including beginning work for atomic icebreakers, and in 1956 AARI established a ship research bureau whose specialists studied the conditions of sailing of ships in heavy ice.[9]

For many years the institute did not have its own fleet, but carried out research on ships rented from the Navy. AARI officials constantly complained that the fleet was inadequate for the effort both in numbers and in quality. Older vessels with wooden hulls could not stand up to the crush of Arctic ice. Scientists hesitated to advocate buying research vessels abroad because of their cost and the pressure to claim Soviet priority in this area; indeed, achievements in space and nuclear research gave good reason to expect results. But they recognized that orders from the Motherland's factories would be met only with great delay.[10] The institute's fleet met the growing demands of being science only in the Brezhnev era (see Table 4.1).

The *Akademic Fedorov* research ship. (Courtesy of the Russian State Museum of the Arctic and Antarctic)

The *Fedorov* was the most modern vessel, sooner a floating institute than a ship, with the newest equipment for observation and collection of data about ocean, ice, and atmospheric conditions, with twelve laboratories, five computers, a helicopter, and bathysphere complexes. Its crew of sixty people could handle an expedition of up to 160 persons.[11]

In the 1960s AARI achieved great stability in research programs and personnel. In Brezhnev-era fashion of gerontocracy, the same director, A. F. Treshnikov, ruled the ice until 1981. By this time the institute was subordinated to Gidrometsluzhba (the Main Hydrometeorological Service). The institute acquired its first computers toward the ends of forecasting weather and ice conditions in 1960, an Ural-2 vacuum tube computer, one of 139 produced. By 1970 the institute's computer center had four computers: two Ural-2s, a Minsk-22, and an M-220. These computers were slow, prone to breakdowns, and based on American machines, yet were a welcome contribution to the speed of analysis. Specialists organized a hydroacoustical laboratory. The recently launched *Lenin* icebreaker assisted in the launching of the SP-10 in 1961.[12] Researchers prided themselves on extensive publications in the Brezhnev era. From 1981 to 1988 they published thirty-one monographs, forty-nine volumes of *Trudy AANII,* and thirteen volumes of *Trudy SAE* ("Works").[13]

In the 1970s a large-scale research program on atmosphere and oceans in polar regions through POLEKS-North (Soviet) and POLEKS-South (international programs) unfolded that focused on pollution and other such issues. The Soviets significantly expanded drift and observation stations, from 1973 that carried out observations on pollution of water and ice by petrochemicals, and from 1979 on SP-22 research on aerosol-optical parameters of the atmosphere in the Program on Global Atmospheric Processes.[14]

In the years 1972–75 the laboratory of ice worthiness of ships together with the Leningrad Shipbuilding Institute developed rules for classification and construction of icebreakers, with D. E. Kheisin and Iu. N. Popov realizing the idea of the creation of an "ice passport": a collection of recommendations for standard operating procedures in Arctic waters.[15] In the 1980s floating ice stations and icebreakers joined hulls in extensive research. The institute launched seven SPs (25–31) with SP-22 working in tandem with satellite systems. The SP were launched, refreshed with personal and supplies, and captured with atomic and diesel icebreakers, occasionally with great difficulty. For example, the evacuation of SP-28 (under the command of N. A. Kornilov) by the nuclear icebreaker *Rossiia* occurred entirely in the polar night. In May–June 1987 the First Complex Scientific Expedition in a near polar region on the icebreaker *Siberia* (under Artur Chilingarov, with the deputy head of the expedition AARI director B. A. Krutskikh) led to several important discoveries on high-latitude dynamics of ice cover, on the ozone hole, and on geologic-geodesic peculiarities of hard-to-reach regions of the Arctic Ocean. On May 25, 1987, for the second time in history, a ship reached the North Pole—yet this was the first time in a period of maximum ice cover, on top of which the mission solved several practical problems: the evacuation of SP-27 and the organization of SP-29. And the scientific fleet of AARI expanded significantly: *Akademik Shuleikin* (1982), *Professor Mul'tanovskii* (1983), and then the *Akademik Fedorov* set sail. In 1986 the institute moved to a new building at 38 Bering Street on Vasilevskii Island in Leningrad from old, cramped quarters in a former townhouse on the Fontanka Canal in the center of the city.[16]

All of these achievements in science and technology and all of the organizational developments reflected the applied science profile of the institute to maximize operations of the Northern Sea Route. AARI research helped extend the time of operation of the route from thirty-five to fifty days to 100 to 105 days. With the introduction of powerful atomic icebreakers in the 1960s and increasingly well-understood regularities of natural processes in the Arctic, the group succeeded in establishing year-round shipping in the Kara Sea to the western shore of the Yamal Peninsula and Dudinka from 1976 and on the Enisei River from 1978. Speed increased; the frequency of problems with ice decreased.[17]

Drifting Research Stations

What were these North Pole (SP) floating research stations? When did they commence? Crucial to the task of exploiting Arctic resources was an extensive effort to understand the central basin of the Arctic Ocean, the coastal regions, the islands, the currents, and the ability of humans to fill those areas. Drifting research stations in which polar explorers literally occupied the ice for months and years were the major vehicle of research. As a first step, in 1935–37 institute polar explorers carried out three high-latitude expeditions on the *Sadko* in northern parts of the Kara, Greenland, Laptev, and East Siberian seas, which gathered great deals of oceanographic, ice, and other data and confirmed that with contemporary icebreakers it would be impossible to penetrate the central Arctic basin systematically. They recognized the importance of aerial observations to support the expansion of the Northern Sea Route, and they began the first of "drifting polar stations," which increased in frequency significantly in the 1950s. The early drift stations offered great assistance to the trans-Arctic flights of Chkalov and Gromov.[18]

But the most important first step was the SP-1 under Ivan Papanin in 1937. The impetus behind Papanin's expedition was the realization that air reconnaissance and research went hand in hand. Glavsevmorput head Shmidt and Vize supported Fridtjof Nansen's notion of creating drift stations in the central Arctic basin. When Shmidt and Petr Shirshov were trapped in the fiasco of the *Cheliuskin* (1934), they discussed in greater detail how to create a constant research presence with aerial support. By 1935 the pilot Vodopianov had been ordered to consider how to pursue this idea and with others spent 1935 contemplating how to support drift stations. In February 1936, while discussing Chkalov's flight across the North Pole to America on an ANT-25, the Politburo approved final plans for a drift station in 1937: the Papanin expedition.[19]

Government officials and scientists in Glavsevmorput recognized immediately how a constant presence in the central Arctic basin would provide extensive information about weather and climate and permit further assimilation of the Northern Sea Route. Already "jumpers" who spent hours or days on the ice undertook observations. But to make them systematic, a long-term presence was required. The first drift stations of the late 1930s and late 1940s facilitated in-depth comprehension of oceanographic processes, ice regime, stratospheric phenomena, the physical nature of the troposphere and ionosphere, and study of magnetic fields of the earth and contributed to significant geographical discoveries (the discovery of the Mendeleev, Lomonosov, and Haekel ridges).[20]

From June 1937 through February 1938 Ivan Papanin, Evgenii Konstantinovich Fedorov, Ernst Teodorovich Krenkel, and Petr Petrovich Shirshov drifted "heroically" through the Arctic for 274 days, during which time the station traveled more than 2,600 kilometers from the North Pole almost due south along the coast of Greenland. The researchers carried out systematic gravitational, hydrochemical, oceanographic, sounding, and astronomical observations, augmented with some zoological research. The crew regularly measured ocean depth, took bottom soil samples, measured water temperature, examined water samples from different levels, and carried out meteorological observations. In February 1938 the "North Pole" station drifted out to the Greenland Sea, and after several attempts, the camp was evacuated successfully with the help of an icebreaker.[21]

Papanin (1894–1986), the leader of the expedition, was a self-made man and true Soviet hero whose career was forged in the revolution.[22] Like many other polar explorers, his path to the Arctic was unexpected, and yet his lifelong tie to the seas ensured that he would explore distant waters. He came from Crimea and a sailor's family, beginning to work in the fourth grade. In 1914 he joined the Navy and served in Crimea during the civil war, joining the Communist Party in 1919 and then the Cheka (the forerunner of the NKVD). He worked as secretary of the Revolutionary Military Council of the Black Sea Fleet when his responsibility was to destroy the remnants of White power including the Tsarist officers' corps. Commander Mikhail Frunze, a leading civil war hero, promised the Whites amnesty of sorts on surrender. Instead more than 60,000 officers were shot or imprisoned. As Cheka commandant, Papanin witnessed these bloody events, but he never wrote or spoke about them. According to several sources, this may be why he left politics and revolution to focus on science—and he was not alone among many leading Soviet scientists throughout history who sought respite from the incessant political maneuverings of Stalinism in academia and research.

From 1922 Papanin served in Moscow as commissar of the Administration of Technical Support of the Navy and in the Commissariat of Post and Telegraph as director of the Central Administration of Military Preparedness for which he completed university courses on communications. He took a position in Yakutia as director of a construction project for a polar radio station. His career in polar regions had begun, perhaps as part of his own internal search for peace and quiet away from the turmoil of Stalinist politics. In 1932 he served on an expedition of the icebreaker *Malygin* in Tikhaia Cove where he directed the establishment of a polar station. In 1934 he took over direction of a polar station at Cape Cheliuskin at the end of the Taimyr Peninsula, a place of snow and wind where high temperatures in summer average 0°C, wintering in 1934 and broadening the station's activities into an observatory.[23]

Papanin's drift earned him accolades and medals. As a Hero of Socialist Labor and laureate of a "Gold Star," Papanin was the logical choice to succeed Otto Shmidt as director of Glavsevmorput in 1939, from which position he strove to increase the fleet of icebreakers available for Soviet power. But health did not permit him to continue at Glavsevmorput. From the beginning of World War II, Papanin supervised the construction of a new Arkhangelsk port, followed by others in Murmansk and in the Far East capable of handling U.S. lend-lease supplies. From 1943 Papanin was Rear Admiral, from 1937–50 a deputy of the Supreme Soviet, from 1948 the deputy director of the Institute of Oceanography of the Academy of Sciences where he worked to expand the scientific fleet of the Academy, from 1951 head of the department of expeditions of the Academy, and from 1951 to 1972 director of the Institute of Biology of Internal Bodies of Water.

Drifting stations served both a scientific and a cultural function. Like polar aviation exploits, they demonstrated before the nation and the world

Hydrological research on drift station, ca. 1937. (Courtesy of the Russian State Museum of the Arctic and Antarctic)

that the USSR was a scientific and military power. They enabled Stalin and his cohort to claim that the socialist citizen could conquer the world, while the capitalist citizen stood in bread lines during the Great Depression. They enhanced the legitimacy and stability of the regime during the Great Terror of the 1930s when so many innocent people, including Old Bolsheviks who had served with Lenin, confessed to crimes against the state and perished in the camps. While Soviet claims that precisely the leadership, brilliance, and certitude of Stalin enabled polar exploits ring of empty propaganda, precisely Stalin's leadership was central to the technological feats of the 1930s, for he insisted upon them at all costs—and often there were great costs.

Using wind-powered generators, the four drift explorers were in constant contact with Moscow by radio, patched through from Rudolf Island to Moscow. This enabled frequent contact with *Pravda*, *Izvestiia*, and other central and regional newspapers so that the nation could follow their adventure. From those reports—and from Papanin's diary published in 1938—we learn that the men were heroic, absolutely modest, extremely hard working, courageous, dogged, and unwavering in their daily work; sleep was not a concern. They men displayed not only courage, but "unruffled courage" in the face of blizzards, deep freezes, drift snow, drift ice, and endless polar

Unidentified, Krenkel, Papinin, and Shmidt at polar research station, ca. 1937. (Courtesy of the Russian State Museum of the Arctic and Antarctic)

nights. Ordinary citizens embraced the men, sending letters to them in care of the Kremlin to be forwarded to the Arctic.

These "true titans" were representatives of "proletarian internationalism" whose achievements were followed by ham radio buffs in the United States, Europe, and elsewhere. Judging by Papanin's diary, they showed their true Soviet patriotism during the drift time and again. Papanin and the others rejoiced at news from the mainland that Marshall Mikhail Tukhachevskii and other "dogs, scum and scoundrels" of the Soviet officers' corps had received their just deserts after confessing for spying for Germany. Tukhachevskii was executed in 1937 after a public trial; the paranoid Stalin ordered the execution of at least 40,000 officers. Did Papanin and his crew not suspect that the officers were innocent? After the war, did they not suspect that the USSR's great losses in the initial weeks of the Nazi attack were partly the result of Stalin's murderous destruction of his best officers? Out of solidarity with "proletarian justice," the foursome determined that from

September 1 onward they would hold a circle to discuss the history of the Communist Party, right there on the ice, every ten days, lest they forget it. On "Constitution Day" (December 5), they respectfully raised the Soviet flag. And so on. Still, it must be an exaggeration to say that Papanin, Krenkel', Fedorov, and Shirshov "felt Stalin's gaze" upon them as their father, teacher, and friend.[24]

In some ways, the goal of Soviet polar research—and of all scientific research—is to make the unexpected routine, to become accustomed to challenges and how to overcome them, and more easily to accumulate data than during initial discoveries. On May 21, the four men were dropped on the ice near the North Pole on an "N-170" airplane commanded by Glavsevmorput director Shmidt and piloted by Vodopianov with a crew to set up camp, with tents for laboratories, mess, and sleep. Backpacks with equipment, apparatuses, hand-cranked and wind-powered generators, batteries, and incidentals were included. They erected the tents on top of insulated plastic platforms on ice at least three meters thick. On June 6, the others flew back to Moscow to let the drift begin.[25]

On August 21, Papanin and his crew celebrated their three-month anniversary on the ice with sweets and a few snifters of cognac.[26] On January 21, 1938, Papanin wrote in his diary, "Today we observe exactly 8 months that we are living and working on the ice. We have become so accustomed to it that we have forgotten about the ocean that roars beneath us." They had made it a custom to celebrate the 21st of each month with cognac. On New Year's Eve, just south of 79° N latitude, they celebrated with caviar, hot dogs, smoked brisket, cheese, nuts, chocolate, and thirty-five pieces of candy each.[27]

It was men's work on the ice. The men joked, why not bring wives along? Krenkel' pointed out that it would not turn out well, for they would have to build ice nurseries and snow maternity wards and bring along teachers and nannies.[28]

They handled the changeable weather with aplomb. In mid-June the ice began to break up and move, so they had to be vigilant to notice fissures and cracks. When this happened, they quickly emptied the tents, brought in fresh snow to fill the cracks, and tramped it down under foot. It many ways, the weather was worse in the summer as fog, freezing rain, and melting ice put the men and their tents at risk. Indeed, it was hard to keep dry in the summer with the snow melting on the tents. When their southward drift slowed in August, they were relieved because this slowed melting ice in and around camp. Sun could penetrate even the "ice" refrigerator, so that many of their stores of meat and fish spoiled in August. But winter was no treat. Often snow blew so hard that nothing was visible at ten paces. When storms hit, especially those with wet snow and strong winds, the men hated to leave the comfort of the tent, but they left for their dangerous chores in any event. Winds to 100 km/hour in January over a three-day snowstorm covered the entrance to the tent that made digging out to establish radio contact with Murmansk a great chore. If in February it was –18°C on the ice, then it was only –12°C in the tent. The ice continued to spread, and in the sea of uniformity nothing was visible to the horizon.[29]

The explorers learned about the ebb and flow of drift ice and about the surprising speed of their drift, often moving about on skis for study. By December they had drifted to the northern shore of Greenland, by mid-December south of Rudolf Island, and on December 22 they reached 81° 7'; in four days the ice moved them thirty-seven nautical miles. Drift accelerated in late January as they moved at twenty miles per day southward toward 74° north latitude.[30] Shadowing them in the last frustrating days as they drifted in the Greenland Sea, the *Murman* and *Taimyr* icebreakers sailed nearby. The explorers and vessels could not immediately locate each other, even with flares. By this time they had drifted twenty degrees in all, and they were ready to return to the mainland. They finally saw the steamships on February 18, and on February 19 they left the ice.[31]

The expedition proved that there were no large land masses or small islands near the North Pole. They learned that warm Atlantic water reached the pole in deeper waters and that Arctic Ocean water temperature increased due to the heat of the Earth. Papanin and his comrades discovered new regularities in the Arctic climate, for example, that cyclones bringing rain, fog, and unstable weather were typical for the Pole area just like in lower latitudes. Ultimately, the drifting stations made future aerial traffic across the Arctic routine.[32]

Drifting into the 1950s

There were four periods of Soviet Arctic exploration through the lens of drift stations and aerial stations: a foundational period (1937–50); a period of temporary financial setbacks (1950–54); rapid expansion in the Khrushchev era and beyond (1954–70); and the modern period (1970–91) until the fall of the USSR. The drift stations came to be large scale in terms of numbers of personnel, quite expensive, and more and more reliant on aerial and ship support including from the nuclear navy. In the first period, Soviet explorers, most of whom were connected with AARI, opened the Arctic to fairly regular, if not systematic, research. They built programs on aerial reconnaissance and supply. World War II interrupted the efforts, which recovered essentially only in the Khrushchev era.[33] The first, aerial expedition "Sever (North)" in March 1937, had the sole purpose to create a drift station on the ice near the North Pole that was established and supplied by the route Moscow–Naryan-Mar–Rudolf Island–North Pole. This led directly to the Papanin expedition. While there were great hopes for further high-latitude aerial expeditions, for example one in March–April 1941, the war delayed any further attempts until "Sever-2" in March 1948 based on six Li-2 airplanes, one Il-12, and a Pe-8 used for refueling. This latter expedition consisted of twenty-three individuals under the director of Glavsevmorput A. A. Kuznetsov. In all seventy-two individuals supported the expedition. Research reflected the interests of Glavsevmorput, Gidrometsluzhba, and Leningrad State University on oceanography, meteorology, and geophysics. Aerial expeditions "Sever-4" and "Sever-5" followed in spring 1949 and spring 1950 respectively.

Now the head of Glavsevmorput, the explorer and seasoned *zimovchik* Papanin approved an important postwar expedition in late March 1946. Given Papanin's extensive resources as head of Glavsevmorput, and the full endorsement of the Communist Party for drift stations as symbols of Bolshevik verve and the superiority of the socialist system, he pushed ahead quickly. By the end of May all staff and equipment had been requisitioned and were on the way to Vladivostok by train. By June 18 the expedition had been readied with food, clothing, and equipment. One final hurdle, an official final government approval of the Council of Ministers, came only on July 16, 1946, and on July 21 the expedition departed, two weeks later than was optimal.[34]

Many aspects of the research project were rather mundane. Would a diesel icebreaker handle ice of different thicknesses and flow? What were the general oceanographic, meteorological, and geophysical conditions of the northern Chukchi, East Siberian, and Laptev seas? What was the impact of current, depth, and water chemistry on ice floes? While the *Krasin, Malygin, Sadko,* and *Sedov* had all explored parts of these seas in 1935–37, they remained insufficiently studied especially in the northern regions—and everyone remembered the recent fiasco when the *Sedov* and other ships had been trapped.[35] The expedition released buoys, took water samples, and made hundreds of measurements between 170° to 180° longitude with many above at 74° north latitude.[36] In 1948 the polar explorers organized expeditions in regions of the Arctic to the northeast of the Novosibirsk Islands and approached the North Pole and over the next two years discovered the Lomonosov and Mendeleev ridges, whose existence would serve as the basis of a Russian claim in 2008 that the vast majority of the Arctic belonged to Russia.

Perhaps because of extremely tight revenues (available investment was focused on postwar reconstruction and expansion of cold war production of weapons of mass destruction), funding for Arctic research fell sharply in the second period from 1950 to 1954. According to a recent Russian study of drift stations, the AARI leadership at that time was less willing to push this kind of research because of a belief that the central Arctic basin had been sufficiently studied, and hence the focus of limited funds should be on aerial research largely on coastal islands that would likely be underwritten by the military. The drift station SP-2 (April 1950–April 1951) undertook oceanographic, ice, and astronomical research, plus meteorological, gravimetrical, and aerological studies as part of this effort.[37]

In 1954 under new Glavsevmorput leadership (Director V. F. Burkhanov) Arctic drift station research expanded significantly. This expansion in the third period was likely part of general trends in the Khrushchev era during which the entire R and D apparatus grew manyfold in terms of the number of scientists, institutes, and disciplines, and scientists were able to reassert autonomy over their disciplines, which had suffered from ideological interference in the Khrushchev era. As is well known, in genetics, computer science, and relativistic and quantum physics, Stalinist ideologues had attacked various fields in modern science—and many of their Soviet representatives—ostensibly to protect those fields from dangerous "western" influences, for example the Bohr interpretation in quantum mechanics. Partly because of cold war expansion of science that mirrored similar processes in the United States and required the training of tens of thousands of researchers, partly because of the growing autonomy of specialists in the post-Stalin era, and partly because of rediscovery in the Khrushchev era of constructivist visions of the power of science to solve problems of economic production, scientists gained significantly greater independence. They founded "science cities" and expanded branches of the Academy (Akademgorodok in Siberia, the Kola Science Center, and so on), many of which had direct relevance for efforts to exploit Arctic resources.[38]

In the Arctic, drift station and aerial research grew intensively and extensively. The research became round-the-clock and focused on long-term predictions in climatology and oceanography. The efforts were large scale—with six, ten, and even more tents to handle twenty, fifty, and more researchers, whose efforts were facilitated by GS-1000 generators, motors, and GAZ-67 jeeps (produced 1943–53), and whose drift covered hundreds of kilometers through the central Arctic basin.[39] Among the noteworthy expeditions where SP-3 (1954–55) during which scientists encountered significant problems with breaking up ice yet wintered 376 days; SP-4 (1954–57), in which three different shifts of specialists spent more than 1,108 days conducting extensive zoological, oceanographic, and geomagnetic research; after they appeared in the area of Spitsbergen-Greenland SP-5 and SP-6 replaced them, so that from 1954 there were constantly two and sometimes three stations. SP-7 (1957–59), perhaps the most comprehensive drift station to date, contributed significantly to IGY and reflected Khrushchev-era talent at using such achievements as the tokomak design of a fusion reactor, the Obninsk reactor (1954), and Sputnik (1957) to demonstrate the world leadership of Soviet science. During these research initiatives, Soviet polar explorers became adept at coordinating aircraft and helicopter supply and research, in this case a total of 328 air freight trips, for example using the midrange Mi-4 helicopter (designed in 1951 and already in production by 1952 at the Saratov Aviation Factory to drop supplies and change crews.[40]

Soviet specialists sought extensive involvement in the IGY, in space and on the earth, both for the purpose of "science for science's sake" and to reenter the international arena as equal partners. Stalin-era autarky in science had ended; now physicists, geologists, oceanographers, and others might reestablish ties with colleagues abroad and ensure the priority of their work. Preparations for the IGY had begun at a 1951 session of the executive council of the International Council of Scientific Unions. In 1956 Presidium of Soviet Academy of Sciences created a coordinating committee under hero metallurgist and academician I. P. Bardin in which Papanin, E. K. Fedorov, V. G. Kort, and other AARI staffers participated.

The participation of Soviet physicists in the Geneva conferences on the peaceful uses of atomic energy beginning in 1955 had indicated these determinations. For Arctic researchers the main directions of activity were study of the ionosphere and magnetic phenomena and also determination of conditions for the propagation of radio waves. In specific, they introduced a method of and apparatus for tilt sounding of the atmosphere, which permitted them to define the diapason of the most useful frequencies of radio communications for any time of day and season of the year.[41] During IGY, in which forty-four countries took place, the number of Soviet coastal expeditions grew from thirteen to twenty-six, and fifteen oceanographic expeditions were organized from the Barents Sea to the Chukotka Sea.[42] Scientists carried out large-scale research in Arctic observatories at Druzhnaia (a research facility that opened in August 1957 on Kheis Island—part of Franz Josef Island archipelago), Dikson, Pevek, Tiksi, and Barentsburg (on Spitsbergen), while the SP-6 through -8 carried out high-latitude expeditions.[43]

Also during the Khrushchev era, the Communist Youth League (Komsomol) became much more active in leadership roles of enterprises, educational and cultural institutions, and even drift stations. The SP-9 (1959–62) was the first "Komosomol-Soviet Youth Drift Station" that was opened by Aerial Expedition Sever-11 in April 1959. The scientific report of the expedition concluded that "neither blizzards nor deep freezes of the polar night, nor tending to or the breaking up of the ice, nor floods during the summer melt could for one minute bog down scientific observations or violate the given rhythm of work. This was due to real friendship, comradely support, initiative, and the courage of that discipline which is specific to the collective of Soviet people who succeed successfully to overcome all difficulties and come out victors in the struggle with the harsh climate of the Arctic."[44] A great indication of the scale and centrality of drift stations in Soviet Arctic research concerned the SP-10 whose activities were supplied in part by the recently launched atomic icebreaker *Lenin*. From the 1970s onward atomic icebreakers and freighters were central to supply and operation not only of the Northern Sea Route commercial interests but drift station research.[45]

In the fourth period, 1970–91, the Brezhnev era of regular, steady increases to scientific R and D, yet of stagnating economic growth, growing public health concerns, inadequate dynamism in the consumer sector, and increasingly burdensome military expenditures, Arctic research remained large scale and stable, reflecting large subsidies to the economy generally to exploit Siberian and Arctic resources, especially fossil fuels, nickel and copper, and rare earth metals. In addition to standard research, observations on special programs on global atmospheric processes, underwater relief of ice, astronomical observations for large-scale drift of ice, and creation of data systems to serve a variety of institutions (AARI, Gidromettsent, the Arctic Weather Bureau, and others) expanded dramatically. Researchers could now rely on regular deliveries of supplies and services as they built runways to accept large transport planes near camps. They lived and worked in prefabricated house designed specifically for Arctic research in longer shifts.

For example, SP-19 (1969–73) lasted 1,235 days, drifted an average of 6.3 km/day and a total of 453, 1,200, and 890 kilometers respectively in three successive shifts. It occupied a "platform" on the ice of roughly seventy-two square kilometers and relied on 738 tons of supplies that were delivered in 705 aerial sortees. In all 369 men took part. Researchers focused on oceanographic, ice, and meteorological research.[46] SP-19 moved ahead under oceanographer Artur Chilingarov. Working in the observatory in Tiksi after graduating from the Makarov State Maritime Academy in Leningrad, Chilingarov was elected secretary of the Bulunski region Komsomol organization of the Yakut Autonomous Republic; at Chilingarov's initiative the SP-19 was designated a "Komsomol" station. (See Chapter 7 for discussion of the resumption of drift research in the twenty-first century and the role of Chilingarov, who became a parliamentary deputy, in using bold, very public research endeavors to support Russia's claims to ownership of much of the underwater region of the Arctic Ocean.)

SP-20 (1970–72, 749 days) drifted 3,135 kilometers north-northeast from Wrangel Island toward the North Pole and, as usual, conducted astronomical, meteorological, oceanographic, ice regime, and radiophysical research. The researchers lived in relative comfort in prefabricated PDKO-16 housing units, powered by two ADPS-20 diesel generators, and used two T-38 bulldozers for drift camp construction, transport, and other operations.[47]

The high-latitude aerial station North-25 (1973) was one of the largest expeditions of the postwar years whose research activities spread across the central Arctic basin like dots on a Cartesian grid. The goal was collection and analysis of oceanographic and atmospheric data for the Polar Experiment POLEKS. Researchers plotted 182 points that were 150 kilometers from each other. Transport planes based in Leningrad and other airports passed through Zhokhov Island in the effort to establish North-25. Work was constant and tense. As the director of the expedition, N. I. Blinov, wrote in his final report, "In spite of all the effort, all the same we did not succeed in avoiding either the rupture of cables and the loss of instruments or the crash of airplanes. This was quite natural for both oceanographic and aviation technology are exploited in difficult climatic conditions and would not tolerate an increase in loads."[48] At least twice, airplanes that landed with personnel and freight were consumed in ice likely because the surface of the ice was lower than the level of the ocean. North-25 was involved in the evacuation of SP-19 when it became clear that the latter was being carried to the Greenland Sea to remove personnel, equipment, housing, and so on to the extent possible.[49]

"SP-22" was the longest of all drift stations in Soviet history. It commenced on September 13, 1973, totaled 873 men over a series of shifts, and lasted 3,120 days with a total drift of 17,069 kilometers and average speed of 5.5 km/day. It followed an anticyclone drift and then went through the central region of the North Pole and through the Fram Straits into the Atlantic Ocean. The station took on nearly 3,000 tons of supplies put on the ice by the icebreakers *Vladivostok* and *Kapitan Kondrat'ev* and airplanes. Nearly 1,500 sortees arrived—essentially every other day—that included An-2 to Il-18s airplanes and Mi-4 and Mi-8 helicopters. The research program involved the standard research on hydrometeorological issues connected with astronomy, oceanography, biology, meteorology, and medicine, with aerological, actinometric, ionospheric, and other research components, and on construction of apartment buildings, airports, and other important facilities. It took more than a month to build the first runway at 900 meters in length and 40 meters wide. Frogmen undertook underwater research on ice regime. The station used three 12T-400-AIP and

E-8-P diesel generators, several gasoline generators, and T-100M, DT-75, and T-38M tractors.[50]

On the eve of the breakup of the USSR, AARI researchers embarked on one last drift station (SP-31, 1988–91). It had to be closed because of tight budgets, rampant inflation, and political uncertainty in the last years of perestroika. SP-31 had three crews fed with 1,385 tons of supplies who inhabited prefabricated housing put together by workers of Sevmorgeo, a geo-prospecting organization that evaluated the oil and gas potential of the Arctic shelf.[51] SP-31 included a runway at 1,000 by 60 meters that by March 1990 received regular flights from Apapelkhino and Cape Shmid on the Chukchi Sea. In all there were forty house and work structures and an army of tractors, bulldozers, and generators to shape the ice according to the needs of science.[52]

All of these stations carried out meteorological, magnetic, and ionospheric observations by the hundreds of thousands; tens of thousands of radiosonde readings; ocean soundings in hundreds of places; and so on. They learned about water exchange, relief, the ocean floor, and microbiology.[53] From 1948 (North-2) through 2009–10 (SP-37) the USSR launched and maintained a virtually constant presence in the Arctic on ninety-four drift stations with a total time of 31,680 days and with a total distance of 183,450 kilometers, in which 2,111 researchers participated. (The exception was the difficult political and financial years of a newly independent Russia from 1994 to 2001 when researchers could not afford a research presence.) The drift stations received 24,500 tons of fuel, food, material, and scientific equipment. Icebreakers supplied roughly 5,000 tons of this support, and 15,200 flights also contributed to the research effort.[54] AARI researchers remain steadfastly in the Arctic to support resource development and study climate change in the twenty-first century.

Science to the Tundra

A crucial new institutional foundation for scientific assimilation of the Arctic was the Kola Scientific Center (founded in 1930 and in its first years called variously "Tietta" [tietta means science in Saami], the Mining Khibiny Station of the Academy of Sciences, the Khibiny Science Base, and the Kola Branch of the Academy of Sciences). It grew out of the surveying and study activities of the geologist Aleksandr Fersman on the Kola Peninsula with the support of the head of the Leningrad Party Committee, Sergei Kirov, and benefitted from the national five-year plans for rapid industrialization with infusions of funding and growth of staff.

A moving force behind geological expeditions in the early USSR in the Arctic and Urals regions, Fersman served as director of the station, with E. P. Kessler, deputy director, and five other staff members, of whom two were scientists. Fersman (1883–1945) was born in Petersburg, although he spent much of his youth in Crimea at the family estate Fersmanovo, where as a little boy he became interested in valuable stones.[55] He studied at the Odessa Gymnasium and graduated in 1901, at the same time putting together a mineral collection. He transferred from Novorossiisk University to Moscow University when he learned about its geology department; on arrival he gave its museum his collection. In 1912 Fersman became a professor at the university and commenced extensive geological expeditions to Central Asia, Northern Russia, and the Ural Mountains. He became a member of KEPS, the Committee for the Study of the Productive Forces, in 1915. Fersman was a lyrical writer and popularizer of modern geological science and one of the founders of the journal *Priroda* (Nature) which is published in Russia to this day. Fersman became interested in igneous rocks, in particular pegmatites, in 1931 publishing a law of the distribution of valuable minerals in different kinds of pegmatite deposits. For his four-volume *Geokhimiia* (1933–39) Fersman won the London Geological Society Wollaston Medal in 1943.

Fersman's Arctic research effort and salesmanship went hand in hand. He spent a great deal of time on the Kola Peninsula exploring in the 1920s, while being a full member of the Academy of Sciences, director of the Radium Institute, and continuing member of the Soviet incarnation of KEPS, among other administrative responsibilities. Having discovered apatite ores in the Khibiny tundra, he also developed a new method of producing phosphorus fertilizers from them, and in the Monche tundra he identified copper and nickel ore. In his later career Fersman was director of the Mineralogical Museum of the Academy of Sciences and also a vice president of the Academy and member of its presidium. From 1930 until his death in 1945 Fersman was chairman of the Kola Base of the Academy and from 1932–38 simultaneously of the Ural Branch of the Academy. Unlike other leading specialists with Tsarist roots, and in this case a Jewish specialist, Fersman seems never to have had his loyalty questioned in the Stalin era.

Sergei Kirov, the Leningrad party secretary, directly participated in the study of Arctic resources of the Northern Region. He actively joined Fersman and others in surveying the resources of the Kola peninsula and setting the foundation for the construction of the "hero" metallurgical cities of Khibinogorsk (now Kirovsk) and Monchegorsk, the latter's copper and nickel smelters still functioning in a barren landscape owing to the deadly, virtually unfiltered heavy metal emissions from its furnaces.[56] He supported the construction of large contemporary hydroelectric power stations on the Niva and Tuloma rivers in Murmansk province, and also the "Kirov" railroad to Murmansk, to kick-start the development of Arctic commerce.[57]

Building on scattered and often anecdotal evidence of small expeditions to the Kola Peninsula sponsored by the Russian Geological Society, scientists charted mineral deposits and were astounded by their extent and diversity—from apatite (a phosphorus ore that can be used to manufacture fertilizers) to copper, nickel, and rare metals. They conducted work on limestone on Kildin Island off the coast from the Murmansk fjord. In 1933 "the problem of surveying" on the Kola Peninsula remained to be solved owing to the climate, the vast region to be explored, and the very few scientists and surveyors available for the task. But the scientists tirelessly covered the tundra when weather, their strength, and their funding permitted and found valuable ores in the southern part of Khibiny Tundra and also "extensive deposits of diatomite after study of Niudozero Lake, the Monche tundra and Vesennee Lake." The ever-growing awareness of the ever-larger deposits indicated that the scientific base required evermore "geochemists, mineralogists, geologists, prospectors, engineers and economists" to ensure the development of nickel, iron, apatite, and titanite deposits.[58] While difficult to determine the extent, a major share of the budget of the Kola Center was associated with expeditions in the summer and early fall months. The station grew slowly but steadily from a handful of employees at the beginning of the 1930s into a major branch. By the end of the decade the Khibiny, Monche, and other tundra had been surveyed, prospected, and in some cases opened to exploitation.[59]

The most important artifact of early industrialization on the Kola Peninsula was the Northern Chemical Combine (or Apatittrest, hereafter "Apatit" or Apatit Trust). On October 16, 1929, the Presidium of the Soviet Academy of Sciences approved the charter for the Khibiny Station to enrich the trust with research. In view of financial problems and challenges of building quickly in the tundra, the construction of the station was pushed back to 1930–31; even as small expeditions scoured the tundra for deposits, scientists housed initial activities for the analysis of samples in a one-story building. This building burned down. In April 1930 the scientists selected a site on the Little Bud'iavro Lake for Tietta—a permanent home for a research institute—and by the middle of June construction had been completed. Supplies arrived with some delay by train from Leningrad and Murmansk. On July 20, 1930, the facility officially opened. It became the base for twenty-five expeditionary groups working the region, several of which headed up the Loparskaia and Iuksporoika river valleys; the groups encountered apatite-nepheline deposits frequently. By late summer the builders finished the heating system and the scientists opened Tietta's laboratories.[60]

Quantitative and qualitative indices reflected a modest but important research program. The initial budget of roughly 20,000 rubles covered the building, administrative costs, and some equipment and research. Apatit Trust, the Committee on Chemicalization, and Narkomtiazprom (the Commissariat of Heavy Industry) provided an additional 20,000 rubles for research and 30,000 for expeditions.[61] By 1934 the facility had three departments (geochemistry, botany, and meteorology) that employed four geochemists, four botanists, four chemists, and three climatologists. They had published several articles. The library grew from 2,000 volumes at the beginning of 1934 to 10,000 one year later, including 2,500 foreign language books, as well as seven journal subscriptions, of which three were foreign. Six standing expeditions had been established: geology, radiology, mineralogy, climatology, zoology, and botany, although with no more than three individuals in any expedition.[62] One year later, the budget had grown to 77,580 rubles from the Academy of Sciences, 6,500 rubles from the government, and another 75,000 rubles under contract with Apatit Trust, of which 25,000 rubles went to the geochemical laboratory, 45,000 to support expeditions, and 5,000 to the botanical garden. Researchers had published four articles on lovchorrite (a silicate) and the geochemical study of various sites. The staff had grown to twelve, with four individuals working under contract in expeditions.[63] According to another document, the number and productivity of scientific expeditions expanded rapidly in the 1920s and 1930s, as reflected in the fact that forty-three works were published between 1924 and 1934 in English, German, and French in such journals as the *Bulletin* of the Society of Exotic Pathology, one article in German on parasites and infections, and another in the *Annales de l'Institut Pasteur.*[64]

Underlining the critical connection between Soviet "big business" and applied science in the USSR, mining and refining pushed scientific achievement. At the beginning of 1932, Apatit together with Giprokhim (the Chemical Engineering Design Institute) developed a five-year plan to expand mining and chemical production on the Kola Peninsula, with the Institute of Rare Elements and Institute of Metals to provide a plan by May 1 for the industrial assimilation of rare elements. This "industrial assimilation of the Kola Peninsula," in particular its chemical and construction industries, lagged during the first and second Five-Year Plan because of shortfalls in technology.[65] Yet the research program of Tietta had expanded to include chemical analysis of ores; study of sulfides; improvement of analytical methods; continued development of construction techniques and materials; investigation of colloidal properties of silicon dioxides for various applications; and further study of phosphorous, especially in agriculture.[66] To facilitate shipment of ore and the development of fisheries, Kola scientists also joined with specialists of SOPS (the Committee for the Study of Productive Forces, the successor to KEPS) to evaluate the possibility for construction of ports in the White Sea in the region of Soroskaia Buxta, Kem, and Kandalaksha; in the Barents Sea; in the Murmansk fjord; and in a wide variety of fishing villages along the shore. Research involved topological photography, study of tides, and geological surveys.[67]

Local, regional, and national party officials, economic planners, and specialists took great interest in Kola science. The development of the vast regions of Karelia and the Kola Peninsula as integral aspects of the first and second five-year plans, not the least because of the White Sea–Baltic Canal, required scientific research. The members of a huge, multidisciplinary, and intergovernmental committee held a plenary meeting in February 1932 to discuss trends and prospects for the economy. Representatives of the provincial planning organization, the Institute of the Socialist Reconstruction of the

Economy, Apatit Trust, the Geological Society, the Leningrad Department of the Supreme Economic Council, the Central Executive Committee of Murmansk region, two industrial fishing organizations, the Oceanographic Institute, and other economic, scientific, and political organizations discussed the place of the Murmansk resources and economy and confidently anticipated that rapid industrialization on the Kola Peninsula would mirror that in the rest of the nation.[68] A representative of the provincial planning organization noted that, "The correct line of the party for the industrialization of our economy and Bolshevik certainty and persistence will permit us to overcome polar conditions and to create a modern trawler fleet, animal husbandry, mining-metallurgical facilities, enrichment factories, new industrial centers and electrical power generation construction. The successes of construction will guarantee the development in the second five-year plan of Murmansk of the basis of its branch industries."[69]

Modern infrastructure was the foundation of the industrial utilization of the region. The effort to develop mining, enrichment, processing, and other facilities to handle apatite, nepheline, and ferrous and non-ferrous ores required "the large scale reconstruction of ports, electrification, railroad transport, [electrification], and also construction of non-rail transport, air and civil communications, extensive urban and socio-cultural construction, and the development of northern agriculture both of vegetables and animal husbandry," as well as fisheries with the construction of processing factories in Murmansk. The representatives concluded, almost as an afterthought, by noting the need to attract a well-qualified labor force.[70] The gulag provided at least a labor force, if not a well-qualified one.

While World War II slowed the development of Arctic science, it resumed its qualitative and quantitative growth in the late 1940s. In the 1950s the Soviet scientific enterprise expanded generally in terms of numbers of scientific centers, institutes and scientists (regarding the latter in terms of doctors and candidates of science), and staff employees. The most notable in terms of size and breadth and quality of research programs was the Siberian Division of the Academy of Science located in Akademgorodok, south of Novosibirsk, approved in 1957, opened in 1962, and by 1965 with twenty-one research institutes and a university. The Kola base became the Kola Branch of the Academy of Sciences at this time and also underwent rapid growth by all indices. The institutes and scientists continued to be the handmaiden of the

Table 4.2 Growth of Kola Science Center, 1951–57

	1951	1952	1953	1954	1955	1956	1957 (plan)
Scientists	65	68	84	104	111	112	131
Scientific-technical personnel	52	52	81	88	93	102	108
Library	2	2	4	5	5	6	7
Production	1	4	8	31	26	33	40
Administration	14	19	30	31	31	31	31
Facilities	33	33	47	54	53	52	62
Other	34		21	23	27	30	46
Total	166	176	272	328	356	374	425

Source: A KNTs, F.1, op. 6, d. 397, l. 3.

extensive geological resources of the Kola Peninsula. They intended to expand research on mining science and continued to put great emphasis on development of enrichment for fertilizers and other applications. Correspondence between the administration of the Kola branch and the Academy of Sciences leadership in Moscow indicates a constant effort to gain approval to expand laboratories, institutes, and numbers of employees. No different from other centers in this time of expansion of the scientific apparatus, the Kola branch grew rapidly, increasing more than 70 percent in number of scientists and 250 percent in total number of employees (see Table 4.2).

Of course, the budget increased a commensurate amount from 895 million rubles in 1951 to 1,376 million rubles in 1954 and 1,843 in 1957 (planned) from state budget sources and from contracts from 280 million rubles in 1951 to 590 million rubles in 1954 to 800 million rubles (planned) in 1957. As before, researchers worked in four major subject areas, all of which were connected with mining, metallurgy, and geochemistry: mineralogy and geochemistry of rare metals; production of cesium from concentrates; development of methods of working with rare titano-niobium ores; and elaboration of methods of extraction of valuable materials from the wastes of the Apatit combine.[71]

While focused on the mineral wealth of the Kola Peninsula, officials still held a small hope of promoting agriculture—that is, beyond animal husbandry and reindeer. A short, unreliable summer, early frosts, and poor soils made this a challenge. In February 1948, during the full flowering of pseudo-scientific Lamarckian biology under the tsar of the Academy of Agricultural Sciences, Trofim Lysenko, leading specialists of the Kola center discussed "theses" on the agricultural assimilation of the Far North; Lysenko's Lamarckian views of the inheritance of acquired characteristics and the rejection of genetics as "bourgeois" pseudo-science damaged agriculture and genetics from the Arctic to the steppe.[72]

At the turn of the twenty-first century the KNTs still maintained four major directions of research, but with ecology and the social sciences gaining support as a reflection of greater concern over the social and environmental impact of development. The focus of eleven research institutes was wedded to mining, refining, smelting, and energetic, if an ecological tint has been applied to a number of fields. The four major areas of research were: (1) in basic science, the status and evolution of Arctic ecosystems including evaluation of anthropogenic threats to them; (2) in applied science, exploration of natural resources in the Russian northwest toward the ends of the creation of ecologically safe technologies for processing raw materials, the development of waste management techniques, including underground storage (specialists have experimented with "peaceful nuclear explosions" near Apatity to consider their use to create subterranean caverns), ecologically safe power-engineering, and wildlife management; (3) investigation of social and ethnic structure of the society, the dynamics of migration, the distribution of manpower, and the quality of life in high-latitude regions, all including the indigenous population of the North; and (4) creation of regional information systems and the automated systems of teaching.[73] The last two foci, on social, educational, and environmental issues, including the socio-economic consequences of environmental change for indigenous people, seemingly have not only newly found interest but growing, if still modest, financial support.

These directions, including the last, meshed with the general thrust of Russian basic research during the third International Polar Year (IPY, 2007–8).[74] The scientific secretary of the Arctic and Antarctic Research Institute in St. Petersburg, Russia, Dr. Viktor Dmitriev, observed that the IPY brought together specialists from all branches of the sciences and social scientists within the framework of a unitary interdisciplinary scientific program and created the potential for further research through readily accessible information systems that enable, for example, understanding of climate change.[75]

Gidrometsluzhba

In spite of the constant presence of research and specialists in the Arctic basin, the Soviet system had difficulty providing regular, accurate weather information to commercial, educational, and other organizations. Another institution, Gidrometsluzhba (the State Hydrometeorological Service), commenced frequent observations at a larger and larger number of points, several times per day to every several days or once a week with thermometers, anemometers, and barometers, from the 1930s onward, yet had difficulty meeting the growing need for accurate information in the Arctic. Gidrometsluzhba established observation points or stations along the coast, on rivers, and in the seas and oceans. The data included temperature, precipitation, barometric pressure, water chemistry, ice makeup and flow, and observations of meteorological phenomena. Ultimately manned and unmanned stations produced hundreds of thousands of data points, samples of seawater, and the like, all of which had to be analyzed at some great cost. Such glitches in the program as bureaucratic infighting, outdated technology, and poorly qualified personnel that hindered analysis persisted into the 1950s. For example, several ship captains refused to allow their freight or passenger vessels to double as research points, even as they led "episodic freight trips on the White and Barents Seas."[76] Rivers served multiple purposes of freight and passenger transportation, mail delivery, fishing, and the crucial spring float of lumber downstream, for example to mills in Kotlas and elsewhere along the Nothern Dvina River all the way to Arkhangelsk, to the Solumbala pulp and paper works, hence the bureaucratic entanglements.[77]

Still, by 1935 Gidrometsluzhba had established 173 stations that covered seventeen rivers, of which sixty-six predated the Revolution, thirty-nine were set up in the 1920s, and sixty-eight were established in the 1930s.[78] The larger rivers had been charted in the Tsarist era; most of the smaller rivers were subjected to data collection only in the 1930s or after World War II. The stations spread from west to east and from delta to the source. Gidrometsluzhba operated sampling points to track maximum spring flows and minimum and low flows.[79]

Gidrometsluzhba had immediate practical obligations. Its staff served Arkhangelsk, Vologda, and Syktyvkar forestry concerns; railroad scheduling and operations; the merchant marine; the Northern, Sheksninsk, and

Pechora basin river routes; various steamship authorities; ports and harbors; the cities of Mezen', Indiga, Naryan-Mar, Onega, Belomorsk, Kem, and Kandalaksha; and the Dvina, Komi, Vychegda, and Vologda lumber float organizations with daily and three-day forecasts and storm warnings to ensure safe passage through deltas and to warn of the onset of ice, strong winds and fog, heavy waves, and currents with 80 percent accuracy.[80] As for the fish industry (Sevryba, Arkhangelsk Administration of Fisheries, Sevrybvod, and Sevrybbakolkhozsoiuz), Gidrometsluzhba provided radio warnings for the fish factories and fleets that operated on or near the shores of the White and Barents seas, including operations in Naryan-Mar, Mezen, Onega, Kem, and Kandalsksha. But discussion at a meeting of Sevrabakkolkhozsoiuz (an Arctic organization representing the fish collective farms) indicated this was not very effective because many of the fish organizations did not have radio receivers![81] If a forecast was sent to a fishing operation without a radio, did it make a sound?

Officials of Gidrometsluzhba recognized the need to extend the network of coastal observation and radio stations because of the time and route of sailing and how heavily ships might be loaded depended on up-to-the minute weather information. In 1950 Gidrometsluzhba pilots flew the White Sea and southeastern Barents Sea a total of 16,500 kilometers over seventy-five hours and completed twenty-two trips along rivers a total of 6,600 kilometers and fifty-seven hours. These observations indicated the great potential of aviation operations for shipping, fisheries, and animal harvests but were obviously inadequate: Pilots devoted fewer than 140 hours over the course of a year to a small region of the western Arctic basin.[82]

In the early 1950s Gidrometsluzhba and its regional branches grew rapidly both because of increasing demands of the Northern Sea Route and because the scientific research enterprise expanded at a fast pace during the Khrushchev era. Yet quantitative growth masked qualitative problems. Of some 1,500 employees, roughly 150 worked in the central administration, twenty-seven in the Marine Observatory, 1,166 at observation points, and seventy-two in aerological work. The central administration used these statistics to request more specialists with engineering training in hydrology, meteorology, or some other related field because employee qualifications lagged behind the scientific needs of their enterprise: 830 persons had unfinished higher education, only 351 had finished university, 208 had some kind of high school education, and 576 had completed only some kind of advanced coursework. To raise the qualifications of staff, Gidrometsluzhba encouraged technical study among staff members. But this study seemed to be mostly rote learning in the sciences and had an extensive political component. A continuous re-attestation of employees was carried out to ensure the political reliability of staff.[83] In 1949, 636 staff members worked to improve their qualifications, nine at the Evening University of Marxism-Leninism, 335 in circles of party history, and 159 in circles of the study of the lives of Lenin and Stalin. The number grew to 733 in 1950.[84] What Lenin and Stalin could add to weather forecasts was unclear.

Largely as a result of low qualifications and limited experience, the quality and timeliness of forecasts and warnings was rather low, and in fact hydrological study of small rivers was carried out quite slowly. Inaccurate storm forecasts led to interruption of flights. Forecasts were especially bad in the northeastern regions of the Nenets region and the northern Urals.[85] Still, Gidrometsluzhba researchers contributed greatly to the hydrometeorology in service of aviation and the economy. In the 1950s they composed a series of maps with 623 indexes, micromaps with 131 indexes, a long term map with 431 indexes, and a baric topography map. They produced a total of 14,164 weather forecasts in 1950: 8,901 short-term, 86 percent of which they asserted were correct; they issued 3,575 three-day forecasts, 89 percent of which were accurate; and they provided 7,930 storm warnings, 87 percent of which were correct. Forecasts were sent out by radio every day at 8:30 and 18:30 in Arkhangelsk and on Komi daily radio "Syktyvkar" at 18:00. Summer forecasts were indicated more for fisheries and farms, winter for lumbering operations, and extended to the Nenets region and well beyond the White Sea into the Barents Sea.[86]

Yet a lag in data accumulation persisted until the late 1950s for three reasons. First, the huge expanse of the Arctic from the border with Norway to Alaska made the establishment of observation points and the collection of data a long, expensive, difficult process. Second, World War II had waylaid the effort, which was slow to resume. Third, government officials were less concerned with data gathering, which they viewed as incidental or at least not crucial to the production process. Paradoxically, they often put the observation points under the jurisdiction of the Soviet Navy and therefore viewed the data as top secret, which also prevented their systematic use and dissemination.

In 1954 the directors of Gidrometsluzhba adopted a series of tactics to improve performance. They changed leadership of regional branches and exhorted workers to approach data gathering and analysis with greater efficiency; they strove to increase the qualifications of employees; and they cut down on the amount of paper being passed around offices. Seventy-eight stations now served agriculture directly, even if long-term forecasts to assist in the production of animal feed lagged. Even these efforts fell short in a number of ways. Poor forecasts led a large number of flights to be interrupted or canceled, although many fewer than in the previous years. Publication of statistical yearbooks lagged significantly. The accuracy of weather forecasts had not improved sufficiently, even if new meteorological devices had been acquired.[87]

Treating the Polar Patient

Lags in meteorological forecasting were tied to lags in understanding the impact of harsh Arctic weather on residents of the polar USSR. Arctic medicine developed slowly and became recognized as an independent field only by the mid-twentieth century. In the nineteenth century many explorers saw it as incidental to expeditionary activity, even if ethnographers, geographers, and doctors who first contacted indigenous people, fishermen, hunters, and their families worried about their precarious lifestyle. A major source of information for nascent Arctic medicine was the Army and Navy, for soldiers and sailors, at great risk at northern latitudes with poor food, clothing, and equipment, could be surveyed relatively easily for data. Scurvy, in particular, remained a serious problem into the 1920s; Georgii Sedov had died from scurvy and exposure. Even in the Soviet period Arctic medicine lagged until the post-war years, with AARI systematically contributing to its development in the 1950s in connection with Antarctic research and eventually in the 1960s with full institutional support.

In the nineteenth century illness and mortality in the Russian Navy was at rather high levels and in Arctic regions, had there been more expeditions and sailors, would have been higher still. D. V. Mertsalov, a naval doctor, demonstrated that disease and illness were significantly higher than in the general population and compared to the Army. In 1863 in the Navy the disease rate was 431.2 and mortality 19.8 per 1,000 individuals, while in the Army the disease rate was 285.5 and mortality 10.5 per 1,000 individuals, which Mertsalov attributed to the behaviors of sailors, the long tours of duty, and the difficult lifestyle of enlisted men.[88]

A. A. Bunge, a well-known natural historian, polar explorer, and medical doctor (1851–1930), participated in Russian Geographic Society expeditions, for example one of the first polar stations on Sagastyr Island in the Lena River delta, established as part of first International Polar Year (1882–84). In the Lena Delta region he offered assistance to local residents who came from afar for help with free medical care; he visited Yakuts and became familiar with their life, activities, beliefs, and habits; and he studied the diseases characteristic for the region. He observed what later became clear—the lowered immunobiological activity of sailors (and others) who were away from land or home for a long time and especially a high frequency of scurvy among participants in zimovka.[89] Scurvy continued to be a problem into the twentieth century, and in the Soviet period as well, with researchers, as well as sailors, dying. A Scottish surgeon in the British Navy, James Lint, had identified some nutrient in citrus foods that prevented scurvy in his *A Treatise of the Scurvy* (1753). Yet most vitamins were identified only in the early twentieth century, and treatment of scurvy with vitamin C was confirmed only in 1932.

Soviet doctors and officials connected with the Commissariat of Health (Narkomzdrav, later Minzdrav) faced wide-ranging challenges after the revolution to improve public health; deliver health care; fight high infant mortality; deal with epidemics of typhus and venereal and other infectious diseases; and combat mistaken folk beliefs about illnesses, their origins, and treatments.[90] During the civil war millions of people took ill and large numbers of them died in epidemics. They achieved great successes in the 1920s in all of these areas, in reducing mortality and increasing life expectancy, lowering disease rates, and eventually eliminating epidemics. Yet health care delivery in the provinces remained a sore point. Doctors, nurses, hospitals, dispensaries, and even feldshers (a kind of rural medical assistant without full qualifications) remained few and far between the further one moved into the tundra.

In the taiga and tundra things were worse still. The number of medical personnel per square kilometer and per person was too low to ensure proper care. As in other regions of polar science, the authorities found it difficult to train and retain qualified personnel. Studies on the impact of Arctic climate and lifestyle lagged until the 1940s and 1950s, so that the few polar physicians available had to struggle with determining the proper protocols, not to mention how to offer advice on nutrition, exercise, clothing, and housing. Several examples from the Nenets Autonomous Region from the 1930s will give an idea of this persistent problem.

Whether Nentsy, Khanty, or Saami, all of these indigenous people suffered illnesses owing to frequent contact with their civilizing Russians. This is usually the case with first encounters between "civilized people" and those they intend to civilize. The civilizers bring microbes; in North America millions of Indians perished from small pox, syphilis, and other diseases.[91] The same held in the Soviet empire, although not on such a dreadful scale as in the Americas. Also, life expectancy among indigenous people increased after Sovietization. When they moved into Arctic regions, the Russian colonizers also fell prey to their share of diseases, to TB and influenza in particular. The illnesses that befell the settlers and the settled unsettled them both.

The polar public health network grew slowly but steadily against challenges of limited budgets, low population densities, and personnel. Officials of the Northern Region Health Department (Sevkraizdrav) related their progress during the Second Five-Year Plan: a hospital with fifteen beds and two beds in a kind of maternity ward and outpatient office opened in Naryan-Mar, with a feldsher point on Novaia Zemlia to serve Russian Harbor in 1932. In 1933 medical personnel organized a feldsher point without a hospital on Vaigach Island, no doubt connected with the gulag camp mines, in 1934 a feldsher point on Franz Josef Land, and in 1935 two feldsher points at Admiralteistva Peninsula.[92] In 1936 145 patients used the Naryan-Mar hospital of whom 117 were Nentsy. There were thousands of outpatient visits. Nurse Mirnova and assistant Valevskaia served as a "medical brigade" with a mobile drug store that brought modern medicine to the served Nentsy in the tundra.[93] But occasionally representatives of the "modern world" themselves had to be evacuated quickly; when the political enlightenment worker at the cultural bases on Novaia Zemlia, Uliana Dekova, was stricken with TB, local officials requested 650 rubles, including an advance of 250 rubles, to send her to a sanitarium in Yalta.[94]

Another ambulatory point in the Bolshezemelskaia tundra consisted of a one-room structure with drug store, hospital, and one bed. It was overburdened with 1,682 visits in 1938, of whom 793 were Nentsy. Medical personnel noted that Russians tended to be more sick with flu, trauma, and toothaches, while the Nentsy suffered from rheumatism, bronchitis, and TB. This facility, still not fully operational, was short of linens, instruments, and equipment, and the doctor, one Matiushev, stood out for his rudeness. He often responded to patients: "There's nothing wrong with you," "You won't die," "It will pass," and "It will heal before you know it ['You will live to your wedding']," and he refused to take responsibility for the sanitary conditions of the hospital.[95]

Yet Soviet power brought medicine to the tundra, something that the Nenets writer Vasilii Ledkov saw almost exclusively in a positive light. In "Month of Little Darkness," Ledkov notes that the Red Tepee, the propaganda point among indigenous people, brought both the teacher and medic into contact with the Nentsy. The herders did not want to study, were afraid of paper, and the doctor for them was more frightening than an evil spirit.[96] But change had come. Women learned to go to the hospital where it is warm and clean to give birth, not in the filthy cold teepees as before.[97] In "Dialogues," a village doctor, a Russian woman, Marina, talks about the challenges but advantages of converting nomads to a sedentary lifestyle. Of course, as a doctor, she welcomes all comforts of life including access to emergency medical care. But she does not ignore the advantages of the Nenets way of life, for example, the chum (teepee) was easy to move about, and when taken down, the wind carried away all of the dust and dirt, while public health officials struggled to teach newly sedentary people how to sweep up dust and dirt in wooden houses. Some Nenets women got the drift quickly, but others in no way wanted to put their houses in order. The doctor said, "There was this Marya . . . how I struggled. She didn't want to clean up in any way. And then dysentery began in the area and her son, Van'ka, got sick, and then it spread to others." They cleaned their house and disinfected with chlorine. Finally, she allowed her son to go to the hospital, and upon recovery the boy refused to return home to his mother.[98]

When medical care failed to deliver on the tundra, it was in part because of poorly trained and motivated personnel like Matiushev who were intolerant of the patients and because of inadequate equipment, poor food, and the absence of clean linens and mattresses. Making matters worse, medical personnel in one facility had commandeered part of the hospital as their living quarters. It did not help that the doctors mistreated patients—and might come to work drunk. Officials "categorically demand from all workers, and in first place from Doctor Gukov, in the future not to permit drunkenness, gossip, and pettiness" in the hospital and no longer "to tolerate from his side rudeness and tactlessness." Thirty-five percent of the hospital workers themselves were illiterate or poorly literate, were rude and inconsiderate, and had taken to throwing their garbage out their front doors,[99] which did not contribute to sanitation.

The growing number of settlers in polar regions put another strain on the health care system, both because of its limited reach and because the impact of harsh climate on physiology that remained poorly studied. The first extensive study of acclimatization of new settlers to Arctic climate that I could find dated to the immediate postwar years. Somewhat belatedly turning to examination of the question of acclimatization of expeditionary parties to conditions of the Far North, the Academy of Medical Sciences and AARI reported in 1949 on studies conducted in the villages of Tiksi, Dudinka, and Sogo and at a polar station from December 11, 1948, to March 27, 1949. Five medical institute laboratories took part: physiology, immunobiology of children's infections, microbiology, infectious diseases, and questions of frostbite. Focusing on physiology, five specialists measured levels of metabolism, electrocardiography, tonometry, the impact of ultraviolet irradiation of children, and thermal regulation and temperature in different parts of the body with help of electric thermometers.[100] In all they conducted 120 electrocardiograms, followed blood pressure on 407 individuals, collected data on caloric needs and consumption of 547 people, and measured breathing among 140 soldiers, seasoned polar explorers, and recent arrivals. They considered the effect of ultraviolet rays on thirty children roughly in two of the first classes of school and then measured their erythrocyte, leukocyte, and hemoglobin levels.[101]

The problem of acclimatization of children to the Arctic had become critical in connection with plans to push construction of large towns and cities, for how to populate the north without making it possible to bring along family members or start families above the Arctic Circle just as in comfortable Leningrad or Moscow? The presence of 322 newborns from 1945–48 permitted researchers to gather data on weight, length, and other vital statistics.[102] Immunity was the crucial issue. Specialists studied phagocytes in children that characterized the nonspecific resistance to infectious diseases. They considered also diphtheria, scarlet fever, and therefore the danger of streptococci. The low phagocyte indices among the large number of children indicated weakened immunity and might explain the fact that 33 percent of the children had scarlet fever, while 41 percent of children carried hemolytic streptococci. In 1948 only one case of scarlet fever was registered, while 65 percent of preschool and 34 percent of schoolchildren were susceptible to diphtheria, and nine carriers existed among 268 children although they could not explain the absence of cases.[103]

A surgical group studied trauma as well, especially appendicitis and mastitis. The frequency of appendicitis exceeded that in other Soviet cities three to five times and also in acuity and complications with more than 385 cases in five years.[104] Regarding mastitis, sixty cases were observed in 168 women who gave birth in 1948, a very large percentage which the specialists deemed due perhaps to unclean breasts, a particular interest of Soviet medical personnel judging by posters urging women to "clean their breasts," lower immunity, and a belated turn to medical care.[105]

Based on all of these studies, the specialists concluded that polar conditions led to an increase in diseases of the cardiovascular system, but not of the gastrointestinal system. They called for more attention to questions of the physical development of children to arrive at better ways to improve their health, for example how better to organize health in schools, improve sanitation at home, and apply UV treatment, vitamins, and fish oil. Further study was needed of acclimatization with the goal to make concrete recommendations to the central authorities.[106]

Another consideration that slowed the development of polar medicine was the fact that medical research fell under the auspices of the Ministry of Health. This served as a barrier to dissemination of information gathered systematically or anecdotally in AARI, Gidrometsluzhba, and other institutions, and it hindered a systematic approach to study of the wide range of issues facing Arctic exploration and assimilation. In 1966 AARI specialists began systematic studies in this field and in 1968 created the department of polar medicine that began to bridge this bureaucratic chasm. Research involved examination by doctors on polar expeditions of medico-physiological, psychological, and public health aspects of research and daily life activities that resulted in the publication of a number of practical recommendations directed toward solving the problem of adaptation of man to life and work in extreme climatic conditions. During the years 1985–89 ten such "recommendations" were introduced.[107]

G. M. Danishevskii,[108] who published *Trud i Zdorov'e Cheloveka na Krainem Severe* (Labor and Health of Man in the Far North, 1970), played a role in organizing the first expeditions for Antarctica[109] and then in promoting systematic medical research in the department of polar medicine. He sought to offer highly qualified medical help to polar researchers in expeditions to prevent and lower illness rates based on the study of long-term acclimatization to harsh climate. From 1966–70 the division equipped all Antarctic stations with medical instruments and apparatuses to support medical treatment and carry out prophylactic work and scientific research. Researchers worked closely with colleagues in the Miasnikov Institute of Cardiology. In fifteen Antarctic expeditions (1956–70) doctors accumulated great experience and approved protocols for treatment of illness and disease. They knew, of course, that success depended first of all on the initial state of health of the participants—so medical selection was crucial. Starting from evaluation of the record of previous expeditions, sector researchers collaborated with specialists of major Leningrad clinical hospitals to produce "Instructions for Medical Examination of New Participants and Long-Serving Participants . . . in Expeditions" of the Institute. Efforts focused, of course, on improvement in the mix and quality of shelter, clothing, and food. The researchers evaluated mechanisms of adaptation of man to Arctic conditions with physiologists, public health specialists, microbiologists, geophysicists, and bioclimatologists taking part. Because many of the changes during acclimatization were transitory, they sought to develop means to limit them by means of "scientific organization of labor and daily life."[110]

The Marxist Veterinarian

Reindeer have economic and cultural importance to indigenous people. Indigenous people have herded them for centuries. The Inuit, First Nations people, Alaska natives, Nentsy, Saami, and Eveny raised reindeer for meat, hides for clothing and shelter, and antlers for tools (others prize them for their supposed properties as aphrodisiacs), sometimes for their milk, and for transportation. Without the animals, the people would perish in the cold. Their languages are rich in terms that reflect their understanding of reindeer life and its importance to them. The reindeer are not fully domesticated, but migrate through the taiga and tundra, with reindeer herders migrating with them annually; this nomadic lifestyle was especially troubling to Soviet party officials and economic planners. Once the people and their reindeer were collectivized, then they could be put to other uses, and not only to feed the cities, for example, on the Karelian front in World War II. Indigenous people have lost their land and lifestyle owing to colonization and industrialization, and in the USSR to collectivization. Somehow, these people survived Stalin and the Cold War and attacks on their shamans and domestication of their reindeer. In the twenty-first century, because of their economic and cultural dependence on the animals and their environment, indigenous people faced direct and dire consequences from climate change.[111] One hope is that specialists and indigenous people can find common understanding that lessen the risk.

In the USSR specialists first brought modern medicine to the reindeer. But the application of modern science, technology, and medicine to the herds and the belief system that accompanied Soviet veterinary medicine into the Russian north often had unanticipated consequences that left officials and specialists at a loss for what to do. But they maintained their faith in modern science; even if their programs and approaches had created problems in the maintenance of herds, they turned not to reindeer herders for help, but to other specialists. Hence they were caught in a cycle of using science to solve problems that originated paradoxically in veterinary medicine. The modernization of food production practices in the twentieth century on the basis of engineered monocultures of plants and animals in viciously industrial settings has led, according to its proponents, to ever-declining food costs and greater certainty about yields. Yet the intractable problems of vast quantities of animal wastes, the feeding of offal and other animal byproducts to the very animals, the forcing of "downer" animals into the food supply, and the increasing frequency of food-borne epidemics in meat, fowl, and vegetables (sickening hundreds of thousands of people annually by *salmonella, E. coli,* and other pathogens and "mad cow disease"), not to mention mediocratization of taste, quality, and diversity of foods, indicate that the costs of industrial food production have not been accurately measured. Soviet veterinary science increased the number of head of reindeer for food and clothing manyfold and encountered some of these problems.

The collectivization of reindeer herding in the tundra was intended to increase and stabilize year-to-year herd numbers and tie Nenets, Komi, and other indigenous economies into the five-year plans. This led to costly and unforeseen consequences that specialists were unable to solve easily. In 1955 an epidemic of foot and mouth disease decimated reindeer in the Bolshezemelskaia tundra, moving so quickly beginning in March that entire herds fell. They encountered the same problems Kerstilli had reported on in the region in 1908–9. Teams of veterinarians equipped with vaccines were sent into the tundra to establish quarantine. Yet the disease spread, even among vaccinated animals. The vets posited that they had applied the vaccines too late, and in any events the origins, pathways, and virulence of the epidemic remained unclear to them. Of 81,817 reindeer in eight collective farms on July 1, approximately 68,000 were infected, 27,000 died, and 12,000 would be too weak to make it through the next winter. Comrade Krupin, first party secretary of the Nenets region, requested 10,500 animals to replenish the herds and informed his superiors not to expect significant production for three or four years.[112] The establishment of industrial reindeer herding had led to its own destruction.

Epidemic problems commenced with the forced collectivization of reindeer herders and the reindeer in the 1930s. On August 10, 1932, a senior veterinarian and the acting deputy director of the sector of veterinary medicine of the Commissariat of Agriculture of the Russian Republic, Folomeev, sent an airmail missive to the directors and veterinarians of Reindeer State Farms of the Reindeer Trust of the Commissariat from the Ural, Leningrad, Northern Region, West Siberian, Far East, and Yakutsk regions. He noted, in view of the growing demand of the working class for high-quality food and the difficulties experienced in supply to several industrial centers of foodstuffs, that more attention to the care and processing of meat in conditions of the Far North was required, including veterinary-public health measures, research, and even rejection (!) of spoiled food. This was all the more important because these products earned hard currency through foreign trade. The sector therefore ordered that examinations be carried out to prevent premature slaughter of even diseased animals, but somehow to provide all meat products with certification of health and safety.[113]

Bolshevik administrators were quite wary of nomadic herders. They could not permit individuals they believed to be illiterate (and therefore also incapable of modern accounting techniques so central to Leninist planning methods) to tell them how many animals were in the herds or permit them to thin the herds to meet pressing individual needs when the requirements of the state predominated. To support Bolshevik industrialization the size of the herds had to increase rapidly to feed and clothe the working masses. Thus the authorities pursued the "passportization" of reindeer. Every individual in the USSR had to carry an individual, internal passport. At the same time, the government pursued passportization of ordinary citizens to deal with rural and urban crises connected with food shortages in the 1930s. New passport laws enabled the police to carry out arrests on a mass scale to exile people or send them to labor camps.[114] In Russia to this day every adult must carry an internal passport; another is required for foreign travel. So the passportization of reindeer was not extraordinary, but a logical step intended to introduce individual accounting into animal husbandry—just as it had been for people. Of course, given limited human, temporal, and financial resources, passportization had to be carried out in stages and only later might scientific selection begin. Making things more difficult was the fact that Lysenkoism had gained strength in Soviet agriculture, so that environmental factors more than genetic ones had acquired unusual influence in decision making.[115] How polar climate might help Lysenkoist agriculture was unclear.

In the first year or so, passportization functioned poorly. The authorities indicated progress in 1936 in comparison with 1935 through shock work, socialist competitions, and Stakhanovism in fulfilling plans for the herds in the Nenets and Pechora districts. To accelerate the process, they asked for budgets to provide such prizes for the best reindeer herders as clothing, animal shears, shaving equipment, pocket watches, record players with twenty records, 1,000 pieces of coal, and guns.[116] Even with these magnificent prizes, passportization was only partially achieved in Nenets region reindeer collective farms.[117]

On August 31, 1937, the authorities announced a "socialist competition" that had become de rigueur in all branches of the economy to achieve the highest indices of reindeer production at the Laiskii, Ust-Usinski, and Petrunskii reindeer state farms in the Pechersk region of the Komi Autonomous Republic. One archival document indicates that Nenets and Komi workers at the Indigskii, Nenetskii, Shapkinskii, Ust-Usinskii, and Petrunskii state farms had met Stakhanovite norms for increasing herds up to 20 percent, although many of them were named Stakhanovites simply for preserving herd levels.

Several individuals who worked as reindeer specialists in the bleak tundra would do anything to get out. In September 1937 the zoologist Philipp Iakovlevich Gulchak wrote the heads of the Arkhangelsk territorial administration of Glavsemorput to ask for a transfer, in the process of which he called everyone around him an enemy of one sort or another. After working for six years in reindeer husbandry, Gulchak had determined his future career was in research. But he asserted that scientific research institutes connected to reindeer were filled "with hostile class elements and other incidental persons who either sabotaged their work or had a vile relation to it." Gulchak hence came to the conclusion that the institutes would benefit if practical workers entered graduate work on the basis of their rich field experience. Gulchak touted his own work on passportization of reindeer at the Nenetsk State Farm. Gulchak also claimed that, during his visit to Naryan-Mar, he had met Otto Shmidt, who learned about his achievements and promised a half-year sabbatical to Alaska. Gulchak promised to learn English fluently. The real reason for Gulchak's letter was that he had not found the time to prepare for his trip, so he requested an additional six-month leave from administration for research and writing.[118]

Gulchak never made it to Alaska, but was sent to the terribly polluted nickel town Norilsk in 1937 as part of the Glavsevmorput effort to develop the local economy, in this case to establish the Agricultural Scientific Research Institute of the Far North—with Gulchak as director. The institute focused on polar farming, animal husbandry, and industrial food production to create to the extent possible a local food supply for burgeoning Arctic industrial centers; it cost millions of rubles to feed the polar latitudes with goods shipped huge distances from agricultural regions. Polar residents quickly tired of tinned fish—when they could get cans. Many of the first personnel in Gulchak's institute were life science specialists from AARI in

Leningrad and others were connected with the Lenin All-Union Academy of Agricultural Sciences—the bastion of Trofim Lysenko. Gulchak continued to focus on reindeer; in 1951 his research group got a Stalin prize for its recommendations on how to rationalize herding practices.[119]

Another reason for the slow pace of passportization was that reindeer collective farms were of massive size owing to the expanse and nature of the tundra, with nomadic hunters, fishermen, and herders traveling scores of kilometers in search of prey and of food for reindeer. The "Red Tent" collective farm had a radius of 700 kilometers. On top of this, if there would be permanent farms, then there would have to be permanent housing, stores, clubs, and schools—and a radical transformation of the nomads' lifestyle and attitudes. But housing for the settlements lagged; herders and their families did not appreciate prefabricated housing, nor did they understand sedentary lifestyle.[120]

The authorities organized another "production-hunting station" of roughly one million hectares of land (3,861 square miles) on the Yamal Peninsula on the Baidaratskaia Bay about 180 kilometers from Amderma in 1933. Only 100 Nentsy wintered there, with another 900 individuals arriving for the summer (June–October) who were joined by twenty-five to thirty Russians and their families, no doubt in a supervisory capacity. To make the station function, the authorities pursued a cultural, education, and medical program to create Soviet Nentsy. Archival materials reveal that they had to import everything from medical and dental supplies to textbooks and pencils, boots, food, and fuel.[121]

In his *Month of Little Darkness,* the Nenets writer V. N. Ledkov explored the conflicts that arose in deer herding between two value systems, norms, and generations. The old people did not understand the changes wrought by Soviet modernization and were afraid of new technology. One older man said, "What did those 'smarties' from the council think of . . . collective farm herds? My reindeer . . . I cannot caress my reindeer when I want, or talk with them."[122] New scientific methods of herding had penetrated the tundra—technology and medicine were in the pastures. Yet not all collective farm chairmen used them, but held on to "grandfatherly methods. At councils they just dismiss what the geobotanists have to say." The rational methods, according to Ledkov's story, overused pastures and destroyed moss and lichen.[123]

Even on paper the goal of modern animal husbandry in the tundra was no easy task. There could be no question of fences and barbed wire to control animals on the tundra. Party and Komsomol workers sent from the center to the farms and stations were instructed to introduce a modern, intensive, scientifically based business. But all too often they arrived without any sense of what they faced, let alone knowledge of reindeer husbandry. They did not understand nomadism, could not fathom why the forced application of veterinary science on the tundra went slowly, and created a new series of problems. The problems included Nenets people who did not comprehend the tenor, direction, haste, or purpose of collectivization and the spread of disease epidemics of animals once they were put in close quarters.[124]

By bringing veterinary medicine to the herds and establishing environments that were more controlled for feeding, mating, culling, and so on, agricultural specialists and party officials rapidly increased the number of head and put reindeer herding on a scientific basis. Yet in gathering so many animals into industrial proximity they created conditions where epidemic diseases beyond foot and mouth disease might—and often did—spread rapidly. Local economists and veterinary specialists created a cumbersome reporting system to follow such diseases as plague, ulcers, emphysema, carbuncles, pneumonias, diarrhea, enzootic pneumonia, influenza, meningitis, small pox, mange, paratyphoid, and various parasites. The authorities hoped to quarantine sick animals and determine which pastures were connected with diseases, but early on were relegated to suggesting that such harsh disinfectants as Lysol in a creolin base be used.[125] An additional important step was to deal with the remains of dead animals, toward which ends the local authorities ordered that "the carcasses of reindeers must not be thrown into rivers, lakes and other bodies of water,"[126] a practice that apparently was widespread. In order more efficiently to prevent epidemics among animals, the economists called for construction of corrals in the summer to facilitate examination of animals, proper maintenance of pastures, careful thinning, and castration.

Ultimately, collectivization failed on many levels. Only time and force made collectivization work by the 1950s. The authorities believed that "planning" would overcome the vagaries of the tundra and relieve the "backwardness" of the aboriginals for whom the process of collectivization would always be a violent and confused process. Specialists and officials had hoped to avoid this outcome through the healthy application of Marxist social science to the little peoples.

Social Sciences for the Arctic

Social science also served the state. Just after the revolution, Marxist scholars created a series of institutes—the Communist Academy of Sciences, the Institute of the Red Professoriat, the Sverdlov University—in which Marxian historical materialist ideas of class struggle, the labor theory of value, political revolution, and so on served as major categories and concepts of analysis. At the end of the 1920s and early 1930s these institutions and approaches were subsumed in state universities and the Soviet Academy of Sciences.[127] But historical materialist ideas remained central to understandings of the evolving classless society, the "new" Soviet man and woman, and Stalin's emendation of Marx with "socialism in one country" (the creation of an unassailable socialist fortress surrounded by capitalist nations, rather than proletarian revolution throughout the world). This meant, among other things, that indigenous people came to be seen as backward and even hostile to Soviet power in their physical and psychological distance from Moscow, and they would be forced to modernize, while "socialism in one country" indicated that the drive to secure and exploit Arctic regions acquired an immediacy and impatience to industrialize, urbanize, collectivize, and militarize the taiga and tundra. The social sciences—ethnography, anthropology, linguistics, and other disciplines—thereby supplemented the physical sciences in Arctic conquest.

In 1926 the "north" consisted of roughly half the territory of the USSR with a population of 645,000 individuals of whom 366,000 were minority nationalities. Specialists counted fourteen minority nationalities across the Arctic Circle. About 2,000 Saami lived in the USSR, mostly on the Kola Peninsula (with 52,000 Saami in Scandinavia). The Nentsy lived over a great region from the White Sea to the Barents Sea to the western Taimyr Peninsula. In 1959 there were 23,000 of them, 5,000 in the Nenets Autonomous Region, and 14,000 in the Yamalo-Nenets Region, and in 1989 34,000 people, of whom 77 percent now spoke Nenets as their first language versus 90 percent earlier.

In USSR at first the authorities pursued "indigenization" to respect and foster small people traditions. Ethnographers in the Committee for Assistance to Small Peoples of the North (Committee of the North) provided guidance to uplift the little people through modern science, education, and written languages on the way to the modern way of life. Party officials and planners sought "de-nomadization" with trading posts, cultural bases, and introduction of modern herds. But by the late 1920s and especially under Stalin collectivization began; it was horrible for people and herds alike. The number of reindeer dropped from 2.2 million in 1925 to 1.4 million in 1937.[128] In Canada and Scandinavia, too, infrastructure developed unevenly—trading posts, mining towns, and sea and river ports functioned relatively well, but never the distant "interior" that was devoid of urban culture and had few roads to enable access even in emergencies. In Canada, bush pilots established ties to distant places of cargo, mail, prospecting, forest fire spotting, and even passenger travel. In Scandinavia iron ore, nickel and copper, aluminum and zinc, coal, hydroelectricity, and forestry stimulated great interest in development, especially at the sites of those resources, but led to the "paternalistic pushing aside" of indigenous people; in Sweden the government pursued horrific eugenic policies of sterilization.[129]

Yuri Slezkine explored the relationship between the indigenous people of the northern regions of the USSR and the colonialist modernizers in *Arctic Mirrors*.[130] By mirrors he meant deforming mirrors that the colonialists held up to the minority peoples with images that oscillated between the extremes of "backwardness-as-beastliness and backwardness-as-innocence." Both Tsarist and Soviet peoples saw indigenes as deeply different, explorers saw them as foreigners, Orthodox missionaries saw them as "unbelievers" because of Shamanism, Tsarist officials saw them as wild savages. Occasionally ethnographers and others praised them for their simplicity, generosity, and stoicism. But others soon enough saw them as wild savages again, and the Soviets saw them as alien. They directed their political power toward them with the goal of involving them completely in the Soviet world.

Initially, the Soviets set out to right the wrongs of the empire and of the imperialists who had ignored and mistreated minority nationalities. They created the Committee of the North to administer the fate of those people, Slezkine wrote, who were not nations and to awaken their creativity and protect their interests, yet within a Marxist context. These were small peoples, and small meant primitive and primitive meant classless. The committee set out to create cultural bases—a kind of missionary station—to spread modernity. The ethnographers and party officials and others in the committee encountered their own hardships and confusion: different languages of the small peoples, many of which had no written form; both nomadic and sedentary people, but difficulties understanding either; people who lived on the margins of existence; and a region intolerable to outsiders yet rich in resources.

Until the Stalin period, Soviet policies favored the natives except when they claimed self-rule. The Commissariat of Nationalities (Narkomnats) considered such questions as how much autonomy for these people should be mixed with central intervention to overcome their exploitation and economic backwardness. Some specialists wondered if the state should attempt to bring them culture and transform them into nations or perhaps skip that phase and transform them directly into proletarians. Ultimately, resources and access to resources overwhelmed these questions, for the natives were necessary to feed the burgeoning urban population. The policies of Glavsevmorput put this new vision into play. The expansion of northward expeditions to tame resources put the lifestyle of the small people in a different light. The primitiveness of the small people became a sin against progress and equality. Their "hopeless" attachment to property signified their hostility to revolution. In 1936 party officials commenced violent and coercive collectivization to end the nomadic way of live; the people's shamans were repressed; their animals were brought under state control. Suddenly they were given the "advantages" of nationality: formal recognition as people with languages and territories and cultural identity. Only those practices that Soviet bosses deemed to be quaint—handicrafts and "costumes"—prevailed because of their financial benefit more than their cultural benefit. By the postwar years Soviet writings on the nationalities referred not to the flowering of small peoples, but to the small numbers of minorities, plus the long-term benefits of the presence of Russian settlers and the "heartfelt thanks" of the "little peoples" to their "older brothers"—the Russians.[131] And, as elsewhere in the world, Soviet ethnographers, geographers, and other social scientists became extensions of colonialism and the indigenous people became endangered.

Specialists at the Institute for the Study of the North conducted extensive ethnographic, survey, and other research to chart Arctic lands and peoples and to begin the process of including indigenous people and their economies in the ongoing industrialization program. Its publishing house energetically released a series of fundamental studies. In one large-scale research program, A. G. Podekrat investigated the various animal-based industries of Kolguev Island in the Barents Sea for their potential contribution to industrialization. In the 1930s Kolguev Island had more than fifty small enterprises tied to animal husbandry that annually harvested 70 to 100 poods (one pood = 16.38 kilograms) of goose down, fox and rabbit pelts, and also such sea mammals as beluga, walrus, and seals, all for skins, fat, and meat. While Russian inhabitants of Kolguev Island tended to the shores, the Nenets nomads lived inland near deltas and lakes and hunted reindeer. The goal of Soviet power would be to develop trade to distant markets based on the tens of thousands of reindeer. While sales of fat and other products grew during the 1920s and early 1930s, the Nentsy's income remained very low. Podekrat concluded this was a result of their incomprehension of the modern market, equipment, indebtedness, and economic uncertainty and also a result of the fact that items on the island (knife, locks, binoculars) cost three to five times more than in Arkhangelsk. Podekrat claimed that the Nentsy bought items that were entirely unnecessary for herding (broaches, necklaces, copper belt buckles and medals), while spending little on what he thought that they ought to purchase in order to expand reindeer trade.[132]

Podekrat acknowledged overfishing (the rivers of Kolguev Island had been fished clean) and destruction of the goose, duck, and other bird populations. But he did not draw the conclusion that tying Nenets economic activity to the demands of distant Soviet markets had led to the rapacious harvest.[133] As have many specialists throughout the world, he concluded that the indigenous people in their outmoded ways were the source of the problem and that the solution to it was additional scientific study: If the resources of industrial mammals, birds, and fish were rationally and properly reconstructed, then production would increase two to three times, meeting the needs of the inhabitants and creating conditions for intensive export of reindeer products.[134]

As he pursued his empire of the Northern Sea Route, Glavsevmorput chief Shmidt tried to think of everything with which to arm the indigenes for modernization, including alphabets. How could indigenous peoples contribute to socialist reconstruction of the tundra without written languages (not to mention small-scale and uncertain reindeer herding)? Beginning in the mid-1920s linguists set out to build Latinized alphabets for the minority nationalities. Institute of the People of the North specialists worked with the Institute of Language and Writing of Peoples of the USSR and with the Commmissariat of Enlightenment (Narkompros) of the Russian federation. By 1931 their task was complete, and in 1936, in keeping with the celebration of Big Russian brotherhood and the glorious, new Stalin Constitution, all alphabets were Russianized. Shmidt sent out approved alphabets for the following northern peoples: Nenets, Saami, Khanti, Mansy, Luoravetlan, Even, Nivki, Ude, Inuit, Nanits, Nymylan, and Selkup. The simple goal was to facilitate learning and literacy in Russian and in native languages, to help local workers, especially schoolteachers, master the grammar of local languages, and to enable orders from central, regional, and local party, economic, and cultural organizations to organize and control the Arctic masses.[135]

Urban specialists who journeyed from Leningrad and Moscow into the tundra proudly reported back to the colleagues and party officials how they prepared the Nentsy, Saami, Chuchki, and Goldi to become socialist workers. They had pride of ownership, feelings of patriotism, and a strong sense of scientific achievement in the alphabets they composed and in the attendant schools and books they used to disseminate this knowledge. There is much to praise in bringing medicine, education, and literacy to the little peoples, in increasing life expectancy, lowering infant mortality, and transforming reindeer herding, fishing, and hunting into reliable activities, and in removing them from the edge of starvation and utter collapse. And yet the coercive vectors of industry, technology, science, and written language nearly destroyed traditional cultures. More important to party officials was the fact that industrial Moscow and Leningrad, Perm and Sverdlovsk, Novosibirsk and Magnitogorsk required reindeer meat and fish for their toiling masses.

As they moved forward in creating written languages, specialists and party officials alike sought to share the experiences—and publications—of cultural construction. The secretary of the Murmansk regional Party committee wrote his counterpart in Kamchatka in 1933 to discuss the challenges raising the cultural level of the Saami who spoke three different dialects and had no written tradition. The specialists succeeded as planned by the sixteenth anniversary of the October Revolution (1933) in providing an alphabet for them and early in the autumn had published in Leningrad an ABC book and the first Saami political brochure *What Did the October Revolution Give the Saami Workers?* The absence of Saami schoolteachers hindered any effort actually to use the books.[136] Members of the Murmansk Committee of the New Alphabet often presented a new book or brochure on their activities to other organizations, asking for copies of their publications and local newspapers in return. Some of these materials concerned the effort to bring literacy and a new language to the Even, Itel'men, Nivlhy, Oroki, and other peoples as far away as the Sakhalin peninsula.[137] Another problem was the absence of libraries throughout the region, although the Leningrad Provincial Political Education (Oblpolitprosvet) organization had promised funding to establish twelve libraries and to pay for 300 books for teachers, adults, and schoolchildren.[138]

The Committee of the New Alphabet struggled mightily to overcome indigenous illiteracy. Its social scientists slowly expanded library holdings in native and nonnative languages to assist in the process and published their hopeful papers, but often with delays because of shortages of funds. Its members sponsored expeditions, museums, and other activities, including those connected with a "Red Boat" floating agitation and propaganda center, translations of foreign ethnographic works into Russian, and collection of local data (on reindeer patterns, populations, husbandry; on fishing in internal bodies of water; on industrial animal husbandry in Kola with participation of Saami, marine biological studies, literacy campaigns, and history, ethnography).[139] An achievement, although with unclear contribution to assimilation of minorities, was the appearance of a Saami arithmetic textbook in fifty copies.[140]

Educational specialists reported significant progress in raising the cultural level of Saami people by early 1935. The written language had made its way into print and into schools. There were now twelve such schools and thirty-one Saami students matriculated in the Saami department of the Murmansk Pedagogical Technical School. The committee succeeded in publishing an ever-growing series of brochures and pamphlets: children's books including an illustrated alphabet book in Saami; Ptokhov's *Preserving the Health of the Reindeer Herder;* Mimichev's *The Strength of the Reindeer Kolkhoz;* a collection of Saami folklore; such posters as "Gather Mushrooms and Berries" and "Why Use the Banya," which was apparently intended to encourage modernity through cleanliness; and Iogan Eikhfeld's *Improvement of Cultivation of Potatoes and Vegetables* that had Lysenkoist overtones.[141]

The Murmansk branch for Study of the People of the North, with a staff of three (a director, consultant, and scientific secretary), focused on language, history, economics, and "cultural construction." It organized a scientific office and library, and it encouraged the study of language and culture directly among herders and fishermen even in such locales as graveyards.[142] Yet from graveyards to boats the specialists found it difficult to penetrate the tundra and its rivers and lakes and convert the Saami to Marxist modernity, not the least because of financial and material shortfalls. The chairman of the Murmansk regional executive committee and chair of its party committee wrote the director of the Stalin Machine Building Factory in Voronezh with a request to speed up delivery of one fifty-horsepower boat engine to outfit a "Red Boat" to serve as a propaganda gathering point for the Saami on the shores of the Barents and White seas. They wrote that the engine "has great political significance because of the mission to change the living conditions and economy of one of the lagging nationalities of the far north."[143] It was as hard to acquire a motor as it was to attract Saami leaders, teachers, and especially women to ongoing literacy and medical campaigns. Assimilation of the written language lagged in particular in the Lovozersko region, where the few Saami teachers were poorly trained, textbooks had not even arrived, and most schools had no paper or pencils. The Lovozerskii Club conducted business in Russian; Saamis did not attend.[144]

Cultural Bases and Red Teepees

The Council of People's Commissariats (Sovnarkom) passed a resolution in June 1936 to transform nomadic and seminomadic peoples of the Far North to sedentary lifestyle by 1942 based on collectivization of reindeer herding. In the resolution, officials noted that they considered this transformation a completely voluntary matter, although one that demanded significant preparatory work from party, Soviet, and social organizations.[145] The process was voluntary if one considered the alternatives: resources went to the institutions of sedentary life—collective farms, small towns established near the farms to operate Red Teepees, and other institutions of cultural assimilation.

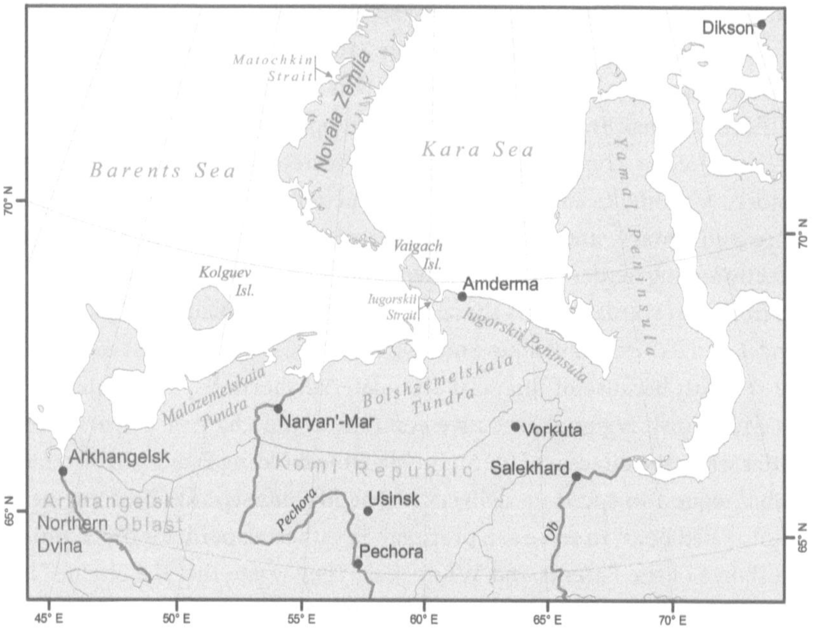

Nenets Autonomous Region and Arkhangelsk Province. (Map by Manny Gimond)

Those nomads refusing to go along would be declared wreckers. Who could decline to become modern under the specter of political coercion, especially when the erection of bakeries, electrical power stations, and cultural institutions accompanied the sedentary life?

Yet many Nentsy refused to join collective farms out of fear and incomprehension, sought refuge in the tundra, and in several cases engaged in violent resistance that was put down more violently by NKVD troops. Apparently "kulak" herders destroyed one-third of livestock between 1926 and 1933. In the winter of 1940 a group of twenty-five to thirty herders escaped from Norilsk and headed south into the Evenki autonomous region using reindeer. They attacked several militia posts to secure weapons. NKVD detachments from Dudinka, Igarka, and Turukhansk pursued them, but turned back exhausted. Eventually the prisoners ran low on supplies, left the Evenki region, and were shot after a being caught by a ski unit.[146]

Indigenous people were also caught up in the Stalinist purges. In the initial stages of the Terror, a total of ninety Saami were repressed (including four women), of whom forty-two were shot, three died in custody, and thirty-nine got prison terms. The first arrests in the early 1930s in Pulozero on the Kola Peninsula saw five individuals—reindeer herders and fishermen—imprisoned for anti-Soviet agitation. Another five people were arrested in April 1935 at the "Lutto" collective farm. Mass arrests began in August 1937—seventeen were arrested over several weeks and sent to Leningrad to be processed. In October–December 1937 another ten Saami—many fishermen from the "Raia Kalastaia" collective—were arrested. In all in 1937–38 fifty-eight Saami, including one woman, fell into the camps.[147]

On Yamal Peninsula in 1931–32 the authorities arrested 488 people, of whom 294 were Nentsy; the others were Russian, Khanty, Zyrian, Kulakis, and Shamans. Occasionally, the indigenous people fought back, for example, the Nentsy on Yamal Peninsula in 1934 and 1943, the Dolgan, Nentsy, and Evenki on the Taimyr Peninsula in 1932; and again in 1948–49 when the authorities descended upon them to confiscate private property, leaving them to die of hunger because of the confiscation of their livelihood, their reindeer, in the name of the party and state. The Great Terror began in 1937—with larger numbers of individuals arrested and shot. Chuckchi, Even, Dolgans, Iukagirs, and others were arrested.[148] After these arrests, the relatives of the arrested were often left to die of hunger and exposure.[149] According to one compilation: Seventeen Dolgan were arrested of whom five died; seven Iganasany with three deaths; twenty-two Nentsy, twenty-eight Sel'kupy, five Chukchi, sixty-seven Evenki, sixteen Eveny, seven Iukagiry, in all 261 people, with twenty deaths.[150]

The authorities had hoped to avoid resistance to collectivization through the establishment of calming outposts of Soviet culture in the Arctic. Party organizers, social scientists, and teachers joined cultural bases, clubs, "houses," Red Teepees, and schools into the tundra. The Red Teepees were a symbolic and physical manifestation of Soviet power in the tundra; reindeer herders traditionally lived in teepees made of animal skins and now were supposed to invite Marx, Lenin, and Stalin to join them for discussion in those teepees. Similar to the Red Corner in industrial enterprises and the Red Boat in fishing cooperatives, the Red Teepee was an important extension of the modern worldview into the tundra.

The Bolshezemelskaia tundra is a hilly morainal plain within the boundary of the Nenets Autonomous Region and the Komi Republic of Russia, with some of the hills reaching 250 meters, but mostly rolling at 100 to 150 meters. It consists of mossy-bush tundra and some forest tundra with fir and birch in sandy, thin soil with rocks and boulders, and the trees do not grow tall. Much of the land is swampy and boggy. Temperatures average −16°C in January and in July 8 to 12°C, although in the summer months frosts are always possible. Precipitation ranges from 450 millimeters per year in the south and 250 millimeters in the north. The major rivers include the Pechora and Usa—whose upper reaches flow from narrow valleys into broader ones that become calm. Here the Nentsy herded reindeer.

All sources indicate the process of conversion of Bolshezemelskaia herders and their families to Soviet lifestyles was unhappy. The intent of Soviet leaders to eliminate illiteracy, combat epidemics, increase life expectancy, and improve economic performance must be welcomed. Yet the coercive process, together with shortfalls of funding, incompetence of personnel responsible for cultural transformation programs, and incoherence of plans made cultural transformation at best an uneven process. Two important sites of cultural reconstruction among Nenets herders were the Kolokolkovskaia (the Bells cultural base, *kul'tbaza*) in the Malozemelskaia Tundra and the Khoseda-Khardskaia base (in Nenets meaning "house on birch bald peak") in the Bolshezemelskaia tundra.

The Kolokolkovskaia Base served six state farms and eighteen collective farms. The collective farms consisted of 131 Nenets groups with 754 people in all. Collectivization led to a fourfold growth of herds from 5,729 head to 22,422 head in two years. To achieve these ends, activists lectured on Soviet culture, read newspapers and books out loud, heard reports from shock workers, offered evenings of chess, checkers, and other board games, studied the speeches of Stalin, and took field trips into the tundra, the last of which must have been like the blind leading people with perfect eyesight. They engaged in a glorious "struggle against horseflies" through mass cultural work in Red Teepees and in a "Nenets House," a kind of parallel to city-based Houses of Culture for workers.[151]

In 1935–36 significant achievements at this cultural base included discussion with the people of the tundra of the magnificent Stalin constitution and other decisions of the party and government, as well as practical assistance to the farms to increase numbers of animals both through the application of veterinary science and through struggle against wreckers. Cultural construction also meant efforts to introduce Soviet norms for cleanliness of the teepee, clothing, and pots and pans. All of these tasks were difficult to achieve given the fact these norms reflected a different system and that among thousands of indigenous people were only five veterinarians, one mechanic, and eight reindeer herder brigadiers. Illiteracy remained widespread; schooling and medical care lagged. But by the standards of Soviet civilization there were positive results. If in 1935 88.5 percent of herding activities were collectivized, in 1936 95.2 percent had been, and nearly 90 percent of the children attended school. Stakhanovism had become widespread among leading herders, although the reindeer surely did not understand the commotion.[152]

Cultural work among the Nentsy picked up in 1936 with thirty-eight trips into tundra by political officers, including a doctor on three occasions. The Nenets herders were thrilled no doubt to discuss the Stalin constitution, the show trials of the so-called Trotskyite-Zinovievite fascist band, and the advantages of Stakhanovism for reindeer with the political officers. They heard twenty-four lectures on medicine, personal health, animal treatment, and veterinary medicine. Cultural enlightenment involved the absurdity of reading out loud sections of Mikhail Sholokhov's *Tikhii Don* ("Quiet Flows the Don," 1934; Sholokhov won the Nobel Prize in literature in 1965), excerpts from Dmitry Furmanov's *Chapaev* (1923, about a legendary Bolshevik Civil War Commander), and Nikolai Ostrovskii's *How the Steel Was Tempered* (written in 1932–36, a socialist realist novel dealing largely with the heroism of the civil war). Still, it is not clear the lectures got anywhere given that only three of fifteen propagandists spoke Nenets and the Red Teepee operated only in the summer among swarms of black flies and mosquitoes.[153]

Erection of a school, dormitory, apartment building with six apartments, hospital, veterinary station, and House of Nenets all moved forward as slowly as might be expected in the Malozemelskaia tundra in the absence of local supplies. A public bath had opened in 1935. An electrical power station was 75 percent complete and would operate in the first quarter of 1937. The generator was crucial to provide light and heat to the school and dormitory, especially because construction was very low quality and the buildings were drafty. The foundation was not plumb or square, doors and windows did not fit, lumber had run out, and reports fill the dossiers that on top of this the contractors were quite rude.[154] Schools were also short of supplies; those on Novaia Zemlia and Kolguev Islands were short 1,000 notebooks, twenty kilograms of writing paper, and 200 large sheets for the bulletin board newspaper.[155]

The even-larger Khoseda-Khardskaia cultural base served the Koseda-Khardsk, Iusharsk, Varaidesk, and Karsk units in which there were seventeen collective reindeers farms with 6,800 toilers, of whom 4,912 were Nentsy who had been liberated from outmoded forms of production to engage in new forms of "Soviet" economic activity.[156] This cultural base served the southeast region and the "Polokha," "Zvezda," "Druzba," "Tet-Iaga-Mal," "Iadei-Segery," and "Udarnik" reindeer collective farms and artels. In March 1929, the cultural base gained the status of a village and became the center of an agricultural council (a local organ of government power) and then of the entire Bolshezemelsk region. It served production, medical-veterinary, cultural-enlightenment, and scientific research activities. In terms of culture, its major goal was a social transformation in the tundra and to raise the "cultural level" of the indigenes.

Ultimately, the Khoseda-Khardskaia cultural center functioned poorly from top to bottom, from personnel to budget, and suffered from endemic misuse of facilities. It lacked good leadership because the director worked two jobs. The school and hospital met expectations if one ignored the fact that part of the hospital building had been taken over by employees as apartments. When an inspector suggested moving the regional division the NKVD and the Chancellery of the Workers' Executive Committee out of the building, the NKVD officials refused. Another problem was huge overspending on feeding schoolchildren. Keep in mind that in the 1930s several states and philanthropies across the globe had begun school lunch programs, but in places like the United States a national school lunch program was not created until 1944. Soviet officials recognized much earlier the importance of providing school lunches both because full stomachs learn better than empty ones and because they needed to attract rural children totally unaccustomed to school and who lacked interest in it.[157]

Party organizers and political educators hoped that schools would reach Nenets children, but here too they encountered problems. At the beginning of the school year in fall 1936, the Khoseda-Khardskaia elementary school had enrolled eighty-two Nentsy and twenty-seven Komi children.[158] The Kololkolkovskaia elementary school had enrolled sixty Nentsy and fifty-three Russian children.[159] As for the most important indices of success of the cultural bases, there was a long way to go. According to an October 1936 report, at the Khoseda-Kharda, Iushar, Kabarov, Vashutkin, and several other cultural bases, 50 percent of the students were illiterate or largely illiterate.[160] The plan to eliminate illiteracy in 1938 had failed miserably: In one school only thirty-six of 100 students had learned to read, in part because the campaign directors shortened the program arbitrarily by forty days. The teachers will ill prepared; they rarely graded notebooks.[161] Pedagogy struggled in the face of uncomfortable desks and chairs, no pencils or notebooks, and few textbooks or maps. The teachers did not help matters when they refused to provide syllabi let alone any other sense of organization. Students in upper classes could hardly speak Russian, while teaching in native languages was unsatisfactory. Nor could students solve mathematics problems. The effort to use art to bring the indigenous people into the fold also failed. No one took part in a district "Olympiad of Artistic Independence."[162] The children wanted to get away—back to the tundra—as fifty-two of them were crowded into five rooms of a dormitory where the temperature rarely exceeded 4°C in the winter and that was poorly furnished to boot. The children were bored with nothing to do; they slept two to a cot; their miserable, monotonous meals did not meet the demands of the tundra nor have the nutrients to fight scurvy. The children themselves were pallid, sickly, and poorly dressed. One can only imagine the quality of the food.[163]

Things moved slowly or not at all. A plan for 1937 revealed only two of three kindergartens had opened and only 115 of 250 children in the region attended, while no kindergartens (versus three planned) had opened in the countryside to serve 100 children in the wide-open winter desert. The authorities had hoped to educate more than 1,000 school-age children, but most of them—and especially the older brothers and sisters—remained illiterate or barely literate because the children were much more interested in herding in the tundra than attending schools in villages. Officials had hoped new schools in Peshsk, Indigsk, and elsewhere, dormitories in Moderks, Indiga, Naryan-Mar, Pesh, and Shoina, and kindergartens in Pesh and Naryan-Mar, as well as houses of culture, red teepees, and libraries, would rectify things. But these facilities were rudimentarily equipped; they were "cultural points" more in name than reality.[164] Everywhere across the region officials told positive stories about cultural reconstruction, building of schools, and ending of illiteracy, while in reality tight budgets and uninterested reindeer people derailed them.[165]

The 1937 annual report of the Khoseda-Khardskaia kultbaza indicated little improvement. Repairs on buildings remained incomplete, those repairs were poor, the wood stove barely functioned, the roof leaked, wood, kerosene, and hay for animals were in short supply, and teachers were poorly prepared, their salaries low and often delayed for two to three months. All of this had the expected result: a sharp reduction in the number of schoolchildren who regularly attended. Further, the officials still did not comprehend that when they brought together children from around the tundra and stuck them in a poorly heated and leaky school, they created a breeding ground for illness. Children suffered from anemia, scurvy, and loss of weight. The school had 116 students, ninety of whom suffered from scurvy and seventeen of whom were quite ill.[166]

Schools were instructed to organize "fundamental teachers' libraries" with the works of Marx, Engels, Lenin, and Stalin plus collections of the classics of literature (Pushkin, Gorky, and so on).[167] The kultbaza augmented Marxist studies with celebrations of the 100th anniversary of the death of Pushkin on February 11, 1937. The authorities recommended performance of songs and competitions among the upper classes for the best bulletin board discussion of Pushkin.[168] What Marx or Stalin or Pushkin could offer for the people of the tundra or their reindeer was not stated.

Correspondence reveals that Glavsevmorput had trouble finding teachers, let alone good teachers, and keeping them in the tundra. To rectify the situation, beginning in September 1936 the Herzen Pedagogical Institute in Leningrad offered a yearlong program for teachers of elementary schools for peoples of the north to raise their qualifications. An effort to publicize the program by radio and newspaper followed. Entrance was open to a teacher with middle, that is, unfinished higher education. Admitted teachers would receive a 200-ruble monthly stipend and a dorm room, although the room was not open to family members.[169] With the other hand, Glavsevmorput officials blamed the teachers themselves for poor preparation and inadequate qualifications.

Modern media were the major tools of cultural transformation, but officials had not fully thought out how to use media creatively. The House of Nenets served mainly the staff and ignored the Nentsy. House directors were replaced quite frequently by impatient party officials. The house had a dorm, but it was open only to men and was poorly equipped. The authorities put emphasis not on comfort and food, but on "cultural" inventory. The house library held 780 books, all in Russian, 274 of which covered political topics, 277 questions of art, and ninety-six scientific-popular themes. Periodical literature included standard Bolshevik fare for political activism, agitation, atheism, and literature.[170]

Kultbaza seem successfully to have employed radio and newspapers, but had great difficulty with the new medium of film. Firsov, who served in the 1930s as director of a base's cultural department, wrote the Archangelsk territorial administration to complain about the small number of auditoriums, projectors, and copies of films. Local authorities had no idea how to use them for propaganda purposes let alone how to run projectors. Firsov requested copies of such films as *Circus,* the legendary *Chapaev, We Are from Kronstadt,* and the infamous *Party Membership Card* to be distributed throughout the tundra bases.[171]

Literary works contributed to the celebration of modern technology and Bolshevism in the tundra. The journalist T. P. Sinitsyn was arrested for counterrevolutionary activity in 1920, then freed, a year later arrested, then freed again. In 1930 he opened a school for the Nentsy on Novaia Zemlia and simultaneously embarked on a writing career using themes of Nenets life under the pseudonym Pelia Punukh. Arrested again in July 1937, he somehow survived the camps.[172] In *From Underneath Five Centuries,* in part written tongue in cheek, Punukh described the glories of modern technology and a new way of life coming to the tundra. He wrote, "Before 1931 the village of Tel'viska, in the center of the Bolshezemelskaia and Malozemelskaia tundra, was considered the countryside. But it had a Nenets District Executive Committee, Nenets school, hospital, cooperative—all this was in Tel'viska. A loudspeaker, telegraph and post office, police station, State Trade Agency, Society of Consumers of 'Rybak', the 'Pechora' Trade Company—all this was in Tel'viska, the countryside which consisted of 40 households."[173]

The Great October Revolution gave the Nentsy—we are told—the opportunity to think, think, and think again about their lives—and to make the conclusion that there was great value for them in Soviet power. Nenets sons and daughters could distinguish themselves from the older generation by going off to school.[174] Yet while school was interesting, and it was necessary to read, write, and do arithmetic, school was a temporary thing. As Punukh noted, writing and arithmetic made it difficult to cheat the Nentsy when they sold reindeer meat and fur. But "the tundra is constant. The tundra is the only thing for the entire life of the Nentsy."[175] To attract Nentsy to a new lifestyle, they showed wonders that "people in the know call 'films.' The films showed pictures of city life. Automobiles, trams, trains—that's what pleased the Nentsy more than anything."[176] The past weighed heavily on the tundra. But the people would free themselves from the past, make the tundra red, make it as Lenin taught, and become literate.[177]

Making light of the Soviet practice of giving acronyms to anything, Punukh offers a scene where the Nentsy give a name to a collective farm. Because the All-Russian Central Executive Committee is abbreviated at VTsIK, the Nentsy joyously decide to name their Parma Nenets Kolkhoz "PNOK." One says, "See how short and easy: PNOK—that's it!" A little boy says, "PNOK, PNOK, PNOK . . . short and pretty."[178]

The story "Novaia Zemlia Everyday Life" put in stark relief the new life of the Soviets versus a past age without technology. The young generation in Punukh's stories was fascinated with technology, with the motors and motorboats, with the boarding school, and with the absence of lice, as one boy tells his grandfather.[179] When the grandfather dies, the young boy kneels at his grandfather's gravestone. He declares he will study, study, and study, and when he knows everything he will return to Novaia Zemlia. He will bring electricity and grow potatoes, and he will not work alone but in a cooperative. He will transform life with electricity and modern reindeer husbandry and by planting potatoes and other crops. "And then no one will sleep long during the Novaia Zemlia night. No one will die of scurvy. And then many people will live there also. And I will love the sun and people."[180]

Cultural Assimilation by Force, Not Science

An outgoing director of a cultural base on the southern shore of Kola Peninsula, near Seven Islands (Sem Ostrovov) and Blessed Nose (Sviatoi Nos) on the brutally isolated north shore of the Kola Peninsula, one Johann

Solomonovich Abramovich, penned a thoughtful, worried report to his replacement that revealed the joys and sorrows of the process of assimilation of the minority peoples of the tundra. Abramovich, an energetic thirty-four-year-old devoted communist, a member of the party since 1921 with higher education from a Marxist higher educational institution, arrived at the base in March 1937. Over the next five months he became intimate with the needs of the base and obstacles to meeting them, as he wrote Pavel Logvinovich Kotov, his replacement. Abramovich warned Kotov to pay attention to three things: supplies; schedule; and the threat of "bourgeois nationalism," although he asked Kotov to keep the latter concern strictly to himself. Regarding supplies, the base required without fail the following items to survive into the next summer: 2,000 lemons for the battle with scurvy, 100 kilograms of dry eggs, 50 kilograms of dry milk, 50 kilograms of cocoa, 100 kilograms of mustard, 100 kilograms of garlic, 5 pairs of rubber boots, 5 pairs of fur boots, 120 pairs of boots, a new typewriter, an arithmometer, pots, pans and other kitchen ware, 120 children's beds, 110 chairs, books, hammer and nails, needles and thread, and medicines.[181]

Abramovich warned that the success of the base depended upon sending a teacher or another worker into the tundra by August 15 to gather up children for school, otherwise they would be lost to herding. Before school itself commenced and before ongoing construction was finished, the workers must finish housing. Toward that end, Abramovich bought two very good tents for temporary workers. Also, the purchase of nonperishable goods for the hospital and school had to be completed and hay delivered. With the remaining funds he suggested building a playground and signing a contract to guarantee deliveries of meat and fish throughout the fall. He strongly advised Kotov, "Never in any case permit the expenditure of operational funds on construction activities."[182]

This fairly straightforward discussion was a prelude to Abramovich expressing his deep concern about the future of the kultbaza, which he believed had been subjugated by "hostile elements." It is difficult to judge if such elements existed or if Abramovich wished to protect himself from the accusation that he himself was such a "hostile element." He wrote, "This, Pavel Logvinovich, is only for you and, I think, at no time should you talk about this with anyone else. For me it is clear: bourgeois nationalist, counterrevolutionary motives lay at the basis of the organization of the kultbazy.

These suspicions of mine were confirmed by a responsible worker of Glavsevmorput whom I met on a steamship, who knows our region well, one Leontovich." Leontovich had informed him that the government had released money in the region in 1928 not to nomads, but to organize a cultural base to serve sedentary Komi people who were not real herders. For Abramovich this meant the kultbaza was "politically and geographically removed" from those it was intended to serve. The schools served Komi children; the hospital treated primarily Russian and Komi.[183] Abramovich based his recommendation on extensive experience in the tundra. He and his staff had spent a total of 1,826 days involved in medical, veterinary, agricultural, and agitational activities. These 1,826 man-days excluded the work of the Red Teepee that itself was nomadic. Abramovich called for the Red Teepee to become more sedentary with the construction of a large club with auditorium, a teepee as a dorm for visitors, and a tent to use as a public bath.[184]

Unlike Abramovich, most emissaries of cultural transformation seem to have preferred the relative warmth of wooden structures to the certain cold of the tundra. Of sixty-three employees at the Malozemelskaia kultbaza, few had actually ventured into the tundra. Most had been in the north for two years, and usually less, and they had not engaged the literacy campaign in over a year. Party investigations revealed significant personnel problems. Some real bad apples stood out.[185] The Red Teepee did not serve its function as a "mobile political-enlightenment institute of the base" because it was a primitive, even wretched place and "absolutely and in no way differed from the very worst teepees of the tundra." The Red Teepees were so "discredited" as an instrument of Soviet culture that during a two-month stay in Khoroi-Vore, not one Nenets stopped by. The local House of Nenets had no inventory of books, newspapers, games, or films for "cultural enlightenment" and no dormitory should a visitor wander by. All propaganda work had superficial character.[186]

In February 1936 the previous director of the Arkhangelsk administration of the cultural base, Stepan Terentev, was "removed from work by a resolution of judicial-investigative organs for poor performance." Terentev had apparently misled officials about the number of employees under his supervision (not forty-four but only thirty, none of whom were doctors, veterinarians, or teachers). Further, the construction of a school, hospital, and veterinary station that should have been completed by November 7—the anniversary of the Great October Socialist Revolution—was only three-quarters complete. "Enlightenment" work at the school in the Kolokolkovskaia base was in "disgracefully" poor condition. Most of the students did not return after vacation, while others tried to run away; the theater, sports, and music programs fell on deaf Nenets children's ears. Terentev was clearly a "swindler" and "embezzler."[187] Fortunately, he survived his incarceration, and when he died at sixty-six years old he was a construction engineer who joyfully built schools throughout the Komi Republic.

Terentev was no wrecker. Like Abramovich, he was devoted to the cultural transformation of indigenous people in the tundra. Under his direction they began to build the Khoseda-Khardskaia cultural base in 1926. Terentev, from Brykalansk, a village of roughly 1,000 people upstream on the Pechora River, was responsible for operations to float lumber down the Khoseda Creek to the tundra to establish the base as an early cultural-educational, scientific research, medical-veterinary, and production outpost of Soviet power. Terentev was exiled after the 1905 revolution with Kliment Voroshilov, a civil war hero, later commissar of defense, and willing participant in Stalin's repressions. After the civil war, Terentev learned the construction trade in Ust'-Usa in the Komi Republic, first building a canning factory, then the cultural base that served a half dozen reindeer collective farms and artels.[188]

Party workers and teachers at the kultbaza plainly were at wits' end trying to do their work. The political turmoil of the Great Terror had clearly spread far into the tundra. They complained to the central cultural department of Glavsevmorput in Moscow about poor leadership that resulted from constant turnover; the workers sent to join in cultural revolution whose qualifications had not been properly vetted; the mixed signals they got; and the way staff members were always picking on each other, giving and countermanding orders, sending telegrams to the center demanding more funding.[189] The workers never adjusted to the seasonal character of their work. They concentrated their efforts in the summer, never in the winter, and the herders never understood why they were there.

Yet by 1949, 500 schools had been established in the national regions of the Far North. Before the war the most important center of higher education for the minorities was in Leningrad, but of course it was closed during the German blockade and did not reopen until 1948 as the Department of Northern Nationalities in Leningrad State University; preparatory courses at the Herzen Pedagogical Institute also ceased until after the blockade. With the tumult of violent collectivization and the war over, pedagogical, medical, agricultural, and technical schools might open in the Arctic Circle. By the 1949–50 academic year students of fifteen northern nationalities were at universities.[190] Yet the authorities acknowledged continuing difficulties in establishing universal elementary education into the 1950s largely because of the nomadic life of the local people and because the older herders opposed Russification; they absconded with their children into the tundra to avoid school.[191]

Delivery of Soviet culture and modern conveniences lagged into the Brezhnev period. Anthropologist Ethel Dunn commented, "I have said that when one attempts to change the culture of a people, humanism is not enough, and that Soviet humanism has failed to prepare the small peoples of the North for the twenty-first century." She noted that roughly 800 indigenous persons were studying medicine in late 1960s and "this is considerably less than the norm set by the government," and "does not satisfy the need for feldshers, midwives, nurses and pharmacists in the RSFSR." In spite of large expenditures to improve the health and well-being of the northern peoples and new settlers, with fully equipped floating hospital ships and floating cultural bases that provided aid, as well as air transport for the seriously ill, "these measures leave much to be desired." Dunn concluded that the lag in medical service and electrification to the indigenous people was the result of the fact that the authorities found it not cost-effective. The facilities were few and far from population centers where they would have preferred to provide services.[192]

No matter the hard work of emissaries of modernity or the imperatives of Marxist science, no matter the exhortations and Stakhanovism, the cultural bases throughout the tundra did not work as hoped. As scientists moved from the rock of the Kola Peninsula to the forests of Arkhangelsk and Karelia and to the rivers, deltas, bays, and seas of the Arctic Circle, they determined interesting regularities. They slowly gained the confidence they could use these understandings to develop Arctic resources of fossil fuels, ore, lumber, and fish. Through extensive, long-term exploration of the Arctic basin, using icebreakers, research vessels, airplanes, and drifting camps, they facilitated the expansion and regular operation of the Northern Sea Route, at the same time learning more about climate and weather, currents and ice regime, fisheries, animal husbandry, and other resources.

Yet it may be that the scientists' successes in accumulating data about the nature of Arctic regularities masked true understanding of qualitative issues,

Red Teepee, Bolshezemelskaia tundra, 1960s. (Courtesy Ekaterina Boikova)

especially those raised by the social sciences. Specialists at the Kola Science Center and AARI tackled the physical world of tundra and the Arctic basin. Trained observers at Gidrometsluzhba developed more and more accurate weather forecasts. Episodically at first, more systematically later, medical personnel began to fathom the great challenges to human physiology of extended periods at higher latitudes. But thorough descriptive and analytical depictions of the physical world and human anatomy did not translate to the world of the settler and indigene in the tundra. When they focused the social sciences on the Arctic landscape they encountered a difficult range of problems. Ethnographers, anthropologists, and linguists excitedly trekked north to assist them in gaining the advantages of modern society, including language. According to Soviet Marxist desiderata, those advantages included sedentary lifestyles, involvement in distant markets, corrals, veterinarians, passports, and the like. Yet the specialists overestimated the ability of science to solve the agenda of modernization across a massive and irregular landmass. Specialists grew frustrated with their inability to fit Nentsy, Komi, and Saami into Marxian categories of class, mutable worldview, and educability.

A close friend from Severodvinsk, Ekaterina Boikova, who grew up in a

Russian school age girl, in Bolshezemelskaia tundra, 1960s. (Courtesy Ekaterina Boikova)

cultural base in the Nenets region, swore by the central importance of the cultural bases in raising the living of literacy and culture among the Nentsy and describes her parents' confidence in the Red Teepee. Her family lived from 1957 in Nizhniaia Pesha on the Pesha River fifty kilometers from its delta on the Barents Sea. Nizhniaia Pesha had 186 residents in 1922 and became a regional government center; in 2010 about 800 residents lived in the village. Ekaterina's mother, a music teacher, and her father, with training in cinematography, met there, married, and had two children. She had the "happiest Soviet childhood" and remembers life as pioneers and Komsomol members.[193] They had many friends including Egor Pankov, a Nenets herder. Her father worked in Red Teepees and showed films to the Nentsy. Ekaterina noted that many Nentsy, Saami, and other indigenes today lament the breakup of the USSR, for it destroyed the calm security of Soviet power and presented them with the challenges of living in a market economy. But she pointed out that they have the opportunity to work through this difficult time precisely because of the educational and other opportunities that Soviet power gave them.

The researchers at AARI have also taken full advantage of the breakup of the USSR. Having navigated the difficult financial shoals of the 1990s, they have resumed basic and applied research. Their work contributes to the continued development of Arctic resources, especially fossil fuels, in the twenty-first century. Soviet power prepared them also for these opportunities.

5

The Nickel That Broke the Reindeer's Back

Oh, beautiful summer! I would love you,
If not for the heat, and the dust, and mosquitoes,
And flies.

—ALEXANDER PUSHKIN

A distinctive feature of Soviet urbanization was the "company town." Owing to a centrally planned economy, forced settlement, and unwavering determination to harness resources no matter climate or geography, Soviet scientists, engineers, and planners and officials joined together in the belief that the best way to exploit natural and mineral resources was through the establishment of towns and cities dedicated, at least initially, to the manufacture of one major product. Soviet citizens might find work in Asbest (Asbestos), a town of 70,000 in the Ural Mountains, or Magnitogorsk, which grew from a few thousand residents to 140,000 inhabitants in the 1930s and to 400,000 people by the end of Soviet power, most of whom were engaged in iron and steel production. In these company towns they faced the constant threat of industrial accidents and pollution-borne illnesses, diseases, and maladies; epidemiological surveys reveal significant life expectancy and infant mortality impacts. Across the Arctic Circle the company towns also often were focused on one major activity: nickel smelting (Monchegorsk), ship and submarine construction (Molotovsk, later Severodvinsk), fluorite mining (Amderma), and apatite ore extracting and refining for phosphorus-based fertilizers (Apatity, the second-largest city on the Kola Peninsula). Or perhaps the city of Nickel.

In addition to these company towns, urbanization in the USSR involved the "socialist reconstruction" of already existing cities into socialist ones and later rebuilding those cities and villages in western regions of the USSR after their devastation during World War II. In Arctic regions, generally speaking, the company-town pattern prevailed. Such diverse activities as the discovery of nickel, copper, platinum, coal, and other valuable ores; the determination to exploit fish or forest resources; and the creation of a railroad, canal, or harbor with new shipbuilding yards and fish processing factories led the authorities, working with engineers, to push the establishment of new industrial centers consisting of a factory, housing for workers, and belatedly stores, hospitals, schools, and clubs. The impetus for resource development came from the search for autarky in the economy and independence from potentially hostile capitalist nations as trade partners for strategic minerals. The government willingly subsidized Arctic urbanization and resource development with cheap labor in the Stalin era (gulag prisoners, exiles, and others). In the Khrushchev and Brezhnev eras, the government had to subsidize Arctic assimilation both with large-scale construction projects, which were far more expensive in the northern latitudes than they were in central and southern regions of the country, and with higher wages for workers including the promise that they might have new housing well before the time promised in other cities.

Arctic urbanization was the outcome of the expansion of the working class that logically accompanied the full embrace of the Marxian imperative of the proletariat as the moving force of history. Largely a peasant society in 1917, by the time of the 1959 census Soviet leaders could claim at last that the nation was more than half working class, with peasants, white-collar employees, and intelligentsia making up the rest of the population. While on the surface and in official histories this process of urbanization followed the plans and dictates of the center, it was messy, irrational, and disorderly. Cities were built in mud by workers housed temporarily in tents and barracks. Even in Molotovsk, important to the nuclear navy, schools and housing lagged years behind the needs of the growing population, and young people with children paid for this inconvenience.

A tour of the urban Arctic through a handful of its cities will provide a sense of the challenges in assimilating resources and in following the Marxian assumption that socialism itself must arise in an advanced industrial society, in urban settings, even if above the Arctic Circle. While chronological in many respects, my approach here is rather to move from one industry to the next, and one city to another, to illustrate the vast problems builders, planners, party officials, workers, and prisoners encountered in their day-to-day lives that were dedicated solely to increasing output. These problems included a dreadful lag in matching progress in factory construction and production with progress in erecting housing and social services on the tundra; a poor industrial safety record; extensive pollution; and the physical and psychological isolation of the people who toiled in Arctic and subarctic regions. Our urban tour moves roughly from west to east from Apatity to Monchegorsk and Pechenganikel on the Kola Peninsula to Molotovsk and Amderma in Arkhangelsk Province.

In many ways urbanization of the Arctic resembled the processes and problems elsewhere in the USSR: Everywhere it had extensive environmental impact and was plagued by irrational planning and shortages, and everywhere its supposed benefits—museums, public transport, access to good medical care and inexpensive housing—reached the residents the least and last.[1] But patterns of Soviet urbanization in the northern latitudes were also distinct from the standard historical experience. Climate and geography imposed great hardships on settlers and great costs on construction and on the provision of necessities. The settlement pattern was unique also in that it included vast numbers of so-called kulaks (ostensibly, the wealthier peasants, but ultimately any farmer who fell afoul of the state or was denounced by angry neighbors), exiles, and political prisoners, in addition to the visionaries and volunteers.

Individuals—scientists, administrators, party members, workers, slave laborers, Saami, Nentsy, and others—make this story, so that I include several short biographies interspersed among description of the smelters, foundries, shipyards, and mills that arose above the Arctic Circle. But because this mammoth task of urbanization dwarfed even Soviet sensibilities and left significant scars on the landscape, I have deliberately left most flesh and blood for other chapters to drive the point home that urbanization of the Arctic was an exceedingly costly—and impersonal.

Beginning in the 1930s, as part of industrialization and collectivization of agriculture throughout the nation, Soviet planners, party officials, scientists, and engineers sought rapidly to subjugate rich Arctic resources. They ordered the construction of mines, enrichment facilities, electrical power stations, and mills as close as possible to rich mineral, forest, and fisheries resources. Teams of explorers combed the vast Arctic plains, the permafrost, the frozen rivers and lakes, they gathered geological samples, they marveled at the change in seasons, and throughout it all they recognized great economic potential in nickel, copper, chrome, phosphates in apatite ore, indeed virtually all elements on the periodic table, and in lumber, fish, and reindeer. They built cities to tame the resources, pull them from the earth, enrich them, and transport them to nearby smelters. From 1939 until 1989 the Soviet Arctic became an urban Arctic, the most urban circumpolar region in the world, with the most cities of any country in the world with 50,000 inhabitants or more, growing in all from a few tens of thousands of inhabitants according to the 1926 census, to perhaps a half million urban residents at the next census in 1939 on the eve of World War II, to 1.3 million Arctic citizens in 1959, 2 million in 1979, and 2.6 million in 1989, before a precipitous decline in population with the breakup of the USSR and loss of political support and economic subsidies that had pushed Arctic resource development. (See Table 5.1.)

Cities of Nickel

Monchegorsk (Monche City, founded in 1938) arises from the Monche Tundra on the Kola Peninsula about 140 kilometers south of Murmansk. Like a dozen other cities anchored in the permafrost to extract, refine, smelt, and process ore, in this case nickel and copper, Monchegorsk grew from tents and barracks, from furious digging, grading, and loading in an amorphous and road-less settlement into a typical planned Soviet city, with industry and mining built smack up against housing, schools, and commerce, and with hospitals, libraries, and other infrastructure added almost as an afterthought, the tree-lined main thoroughfare, Prospekt Metallurgov (Metallurgists Boulevard), only a temporary distraction from the tumult of the massive central factories that employed many of the city's residents.

Monche means "beautiful" in Saamese, and the surrounding mountains, birch forests, and tundra certainly inspired wonder among the first Soviet settlers in the 1930s. Birch forests occupy about a fifth of Murmansk Province, covering plains and low mountain areas and forming a narrow band between coniferous forests and tundra in high mountains. When construction on Monchegorsk began, the Lake Imandra shore was covered with dense coniferous forest, but builders pushed the first streets of Monchegorsk

Table 5.1 Population of circumpolar Soviet cities, 1926–present

City	1926	1939	1959	1979	1989	2002	2010
Amderma			3,598	4,221	5,495	650	556
Apatity	~4,000		1,3958	62,010	88,026	64,405	59,690
Arkhangelsk	77,000	251,000	284,600	256,309	415,921	348,399	356,051
Dikson			3,470	4,045	4,449	1,198	643
Dudinka		13,900	16,300	24,800	32,300	25,132	24,600
Igarka		23,300	14,311	16,335	18,820	8,627	6,141
Kandalaksha	4,195	22,172	38,222	45,430	54,080	40,564	35,659
Kirovsk		22,542	39,047	40,521	43,526	31,593	28,639
Kotlas	4,000	17,300	52,608	61,454	68,021	~65,000	60,562
Lovozero			2,330	3,397	3,638	3,141	2,871
Medvezh'egorsk		12,108	15,824	20,300	20,373	17,283	15,536
Monchegorsk		28,450	45,980	51,401	68,562	52,242	45,381
Murmansk	9,000	110,000	221,874	380,817	468,039	336,137	307,664
Naryan-Mar		13,700	13,222	23,435	20,182	18,611	19,820
Norilsk		14,000	118,000	~160,000	~165,000	110,000	~105,000
Olenogorsk			12,110	27,369	35,584	25,166	32,079
Pechora			30,586	56,361	64,746	48,700	45,543
Petrozavodsk	27,105	~60,000	~140,000	~225,000	~255,000	244,760	~247,000
Poliarnyi	408	7,068	11,354	20,015	27,635	~22,000	17,304
Salekhard			15,567	24,935	32,334	36,827	42,494
Severodvinsk		21,000	78,657	197,232	248,670	201,551	193,519
Syktyvkar	5,068	25,281	64,461	170,980	232,117	230,011	235,006
Tiksi			4,833	9,505	11,649	5,873	5,892
Usinsk			15,000	19,513	47,219	45,358	43,283
Ukhta		3,000	36,154	87,467	110,548	103,340	99,642
Vorkuta		~5,000	~176,000	200,210	~200,000	~100,000	~69,000
Total			1,397,713	2,193,062	2,757,580	2,211,512	2,073,805

Note: Rough estimates in many cases; data from a variety of online and published sources; for Vorkuta, based on data on "non-prisoner" population, taken from Alan Barenberg, personal email concerning his forthcoming book on Vorkuta.

through the forest, felling the trees for lumber. Several mountains, a few at more than 600 meters, rise above the lake and surround the city. In regions not in the path of industrial emissions moss and lichen cover boulders and mushrooms and berries fill the forests. In addition to Lake Imandra, which in places reaches twenty kilometers in width and has a total surface area of 812 square kilometers and holds 150 small islands, another thirty-seven lakes of more than one square kilometer surface area are nearby.

Monchegorsk industrial pollution, June 2010. (Photo by Paul R. Josephson)

Urban planners were blessed by the fact that nearby forests, mountains, and lakes offered recreational opportunities year-round, a time for workers to rest up, reflect on the achievements at the forge, and enjoy the steady if modest improvements in their standard of living. But as usual for a Soviet city, local industry left an indelible impact on the surroundings; the prevailing winds from the smelters denuded the trees from the mountains, killing the forest in some areas entirely, and in the absence of reliable information citizens were left to guess how their lungs must be faring if the trees had vanished. They knew from *Monchegorskii Rabochii*, the daily newspaper, that they had reached glorious norms of production. But the winds had carried uphill to the south-southwest from the smelters, all the trees are gone from the hillside, acid rain plagues the region, and in many places Lake Imandra became a deadzone. Today things are somewhat better because of the economic crisis in the 1990s and operations that rely more on western technology; it will be decades before true remediation occurs.

The industrial cities of Apatity and Khibinogorsk (now Kirovsk) at the southern end of the Imandra and Monchegorsk at the northern end of the Imandra changed the natural environment irreversibly. The goal was to

Closed mine, Kirovsk. (Photo by Paul R. Josephson)

develop strategic resources for the growing Soviet economy. In 1929 an Academy of Sciences expedition under G. D. Rikhter of the Geography Institute and the next summer another expedition under the nation's leading geologist, Aleksandr Fersman, identified exceedingly rich flora, fauna, and mineral resources in the Khibiny tundra. In 1931 the Leningrad Geological Survey Trust undertook extensive prospecting on a series of nickel deposits. Its scientists reported to Leningrad Party Secretary Sergei Kirov. In a major speech in January 1932 Sergei Kirov pushed for the Kola Peninsula to succumb to an industrialization program through the building of mines and cities in the tundra. At Kirov's initiative in the Leningrad Executive Committee a Karelo-Murmasnk Committee for the Study and Utilization of the Wealth of the Northern Region came into existence. By the fall of 1933, a new trust had organized year-round efforts to exploit copper and nickel reserves.[2] Roads and railroads came later.

Fersman claimed that the industrial victory in the Far North was Kirov's grand achievement. He was "victorious over nature, and he knew how to recast people as well." Painted as wise for this policy, Kirov also pursued draconian measures against "kulaks," wealthy peasants, in the northwest region who were forced into exile, at times into Arctic regions. In November 1932 Kirov proudly noted that by the end of the First Five-Year plan Khibinogorsk had mining, chemical, and medical technical schools, thirteen elementary and middle schools, and an evening and party school, while 600 workers and their children studied art. A radio station, a daily newspaper, four clubs, nine "red corners," a theater, a House of Culture, and eight libraries had all opened. By the time of his murder in December 1934 in the Smolny Institute, the party headquarters in Leningrad, no doubt arranged by Stalin, the industrialization of the Murmansk region was in hand: The Apatit Trust, the Kirovsk mines and enrichment facilities, and the foundations of Severonikelikel in Monchegorsk were in place.[3]

Soviet ideologues considered industrial technology to be the highest form of culture. Propagandists began to use film extensively in the 1920s to publicize cultural construction on the Kola Peninsula, and Kirov was in the camera's lens. The topics were as varied as ski racing, launching of icebreakers, catching and salting of Murmansk herring, reindeer herding activities, and so on. Film reports and photos appeared in such publications as *Sovkinozhurnal* (Soviet Film), *Sotsialisticheskaia Derevnia* (The Socialist Countryside), *Soiuzhkinozhurnal* (The All-Union Film Journal), and the children's *Zvezdochka* (Little Star). From 1939 a branch of the Leningrad news studio operated in Murmansk. The films focused on the assimilation of mineral ore, leading workers ("peredoviki") of the trawl fishing fleet, the natural environment, and the life of the indigenous people. Such black and white productions as "The Trawling Fleet of the North," "Eastern Lapland," "Kirovsk," "On the Far Shore," "The 68th Parallel," "City in the Tundra," and "Seven Islands" celebrated the economic and cultural achievements.

One of the more extensive efforts was Vladimir Erofeev's documentary "Sergei Mironovich Kirov," shown on the occasion of the fifth anniversary of his murder. The newspaper *Kino* reported on "the great role of S. M. Kirov in the construction of the Soviet Arctic. The wild nature of the cold north. The distant Kola Peninsula. And throughout flowed the creative thought of Kirov, his economic eye glancing there. The audience sees a modest wooden house where at a meeting on New Year's Eve of 1930 Kirov opened Arctic development. Three [dynamite] explosions follow. He speaks at the podium:

'There is no place which, under the capable hands of Soviet power, cannot be transformed for the good of humanity.'" The film focused on the "gigantic" Khibiny Mining Combine (where dynamite uncovered ore for processing) at Apatity, the Tulomsk hydropower station, the northernmost in world, the reconstruction of old Murmansk, and the founding of Khibinogorsk. At the state farm "Industry" beyond the Arctic Circle even vegetables grew. "All of this was created and built on the orders of the great Stalin under the direction of Kirov."[4]

The violent, sudden, industrial essence of urbanization in the Arctic Circle becomes most clear at the cities of nickel and copper that grew out of Kirov's determination to tame the far reaches of the huge Northern Region (Severnyi Krai) and the Kola Peninsula. The Severnyi Krai was the major administrative unit of the Russian northwest from 1929 to 1936; in 1929 it consisted of Arkhangelsk, Vologda, and the Northern Dvina provinces. From 1930 to 1935 the authorities made constant changes to the region with districts abolished, established, and renamed,[5] but they did not change their determination to extract its resources as rapidly as possible.

As quickly as manpower allowed—they had little equipment and in any event it was hard to bring machinery from the faraway Murmansk-Leningrad railroad—they built cities around new mines and factories. In May 1935 the construction of Severonikel commenced, and amorphously, like in the Wild West, Monchegorsk began to form. The "Rudstroi" Construction Trust dug the Niud and Sopcha mines under Dmitrii Matiushkin who became director of Severonikel in 1960s. The workers and managers lived in tents and later in drafty wooden barracks. They had one compressor, one tractor, and horses. Sometime later, ZIS dump trucks, four more tractors, and one bus appeared; drivers ventured out across the ice and occasionally broke through. In March 1936 they dug the foundation pit for first school. The "city" filled with more wooden houses. Soon concrete mixers, cranes, and three excavators arrived.[6] By 1938 30,000 laborers, bosses, and prisoners crowded into the city's inadequate housing of roughly two square meters (under twenty square feet per person) on average. A hospital, polyclinic, kindergarten, workers' club, bakery, vegetable storage facility, and fire department were built next. Stone construction began on the eve of the war, but waited until the late 1940s.[7] When the Nazis invaded, the city fathers ordered machinery and equipment to be evacuated to Norilsk, Dzhezkazgan, and Orsk, and they filled the mines with water to make them unusable. They removed window frames, doors, and radiators; many of the workers went to the front to fight the fascists, leaving a small group behind. In 1942 operations began again but on a modest level. During the Cold War nickel production accelerated; by 1959 there were 45,200 inhabitants living in 250,000 square meters of apartment buildings (now fifty-five square feet per person).[8]

Vasilii Ivanovich Kondrikov, director of the "Apatit" Trust and "Kol'stroi" (the Kola Construction Trust), pushed the design and construction of "Severonikel" and Monchegorsk until his arrest and execution in 1938. Like many of the innocent people who later became "enemies of the people," Kondrikov was devoted to Bolshevism. Kondrikov served in the civil war on the southern front and was appointed head of the political department of the Bashkir Cavalry. Demobilized, he was sent to Nizhnii Novogorod to head the provincial department of communications, to Tver as head of the administration of post and telegraph, in 1921 to Petrograd as first assistant of the head of the regional administration of communication, and then to administrative work in trade unions, the Industrial Bank, and the Communal Bank, all with but seven years of education. During the New Economic Policy (NEP) of small-scale capitalism, party officials and bureaucrats recognized his intelligence, entrepreneurial spirit, and ability to make good investment decisions with the state's capital. Kirov said, "Give him a ruble and send him to America—in a year he'll return a millionaire." In spite of his talents, in the feverish atmosphere of Stalinist politics Kondrikov was criticized for being unscrupulous in pursuit of his goals and "rude" to subordinates. Recognizing that his career—and perhaps life—was at risk in Leningrad, Kirov sent Kondrikov in November 1929 to the Kola Peninsula as temporary head of what became Apatit Trust. At that point, neither Kirov nor he likely understood the future scale or responsibilities of these operations. Prospecting had just begun and the theoretical questions of enrichment of apatite ore had hardly been addressed. When Stalin came to power, all bets were off in terms of past mistakes and future promise, but in the wild Far North Kondrikov managed to found two cities: Khibinogorsk and Monchegorsk.[9]

At the end of 1929, Kondrikov joined Kirov and Ivan Kadatskii, the chair of the Leningrad Central Executive Committee, on a trip to the Khibiny tundra. He returned full of ideas. After his transfer to Khibinogorsk,

Kondrikov's wife, Inna Tartakovskaia, rarely visited him because of the absence of housing and because she had been trained as a ballerina, not a mining specialist. Yet she was excited by the discussions of industrialization of the tundra that unfolded nightly in their Leningrad apartment. Kondrikov saw Kirov almost every morning when in Leningrad and once complained to him that he did not understand enrichment technology sufficiently. Kirov instructed him to teach his wife, so that she could help him. She took courses at the Mining Enrichment Technical School. If she mastered theory, then she complained to Fersman that she had trouble with practical problems. He introduced her to a young scholar who worked with Fersman in Committee for the Productive Forces who served as her consultant.[10]

Like Trotsky who commandeered an armored train during the Civil War, Kondrikov had a special train at his disposal to move around quickly on the Kola Peninsula from Kirovsk to Kandalaksha, Olenogorsk, Niva, and Apatity, and to Monche on a newly constructed spur. In the summer the train provided refuge from mosquitoes and black flies. Men tramped around in dirty shirts and rumpled jackets, their boots tracking mud from the construction site into the wagon, their hands loaded with briefcases with documents and drawings. Only when Kondrikov went home would the wagon occupy a siding in Kirovsk. Kirovsk itself was quickly transformed from a mine and tent settlement into a real city with an enrichment facility. The recently opened Niva power station sent electricity across the Kola Peninsula to the burgeoning operations that Kondrikov celebrated like a new daughter. In addition to this station, Kirovsk, and Monchegorsk, Kondrikov was responsible for the construction of the Chemical Combine in Kandalaksha, in September 1935 also acquiring the portfolio for the Severonikel factory.[11]

In the fall of 1930, Tartakovskaia moved to Khibiny; she and her husband lived in barracks. Later they moved to a barracks in "19-kilometer," a settlement named for its distance from the railway line, that they shared with Georgii Geberi and Iakov Tsagareli, who were deputies of the trust. Their apartment again served as a meeting place for new arrivals; at all times the copper kettle was hot and tea accompanied discussions about the next day's activities into the late hours. The most frequent guests were representatives of Mekhanobr. Tartakovskaia was also frequently involved in activities of the newly founded Khibiny base of the Academy of Sciences, sometimes as secretary of meetings of scientists.[12]

In the construction of Kirovsk and Mochegorsk, Kondrikov reported to the Commissar of Heavy Industry, Sergo Orzhonikidze. One reason to build Severonikel rapidly was that the Ufa Nickel Factory hardly met national demand and ore had to be purchased abroad to stoke its furnaces. By the summer of 1934 the Communist Party leadership was convinced of the utility and promise of the Monche operations; the Commissariat of Heavy Industry ordered the Leningrad Design Bureau of Nickel Production and Soiuznikelikelolovoproekt (SNOP, which was actually served by prisoners in SLON) to assist the operations in every way. By end of 1934 the reserves of nickel reached 70,000 tons. Another commission arrived and recommended building an enrichment factory. In spring 1935 Ordzhonikidze ordered redoubled pace of construction to produce 10,000 tons of nickel and10,000 tons of copper annually, while setting up Severonikel as independent from Apatit Trust.[13]

Kondrikov's position became more complex with the assassination of Kirov and Andrei Zhdanov's appointment as Leningrad Party Secretary, for if Kondrikov and Kirov understood each other well, Zhdanov kept his distance. Kondrikov sent Zhdanov a ten-page typewritten report that described the great achievements in Monchegorsk, Kirovsk, and Apatity, and he included analysis of factors that slowed efforts to have achieved even more. This narrative may have contained the kernel of Kondrikov's sad fate. He called for the unification of the Kirovsk and Monchegorsk operations in one combine because it was hard for him to direct two separate "economic organizations, groups of employees, construction materials, and so forth." He urged that Kandalaksha's construction be resumed. Kondrikov thus seemed to request that he officially be given power over a huge, rapidly developing region, as if he had political aspirations. Western newspaper referred to him as a "polar prince." As the report lay on Zhdanov's desk, Kondrikov's difficulties began in earnest.[14]

The deputy director of Apatit Trust for technology, Kasparov, with whom Kondrikov received an Order of Lenin in 1934, wrote a complaint against him that he sent directly to Stalin and Zhdanov, accusing Kondrikov of unprofitable operations, profligate expenditures, poor planning, disrespect for party workers, and "fascination" (!) with molybdenum, nepheline, and other incidental minerals. The local newspaper more accurately reported on the situation: "The entire country built gigantic new factories, the children of the Stalinist five-year plans, railroads, bridges, the marvelous Moscow Metro, grandiose canals which joined the White Sea with the Baltic and the Volga with the Moscow [Rivers]." Yet in spite of these marvels, on February 11, 1936, the Council of Ministers resolved to accelerate work and lower construction costs. The newspaper ominously noted that significant challenges remained, including the fact that "enemies of the people in every way possible brake the assimilation" of modern methods, on top of which workers lacked the proper "culture" of labor.[15] A May 17, 1936, party meeting considered Kondrikov's personnel folder in his absence. Hearing wind of this, Kondrikov had journeyed to Moscow to meet Ordzhonikidze to defend himself. But Ordzhonikidze for some reason could not help, and on July 31, 1936, the Leningrad Obkom recorded a severe reprimand. The denunciations of Kondrikov accelerated, and he had few allies: Ordzhonikidze committed suicide in February 1937, likely in protest of the purges, his wife was arrested and shot shortly thereafter, and the entire leadership of the Murmansk regional party committee was arrested.[16]

The NKVD arrested Kondrikov at the height of the Great Terror on March 15, 1937, in Monchegorsk. People in uniform met his train, sat quietly in his carriage, and accompanied him home. They searched the apartment and took old books and documents. They carted him off to Leningrad where at his first interrogation on March 21 they accused him of creating a counterrevolutionary Trotskyite-Zinoviev terrorist diversionary-wrecking organization that he allegedly joined a month after Kirov's murder. Yet another fabricated document indicated he joined this organization in the late 1920s when at the Communal Bank. Over three months, they interrogated Kondrikov constantly in a conveyor; he never confessed nor slandered others, so that this counterrevolutionary organization consisted of one person. On August 25, 1937, Kondrikov was found guilty by a Military Tribunal of the Supreme Court of the USSR, sentenced to death, and shot the next day.[17] He was rehabilitated after the 20th Communist Party Congress in 1956 almost twenty years later.[18]

Scars of Nickel

Metallurgical combines—and their waste—now dot the Arctic. One could explore the history of Norilsk, Vorkuta, or a dozen other cities to understand how the Soviet determination to conquer the tundra had a series of familiar features that included pushing workers and nature and leaving scars behind.[19] To take one example from the Kola Peninsula, in 1939 at the "Kaula Kotselvarra" mine, a Canadian–Finnish firm "Petsamonikel" opened that operated until 1944 with its output shipped to Germany for further processing. After the war, now "PechengaNikel," it fell on Soviet territory in the city Nickel (Nikel) that, having been taken from the Finns, was one of the few such strategic enterprises in the USSR so close to a border. In November 1946, with the help of metallurgists from Severonikel in Monchegorsk and Iuzhuralnikel in Orsk, the first nickel was produced. Local mining expanded with several new deposits opened by 1958, as did an enrichment facility. With the opening of the "Zhdanov" mine, the city "Zapoliarnyi" opened, followed by an enrichment facility that opened in 1965, and the "Vostok" and "Severnyi" mines in 1971.[20]

Although crucial to national postwar reconstruction efforts, tight budgets and other priorities slowed PechengaNikel production. Each region competed with every other region, and of course Moscow, the central industrial region, and Ural factories received the lion's share of investment under Stalin. The director of PechengaNikel, Shchelkunov, wrote the deputy minister for non-ferrous metallurgy, S. A. Raginskii, about the need for twenty million rubles to finish construction and meet targets. Yet the budget for equipping the metallurgical foundry for 1945 was but 900,000 rubles, with only 250,000 allocated; for mine equipment—1,150,000 rubles with only 650,000 rubles designated; and for the transformer substation—100,000 instead of 650,000 rubles. On top of this, the funds for apartments and daily life were completely insufficient at 1,600,000 when 2,700,000 rubles were needed for housing, workers' club, school, and hospital.[21]

In September 1945 industrial officials established a production level for PechengaNikel of 6,000 tons of nickel based on mining of 210,000 tons of ore per year and as quickly as possible to 12,000 tons of nickel a year with the mine producing 420,000 tons of ore per year. To achieve this rapid increase, the factory would need a significant upgrade in electrical service from Kolenergo hydroelectricity. All of this required reopening mines at Kaula and Kamikivtunturi, establishing a new lumber mill on the lake, a quarry for quartz sand, ports at Linakhamar and at Trifonovo, and a road from the port to the combine, as well as power lines, transformers, water supply, sewage, boilers, and still more housing.[22]

The planned economy was anything but planned. A decision to advance production in one sector nearly inevitably meant bottlenecks in another. Each manager of every burgeoning enterprise was pushed to increase production—and pushed himself to meet and exceed targets. This required more regular transportation, more electrical energy, better communications. This never happened easily. Although Kola Peninsula electrical energy was directed toward industry, there was not enough to go around. As of March 10, 1948, work to get the Finnish-built Ianiskoski hydropower station that had been dynamited by the Nazis back on line to serve PechengaNikel was "entirely unsatisfactory," not in the least because of huge cost overruns.[23] A second smelter that came on line at PechengaNikel in 1949 was the nickel that broke the reindeer's back. When there was high water on the Tuloma River, then simultaneously both furnaces could fire because the hydroelectric power station operated at near full capacity. But when low water, from October to May, the grid could not produce enough electricity for the second furnace.[24] For these huge undertakings not even such crucial equipment as trucks was available; they were short dozens of vehicles.[25]

In October 1946, N. V. Trofimov, the director of PechengaNikel, wrote the minister of non-ferrous metallurgy to complain that the combine did not even have the resources to clear snow from the main roads in winter. One road ran 100 kilometers to the port, while another pointed fifty-seven kilometers toward the lumber factory. He claimed that he would need to put at least 600 or perhaps up to 2,000 men to work on snow removal and that this would completely stop construction and close the entire auxiliary facility. The director requested that troops be put to work on this task without delay—already a half meter of snow had fallen.[26] Problems with road maintenance and snow removal persisted. On February 2, 1948, the main engineer of the Main Administration for Nickel and Cobalt, A. Mironov, informed Trofimov that the Ministry of Internal Affairs would be responsible for repair and maintenance of the road to the border with Finland (and protecting the border from enemies), but that PechengaNikel would be responsible for keeping the road clear because it had two powerful "Snogo" snow removal machines.[27] But trucks were still in short supply. Mironov wrote Trofimov in September 1948 about his request for thirty-five trucks. Of the forty diesel vehicles in the combine's autopark, thirteen were in need of repair, and of sixty-seven gasoline trucks, fourteen were in need of repair. Mironov informed Trofimov not to expect additional vehicles until 1949 and to focus on repairs so that each vehicle could complete at least twenty trips per month.[28]

On the Kola Peninsula—and elsewhere across the Soviet Arctic—the worker and his family paid with their health, while trees wilted on the spot. Industrial safety was a great concern in theory, but in reality workers were at great risk because of the absence of emergency plans or equipment. The central government took the position of ordering factories to create fire departments and rescue teams only in the late 1940s, but did not provide them with the funding to do so.[29] The effort to create an emergency rescue team at PechengaNikel and its mines took more than a decade—and this in the worker's paradise of safety and socialism. In 1949 the nonexistent emergency rescue team needed respirators, pumps, oxygen indicators, tanks of oxygen, safety glasses, and walkie-talkies at the Kaula mine. As of May 1952 the combine still did not have a building for the emergency rescue team. While many of Kaula operations were open pit, much of the work was underground. Repeated requests to organize a rescue unit with a full-time staff and facilities for the equipment got nowhere.[30] No one accepted responsibility for taking action. Gipronikel, the All-Union Nickel Design Institute, founded in Leningrad in 1934, was focused on production, not safety, and so offered up incomplete projects for industrial safety only after its engineers were prodded repeatedly to do so. At PechengaNikel parameters for ventilation constantly changed, but never adequately addressed evacuation of mine gases or the pumping in of fresh air, while 500,000 rubles on a project had already been spent. The exposure of workers to heavy metals and particulate far exceeded norms[31]—and their wives and children lived only a few kilometers away.

While the central authorities reasonably held plant officials responsible for safety, they unreasonably did not provide resources or incentives to address safety. The administration of PechengaNikel wrote central government officials in the non-ferrous metal industry, on September 9, 1953, to report on several safety measures: they had built a ventilation shaft, repaired and cleaned special worker clothing, replaced compressors, and provided better illumination both in the mine and at the factory. Yet most other buildings were not in compliance with the law; the garage for rescue equipment was not finished; ventilators at the Kaula mines were not switched on; roads and sidewalks on the territory to provide emergency access had not been built; warning lights did not operate.[32] Two years later the ventilation system of the mine still was not working properly, areas properly indicated for loading and unloading were not demarked, and no emergency evacuation plans had been formulated.[33] In fact, there was no emergency crew! Workers shared equipment, clothing, and lockers. In one and the same locker several laborers kept dirty and clean clothes. There was no "de-dusting" or laundry equipment, let alone an infirmary.[34]

Sometimes workers gave up and tried to lose themselves in hangovers. On February 25, 1946, the brigadier of the PechengaNikel auto park signed out for three liters of brake fluid—glycerin diluted with spirits—which he and two friends drank. He died, the second was in serious condition, and the third managed to crawl away.[35]

Finally, in the Khrushchev era officials took real interest in concrete measures to ensure safety. As in other countries, this required the separation of promotional from regulatory functions of bureaucracies, for example, the breaking up of the U.S. Atomic Energy Commission in the 1970s into the Nuclear Regulatory Commission and the Energy Research and Development Administration. In the USSR, for mining, one step, apparently, was the transformation of the Mining Technical Inspectorate into Gosgortekhnadzor and the requirement that emergency plans be constituted that included evacuation plans and emergency contingencies. These necessitated the determination of minimum width of shafts with or without rails, regular inspections, the presence of fire extinguishers, and operation ventilation systems.[36]

Monchegorsk, Zapoliarnyi, and Nikel continued to grow in the Brezhnev era, producing more and more nickel and copper. Levels of pollution also continued to grow with terrible impacts on public health and the environment. There were a few positive signs, however. A forerunner of an environmental protection agency, the Hydrometeorological Agency gained expanded powers for monitoring background air and water pollution levels in the 1960s. In 1980, the government introduced two laws that attempted to put teeth into previous regulations by coordinating the numerous requirements adopted by various ministries and agencies. One law, "On the Protection of the Atmosphere," set air quality standards for the country as a whole, as opposed to industrial regions alone. The law also stipulated setting up water and air pollution permits and specified the creation of an "environmental cadastre" of all enterprises in the country. A second law, "On the Protection of Animals," followed. The USSR finally also introduced a system of fines for pollution. But it remained cheaper to pay fines than risk punishment for not meeting production targets.[37] Hence, in spite of the creation of a number of agencies with environmental functions, real decision-making power remained in the hands of economic ministries like those connected to nickel production. These were huge organizations with great bureaucratic momentum, and they were hardly interested in pollution control.

Perhaps the nature of technology itself rather than the nature of Soviet socialism led to the extensive environmental degradation of the Arctic landscape near and around the mines and smelters that arose on the tundra. Across the American West large-scale technologies were central to the effort to push nature to give up its water and mineral resources—and to the effort to push Indians aside to enable massive operations to remove coal, copper, and other ore. As on the Kola Peninsula, so in Colorado's mines mineral wealth created tension between managers and workers, party organizers and managers, and the mining cities, nature, and indigenes. The workers suffered at the hands of the bosses, whether the bosses were socialist or capitalist, because the bosses wished to extract wealth from the ground as cheaply as possible. The mines drew in tens of thousands of laborers, coercively under Stalin. In both systems, physical dangers abounded and the bosses were late for one reason or another in developing mine safety regimes, no matter the moral or regulatory impetus to do so. In both, mining operations controlled housing, clubs, schools, and stores—in the U.S. case, at the very least into the twentieth century. In both, crucially, resources served purposes and persons far away from the mines and smelters themselves.[38]

Huge open pit mines poisoned the surrounding lands far and wide. Technology for excavation and processing, what I call brute force technology, was so efficient that it could work with trace deposits while removing entire mountaintops and polluting entire valleys. Enrichment facilities promptly extracted ore from deposits. Vast hills of tailings and riff-raff slowly leached heavy metals into the environment. In the haste and violence to remove overburden and extract ore, the mining concerns destroyed ecosystems and accelerated erosion, processes that Tim LeCain, in a study of copper mining in the United States, calls "mass destruction." These open-pit mines seemingly extended to the horizon as part of efficient removal but thoughtful devastation. Similarly, many Kola Peninsula mines were open pit, leading to the same kind of "mass destruction"—mechanized exploitation of low-grade ores in search of copper for electrical, telephone, and other industries.[39] In a word, large-scale technologies dominated the landscape.[40] In all these ways, the determinist argument that technologies rather than polity or economy are the crucial feature of environmental degradation across the globe seems to carry great weight.

Yet the metallurgical industry—all industries—in the USSR differed significantly from those in the United States. Having stultified the innovation process through heavy-handed planning mechanisms and tying severe penalties—including prison terms—to failure to meet targets, Soviet industry was more labor-intensive and less capital-efficient than western. The replacement of workers by machines in Soviet mines and factory sites was much slower than under capitalism. And the workers for the Soviet mines were requisitioned to the site, rather than drawn by jobs.

In 1998 specialists of PechengaNikel and Severonikel suggested that such conditions of pollution were a "usual and regular phenomenon" in spite of the impact on the health of local residents—and the constant complaints of the residents. Such pollution has continued for decades because of inadequate and outdated technology, while Rostekhnadzor of Murmansk, the regional inspectorate, has refused to take appropriate action and instead raised the allowable limits—with the expected health and environmental consequences. While required by federal law in Russia, improvements in pollution control and monitoring and worker safety have lagged significantly. In June–July 2007 the concentration of sulfur oxides in Zapoliarnyi, Nickel, and Monchegorsk was five times the legal limit. This had an impact not only on the health of the inhabitants in terms of respiratory ailments, but also contributed to acid rain in the region.[41] Instead of openness and legality, the people of Murmansk Province get obfuscation and poor health—as do their Norwegian neighbors who are unfortunate enough to hug the Russian border. In 2012 Zapoliarnyi was part of the largest mining company in Russia, Norilsk Nickel, and still polluted substantially—and its owners without remorse.

Molotov's Shipbuilding City

Stalinist industrialization extended far into Arctic and subarctic regions, and with it the Stalinist terror. Even as party officials and managers and engineers strived honestly to meet targets for building new cities and factories and mines, they faced all around them the constant threat that they themselves might be accused of being "enemies of the people" if they could not against all odds meet targets. Ivan Latunov and Savelii Loginov successfully defended the interests of Arkhangelsk province as secretaries of the provincial Communist Party committee ("Obkom") from 1948 to 1960. They worked with local authorities to support the growth of shipbuilding Severodvinsk, pulp- and paper-producing Novodvinsk, and reindeer- and fish-processing Naryan-Mar. Their predecessors sought to achieve industrial growth and collectivization, too, but then disappeared into the meat grinder of repression. Dmitrii Alekseevich Kontorin, who as an eighteen-year-old joined the Cheka, fought in the civil war, served in the Komsomol, was a party instructor in Arkhangelsk region, moved up through the apparatus in the mid-1930s, became secretary of the Obkom early in 1937, and was arrested and shot as a counterrevolutionary and enemy of the people at the height of the Great Terror.

Kontorin's successor, Aleksandr Fillipovich Nikanorov, served the party in the provinces, was "advanced" as a good communist to the Sverdlov Communist University and the Institute of the Red Professoriat with a specialization in agriculture. As a devoted agitator and propagandist, he entered the Leningrad Party apparatus after the assassination of Sergei Kirov, whose murder served as a pretext to purge it of tens of thousands of devoted party members. He was ordered to Arkhangelsk to clean up Kontorin's mess. As the newly designated secretary of the Okbom, Nikanorov willingly and energetically carried out a purge of the Arkhangelsk party military, trade, and government organizations, finding criminal negligence and espionage everywhere. He seemed to relish rooting out "enemies of the people" and enjoyed dressing down party members publicly during purges. In 1937 alone, reflecting the same pattern throughout the nation, sixty-two leading obkom officials were purged, only ten were left, and roughly 1,300 party members were kicked out. Of course, Nikanorov himself was relieved of duties a year later for unsatisfactory leadership in the lumber industry so important to Arkhangelsk region and was arrested and shot in 1940. Kontorin and Nikanorov were in Arkhangelsk when Molotovsk was established using gulag prisoners.

Viacheslav Mikhailovich Molotov (the "Hammer," Molotov's family name was "Skriabin," but he was not related to the composer and pianist), a sympathizer of the Bolsheviks since the revolution of 1905, a supporter of Stalin from the mid-1920s, Stalin's foreign minister and negotiator of a Non-Aggression Treaty between the USSR and Nazi Germany (the Ribbentropp-Molotov Pact), knew nothing about shipbuilding. But in the authoritarian USSR all leaders would have favorite ships, icebreakers, airplanes, and cities named after them. Molotov got two in one: a shipbuilding city.

The study of the creation of almost any new city in the Soviet Arctic or "socialist reconstruction" of an existing village like Kandalaksha in Karelia gives a sense of the frenzy of urbanization. The experience of the closed shipbuilding city of Severodvinsk indicates that even in a priority area of the Soviet economy—the military sector—industry moved ahead in fits and starts, while social services and housing lagged far behind. This was the result of the physical location of Severodvinsk on swampy lands on the White Sea shore, Arctic winds off the water, and the bottlenecks endemic in the Soviet planning system. But while geography played a role, the Soviet planning system was the real culprit in the lag in planned urbanization.

Since 1955 called Severodvinsk, this closed city of 200,000 people in 2010 (down from its height of 250,000 in 1989) is the home of nuclear shipbuilding in Russia. Since the 1980s the city administration has fulfilled the needs of citizens to provide good housing, museums, and clubs. Three major shipbuilding facilities, Sevmash, Zvezdochka, and Severnyi Reid, dominate employment. The city is laid out on a Cartesian grid with ponds and parks next to many residential areas; some nuclear waste was haphazardly stored nearby. In addition to the center of the city with the municipal building, Lenin Square, a major hotel, and shops nearby, several residential neighborhoods ring the city, including on Iagry Island where prisoners who built the city in the 1930s, 1940s, and 1950s were incarcerated.

Special permission and permits are required to enter Severodvinsk and other "closed" military cities because of their strategic importance, and most foreigners are accompanied by official hosts. Severodvinsk was, until 2008, a kind of second home for me in Russia. I had been American president of the Portsmouth-Severodvinsk Sister City organization, the P/S Connection, developed close friendships and good working relations with Severodvinsk officials, and helped to establish fairly regular exchanges of high school students and some professionals in the areas of environment, business, and the arts. But, although I have no military or other security interests whatsoever, in 2008 the new Russian KGB (the "FSB," the Federal Bureau of Security) decided that I could no longer visit Severodvinsk, even at the request of the mayor of Severodvinsk. The FSB will be an ominous and vigilant police presence in all strategic regions of Russia, which now includes virtually all of the Arctic.[42]

Much of the literature on the "closed" cities focuses on their expansion and proliferation during the Cold War. Many if not most of them were connected with the military-industrial complex, for example Krasnoiarsk-26, Cheliabinsk-40 (now Ozersk), Obninsk, Pushchino, and Dubna. According to one scholar, "It is not known how many cities were closed during the Soviet era, for information about them was highly classified by the Soviet authorities, and has not yet been entirely declassified." The government even created an official designation for these military, industrial, secret, and otherwise "closed" cities, the Closed Administrative-Territorial Formation (ZATO in its Russian acronym). Many of the cities never appeared on maps; many were known only as a post office box number. The people who lived and worked in these cities needed special passes to enter and leave. In his *Memoirs* Andrei Sakharov discusses his experience in the "installation" in Sarov that gives a sense of the borders, boundaries, gates, passes, prisoners, and other standard fare of control and coercion in the closed cities.[43] But these cities, strictly speaking, are not a postwar phenomenon, but date to the first-five year plan and the reintroduction of the Tsarist system of internal passports, tied to the Soviet practices of ubiquitous ID cards and labor books, especially in such strategic cities as the shipbuilding center Molotovsk and because of the creation of gulag camps cum cities that served construction ends.

To go to a "closed" city in Russia today thus cannot help but remind us of 1936 and 1937 when, in Molotovsk, each morning, battalions of emaciated prisoners were marched from Iagrinlag, a camp on gray Iagry Island, out of their thin wooden barracks into the wet, windy cold, across rickety wooden bridges to the shipyards, to build structures and docks and dry docks to expand military shipbuilding capacity. The prisoners continued to march across the bridges until late 1961—when the Khrushchev reforms were well under way and when virtually all other gulag camps had long since been emptied. From this time, nuclear Severodvinsk became entirely a post-Stalin shipbuilding facility of free workers who rarely moved elsewhere in the USSR and with a strong secret police presence.

In 1936 Communist Party officials ordered the construction of a shipbuilding yard about forty kilometers northwest of Arkhangelsk along the Northern Dvina delta and White Sea. Severodvinsk was built by thousands of volunteer enthusiasts and a greater number of prisoners in the Sevdvinlag and Iagrinlag labor camps. By the eve of World War II Severodvinsk shipbuilding facilities had begun to launch ships, and in the late 1950s it became the home of nuclear shipbuilding in the USSR. In Severodvinsk, as in other Arctic cities, families lacked housing, children filled crowded schools, and automobiles and trucks toiled along muddy, nearly impassable roads until well into the 1960s; the main road from Arkhangelsk to Severodvinsk was a substandard, dangerous, gravel road into the 1970s. Supply bottlenecks and worker ennui led to failures to meet targets, and Arctic climate also failed to cooperate.

Severodvinsk grew rapidly as a central facility in the military-industrial complex. Yet everything needed for construction of factories or housing—lumber, bricks, cement, stone, gravel, narrow gauge rail—was in short supply. Three massive shipbuilding factories that dominate the industrial landscape, Number 203 (the Shipbuilding Factory), 402 (today Sevmash, *Severnoe Mashinostroitelnoe Predpriiatie*), and 893 (Zvezdochka, "Little Star") could not themselves meet ambitious targets, although they launched the Soviet Navy into the Cold War an equal to that of the United States. The USSR/Russia manufactured 245 nuclear submarines through 1997, more than the rest of the world combined, most of them in Severodvinsk, most with two reactors, and also with the largest number of radiological accidents, including the sinking of the "Kursk" in August 2000 and the loss of all 118 men on board.

Throughout the postwar period of their expansion, the Sevmash, Zvezdochka, and the Shipbuilding factories lagged behind capital construction and production for all of these reasons: shortages of workers, insufficient construction materials, and lags in delivery and installation of cranes, presses, and other machinery.[44] Expansion of foundries required more than 300,000 m^3 of lumber per year from the Kudemsk forest some forty kilometers away. Lumberjacks could not fell and yard trees fast enough to meet demand because their operations were not mechanized. One Roshchin, the deputy administrator of Factory 203, reported that they had a stable of 250 horses of which 130 were used in forestry operations. But most of the reins, saddles, and bits were ragged, and at least twenty of the horses were destined for the glue factory—or the dining room table.[45] If people were treated like animals, then how were animals treated?

To speed construction, other massive operations had to fall into place. In June 1952 Gushchin, the director of Factory 203, asked the provincial bureaucrats responsible for forestry and for machine building to facilitate construction of forty kilometers of narrow gauge railroad in the Kudemsk forest and construction of seventy-five apartments in a nearby village to house workers along with garages, railroad depot, and repair shop and supply a generator at 75 kW. The lumber operation also required tons, cubic meters, and square meters of bricks, concrete blocks, cement, concrete, stone, gravel, sand, prefabricated concrete forms, lumber, windows, pumps, and a way to transport all of this stuff.[46]

In the 1950s Nikita Khrushchev launched a campaign to mechanize, automate, and introduce industrial methods of production in every sector of the economy to overcome these endemic problems. Automation and mechanization would envelope the nation including Arctic regions. The directors of Severodvinsk's shipbuilding factories hoped that these new methods, for example large-panel concrete construction properly adapted to the climatic conditions of the White Sea, would enable them to move forward with factory expansion and apartment construction. But labor and capital inputs were tight. Iagrinlag prisoners had been spread thin in forestry, brick, and other operations; the city fathers asked for another 7,000 laborers. On top of this, throughout 1952 and 1953 housing lagged at roughly 70 percent of targets, and ten eight-story apartment houses, three dormitories, a kindergarten, and other facilities were far behind schedule.[47]

In the absence of capital and machinery, one of the traditional ways to improve production was to raise the qualifications of workers. The USSR commenced a second nationwide effort to expand education in the Khrushchev era, an effort that had lagged in recent years after tremendous achievements in the early Stalin era to overcome illiteracy and make universal elementary education accessible. Although many workers in the 1920s and 1930s required remedial language, mathematics, and other courses, the labor force was far better educated on the eve of World War II than in 1917. Yet things had stagnated in the late Stalin era. In 1952 the main Severodvinsk construction trust commenced a series of seminars as part of the effort to raise qualifications. One seminar focused on winter construction techniques and garnered thirteen students; another for labor engineers involved the study of technical norms to inculcate notions of efficiency; and another for advanced railway specialists focused on the study of laws of traffic and signals that initially had zero students, but with some cajoling fifty-five students joined in.[48]

At virtually every economic unit, the authorities offered training programs to raise the qualifications of workers to increase the productivity of labor. This was common sense given that it is less expensive to hire and train workers than constantly to seek new workers, ignore the older ones, and deal constantly with turnover—as capitalist and socialist managers alike should know. In the late 1940s and early 1950s in Severodvinsk a training program at Sevmash organized through the Leningrad Institute for Raising of Qualifications of Technical Personnel sought to engage thousands of individuals studying for all positions—brigadiers, mechanics, Taylorist time-and-motion specialists (*normirovshchiki* or rate setters)—all to get more out of the workers. They would take courses at technical and Stakhanovite schools as well. Programs for specialists and white-collar workers also expanded rapidly into the thousands of personnel.[49]

Even among the white-collar workers, however, problems with "labor discipline" were growing. To counter this trend they established mass political agitation with "red corners" in every industrial nook and cranny; in 1952 political officers delivered 218 talks to 11,675 workers on such crucial aspects of port life and shipping as "The International Position of USSR," "Hero Projects of Communism," "Events in Korea," and "On the 19th Communist Party Congress" (in 1951, the first party congress since the 18th congress in 1937.[50] Training programs promised results over time, but there was a lot of work to do. Even among party officials special education lagged. Or to put it more bluntly, archival materials indicate that it was more difficult for almost any organization in the Arctic regions to attract and retain employees, let alone leading personnel, who had finished higher education than for any other region of the country.[51]

Even shipbuilding yards that fed on the Cold War had trouble hiring and keeping personnel. Roughly one-third of shipyard employees left their jobs each year, and only a small number because of illness or injury. High turnover was to be expected; most of the laborers were unskilled, and I suspect many were barely literate and went from job to job and bottle to bottle. Few individuals with higher or special education filled the offices near the docks. Only one-fifth of employees were engineers or white-collar workers with some higher or middle education, and of 190 port workers with engineering or technical responsibilities, only sixty-six had higher and middle education.[52]

Military significance did not guarantee labor or capital inputs—let alone housing. Obkom Secretary Latunov implored the Council of Ministers to assign more workers to Zvezdochka to complete construction in the fourth quarter of 1954. The factory required 6,580 workers, but on October 1 there were only 5,790, of whom 725 were soldiers and 1,091 prisoners. By the end of the year 600 soldiers would be demobilized and 400 prisoners released, which left Zvezdochka with only three-quarters of its labor force—and still without housing for them. Latunov asked that the Ministry of Internal Affairs drag another 1,700 prisoners to Iagrinlag because there were barracks available there.[53] Shipbuilding orders rained down upon Sevmash and Zvezdochka from military ministries, yet the town fathers could not promise additional housing until years later.[54]

Latunov did what he could. Ivan Sergeievich Latunov (1906–1970) was born in a village in Samara province, the son of peasants. He joined the Communist Party in 1930. He graduated from the Ural Forestry Institute in 1935, moved to Syzran in Kuibyshev Province to work in a factory, and in 1938 became secretary of the Syzransk gorkom (city party committee), likely because of the arrest and execution of the previous secretary. He entered the personnel department of the Central Committee in 1938 and became second secretary of the Arkhangelsk obkom until 1946 except for a break from 1941 through 1945 when he served in the Army. From 1946 he was the chair of the Arkhangelsk Executive Committee, and from 1948 through 1955 the secretary of the Arkhangelsk obkom. Latunov supported the Khrushchev reforms and served in the Central Committee of the Communist Party from 1952 through 1970.

Stalin's death triggered uprisings in socialist Germany, Poland, and Hungary and in Vorkuta and elsewhere in the USSR. It may have triggered an assassination attempt on Latunov on May Day in 1954. In spring 1954 the opening of navigation was delayed; the Northern Dvina remained frozen, and the first steamship from Kotlas arrived only May 8. The May Day parade was held on a cool, windy day, and soldiers and officers wore overcoats. Leading figures of the province and city—first secretaries of the obkom, gorkom, and ispolkom and representatives of industry—stood on a tribunal. First, the soldiers passed by, then workers and heads of enterprises, institutes, and organizations. Suddenly a man of average height jumped out from the columns of lumberjacks and workers, calmly took out a pistol, and fired at the tribunal. The deputy chair of the City Executive Committee, a general, and a member of the obkom died. The assailant, a worker demanding "freedom" for the Russian people, managed to shoot a captain who tried to subdue him, but the bullets hit a matchbox in his coat. Latunov escaped unharmed. The newspapers never reported this incident, which was known only by word of mouth until 1990.[55] This kind of incident, which likely reflected protest against miserable living and working conditions, certainly had an impact of Latunov's attitudes, and he struggled to build housing for workers in Arkhangelsk province until his transfer in November 1955 to the position of secretary of the Vologda obkom.[56]

By the mid-1950s the Zvezdochka shipyard needed forty to fifty million rubles annually to expand and modernize machinery and equipment, not to mention hire employees. Yet the need to house existing and new workers demanded an additional twenty to thirty-five million rubles per year.[57] Even these initial projections became much too modest given the creation of a nuclear submarine fleet. As of October 1, 1955, expansion at Zvezdochka was moving at the pace of a steam-powered icebreaker. While capital investment was indicated at nearly 600 million rubles, only 150 million rubles had been allocated. This contributed, as well, to the significant lags in ship repairs carried out at the factory for merchant marine, the Northern Steamship Administration, Sevryba, and military ships that sailed the White, Barents, and North seas. The secretaries of the provincial and the city party committees requested that the Council of Ministers ensure the flow of at least seventy to seventy-five million rubles annually.[58] In response at the end of November 1955, Gosplan awarded 300 million rubles for Zvezdochka for capital investment for 1956–60.[59] But the plans did not include such requested facilities as an electrode manufacturing facility, a 1,500-ton press, an electric arc furnace, and other modern equipment with the justification that these facilities existed elsewhere; once again, the "rational" Soviet planning system could not provide crucial machinery where needed because of bottlenecks and shortfalls. In addition, General Secretary Khrushchev was trying mightily to reign in the costs of the military.[60]

These problems slowed industrial expansion throughout the city. At the Shipbuilding Factory they could neither pour a concrete foundation, nor finish work on railroad spur to receive reinforcing rods, nor install cranes—rated at 50 to 350 tons—nor acquire winches and pumps, let alone heaters and generators. The absence of winches, pumps, and ejectors hampered manufacture of cruisers. The "Komsomolets" Factory was far behind on supplying the heater units, vaporizers, and condensing apparatuses that had been ordered.[61] Construction on nine major projects lagged 30 percent behind schedule. In addition to shortages of supplies and tardy deliveries of machinery and equipment, the factory was also short thousands of employees. In 1952 the Shipbuilding Factory began the year with 3,866 workers, hired 1,456 hired during the year, but 1,499 left or were fired because of family issues, criminal prosecution, being drafted, or moving to other jobs. Labor discipline for first and second quarter was a real problem with 429 cases of shirking, 210 truancies, seventy-three desertions, 249 workers given reprimands, and several hundred other violations.[62]

If Molotovsk had trouble with plans, supplies, and employees, then it is not surprising other smaller shipbuilding facilities in the Northern Dvina River basin also suffered. In the early 1950s the Limenda Shipbuilding and Repair Factory could not fulfill its plans. Of seventy-eight needed engineers, the plant had only thirteen. As the Kotlas port and Limendsk Factory grew, they required more power while generators were decades old—dating to 1914—and poorly operating. The secretary of the Kotlas city party committee, N. Berezin, implored the minister of inland fleets to provide 3,000 kW in generators as soon as possible.[63] To this day one can see rotting ships and boats on dry dock stands along the river.

In the twentieth century housing became a right of decency rather than a privilege of the entitled middle and upper classes. Public housing dates to the nineteenth century, expanded in the interwar years, and became a major aspect of state policy after World War II, both because of the interest in social equity and because rising populations and rural-urban migration put pressure on markets that the private sector could not meet. Weimar Germany was an incubator for ideas about public housing, and its constitution even noted housing as a right, although economic and political turmoil prevented this right from becoming widespread. Postwar France saw major and successful programs come to fruition. In the United States, New York City was a pioneer in the 1930s of public projects, while in the postwar years, together with federal highway projects, housing projects expanded, although nearly always as a means to "cleanse" or alter urban neighborhoods and rarely in suburban regions, with the result that the poor—usually people of color—were pushed into mass-produced, poorly illuminated and ventilated buildings that lacked sufficient green space.[64]

The Cold War expansion of the nuclear arsenal gave impetus to the rapid but extremely uneven expansion of dozens of facilities and cities of scientists devoted to creation of weapons of mass destruction and their families: Oak Ridge, Tennessee, for production of uranium; Hanford, Washington, to manufacture plutonium; Groton, Connecticut, for Electric Boat and nuclear submarines, which doubled in population in the 1950s and then again by 1980s; Cheliabinsk, home of the Mayak Production Facility, and Krasnoiarsk-26 (Zheleznogorsk), both for plutonium production; and Severodvinsk for nuclear shipbuilding. This required new production capacity and in turn energy production, infrastructure, housing, and schools.

Over the course of seventy years, the USSR never met demand for good quality and spacious housing for a variety of reasons. Twice, war or revolution led to the ruination of housing stock. After the 1917 revolution, with inadequate funds to build, but with the need to house the masses, the Bolsheviks confiscated virtually all real property and divided it into smaller and smaller pieces. A prerevolutionary merchant's house became a communal apartment for several families who were forced painfully and uncomfortably to share sights and sounds of daily life—and bathrooms and kitchens. With their interest in creation and resurrection of heavy industry Soviet leaders rarely gave housing the investment it needed or, if so, as an afterthought. World War II led to the ruination of much of the housing stock. Postwar Soviet economic plans focused as resolutely on industry as had the first five-year plans, and when housing arose it often went to the Stalinist elite. On top of this, because of a collectivist ethos and criticism of bourgeois waste, planners adopted simple, even Spartan styles for apartment buildings, with the goal of rapid—and egalitarian—construction. This meant barracks and dormitories, cooperative kitchens, tiny rooms. It meant above all else very tight housing conditions, with people forced in the postwar years to live in rubble or even underground. Khrushchev determined to pay much greater attention to consumer goods, including to housing. Programs for new apartment complexes rapidly expanded. Built hurriedly and with mass production techniques, they acquired the derisive name of *khrushchoby* (a cross between the Russian words for "Khrushchev" and "slum"). Yet in 1957 alone the Soviets built 2.7 million square meters of housing, and most citizens were thrilled to have their own apartments with their own bathroom, sounds, and smells.[65]

Even with Khrushchev's program, however, housing lagged. Newly adopted methods of industrial housing construction worked poorly everywhere in the USSR, but especially in cold climates that affected concrete production and curing. Efforts to open a mine for aggregate and a factory to produce concrete forms for Severodvinsk lagged because of the usual shortages of equipment appropriate to the tasks at hand, and so they had to rely as in the 1930s on Stakhanovism to get any work done.[66]

In autumn of 1955 officials of Sevmash complained about the terrible situation in their housing fund. Only 24.6 percent of funds awarded in the plan had been spent in seven months, and in terms of square meters only 65 percent of the plan had been met. As a result, more than 1,000 newlyweds had no housing whatsoever, 385 of them lived in dorm rooms meant for single individuals, and 1,243 other people had already waited from three to five years for their modest allotment of square meters.[67] Roads, schools, dining halls—even housing for military personnel—lagged through 1955, no matter the high level of attention of the provincial government and party apparatus.[68] How could the workers build ships if their children were unhappy, their stomachs empty, and they waited in line for housing?

The town fathers faced continuous problems with social overhead capital, heat, and electricity. The shells of buildings sprouted rapidly on the White Sea shoreline, in many cases replacing the wooden barracks that had served as home for many residents, but more often in new neighbourhoods dedicated to new housing. Yet whether old or new neighborhoods, linemen and linewomen fell behind in hanging phone and power lines. Provincial officials

Construction of Lenin Square, Severodvinsk. (Courtesy of City of Severodvinsk)

requested an additional 33.6 million rubles in 1954–55 to build eighteen kilometers of roads, replace forty kilometers of sewers, rebuild pump stations, wire the city, and add substations.[69] Streets, lighting, and sewer and water treatment lagged behind the demands of a modern Soviet city. Local and regional officials requested in 1954 nearly 6.5 million rubles for roads, 840,000 rubles for twenty kilometers of streetlights to illuminate the dark streets, and nine million for a vegetable and grains greenhouse.[70] A complete makeover faced the city. The town fathers had budgeted 12.8 million rubles to cover streets with asphalt (Pervomaiskaia, Sovetskaia, Lesnaia, Lomosov, Komsomolskaia, Pionerskaia, and Polarnaia streets) to put an end to rutted, muddy, nineteenth-century thoroughfares.[71]

Both the brick factory in Glinniki, on the western shore of the Northern Dvina River between Arkhangelsk and Severodinsk, and the Concrete Block Factory in Severodvinsk operated poorly, hardly meeting construction targets. In nine months of 1952, the brick factory was baking at 76 percent of 1951 levels, apparently explained by the fact that in March and April the prison labor contingent working at the factory changed twice, and for at least five months the factory was 100 workers short. Concern tempered each positive result. If in 1951 enterprises of the province produced nine million bricks, then in 1952 the demand reached forty million, and within three years it would reach seventy million bricks; Latunov asked Malenkov for 2.7 million rubles to increase production capacity.[72] And for more bricks.

Winter municipal construction, Severodvinsk. (Courtesy of City of Severodvinsk)

The concrete factory was short of cement and could not fulfill its plans, while the wood-working combines failed to meet targets for windows, doors, and other mass-produced items, in part because of shortages of wood.[73] In the early 1950s window-making factories 1 and 2 manufactured only roughly 20,000 m^2 and 13,000 m^2 of door and window frames respectively annually because of shortages of raw materials and of workers.[74] The concrete block factory increased output substantially—in 1952 in nine months spitting out 3,820 blocks versus 3,354 in the same period in 1951. But a shortage of slag that came in any event from more than 300 kilometers away prevented further expansion of production.[75] All of the work of the concrete factory—mechanization of construction, the stockpiling of stone and sand for aggregate, mixing and pouring of concrete, plastering, painting, and transport of forms fell well behind schedule.[76]

Problems extended to the supply of heat and electricity to the city that faced cold, wet wind off the White Sea for much of the year. In the USSR the authorities supplied heat, like everything else, from central boilers usually for cogeneration with electrical power. Then steam heat moved to buildings in pipes buried near to the surface or even above ground. Muddy spring-like pathways might appear through deep snow. In the 1950s construction organizations faced a massive project of replacing 40 kilometers of leaking temporary pipes with 45 kilometers of asbestos-encrusted heating ducts.[77] Three boilers that generated 400 tons of steam per hour and two turbines that produced 25,000 kW of electricity hardly served the masses—or the ships. They needed roughly 50 percent more heat and electricity, but expansion of the cogeneration power station would be complete only in 1955 or 1956, and this delay had an impact on Zvedochka's production. On top of this, everything built to date had been temporary. For example, sewers emptied into a rudimentary water treatment facility that discharged largely untreated sewage into a local river. Waterborne infections were a serious problem, so that all pumping stations required reconstruction and building of new ones. The authorities planned to drain and reclaim 200 hectares of swampy land subject to spring flooding. Construction of schools, a maternity hospital, pediatric department, and public baths lagged. Inside buildings linoleum, paint, carpet, wallpaper, and other building materials were in short supply. Financing was millions of rubles in arrears. In the summer and fall of 1954 another problem arose: In the effort to cut the national budget, Khrushchev had ordered cuts in the size of the Red Army, but the demobilization of soldiers and the release (and eventual amnesty) of prisoners at lagry left the shipbuilding labor force short by 25 percent. The town fathers complained bitterly about the critical situation: Existing structures often were dilapidated, drafty wood structures poorly anchored into the permafrost. Spring floods and ravenous mosquitoes made things worse.[78]

Unclear division of labor among ministries—and limited budgets from one ministry to the next—put additional strains on the unfolding capital reconstruction of Severodvinsk. Admiral Gorshkov, in the headquarters of the Navy in Moscow, wrote Obkom secretary Savelii Loginov in late July 1955 that plans for 1956 included various sewer projects, but no money for schools and other children's institutions; he insisted that these funds were the responsibility of the Ministry of Education and Health of the RSFSR.[79] In late September 1955, Loginov[80] wrote Georgy Zhukov, a war hero and for two years the minister of defense under Khrushchev, on the need for his direct intervention. While construction on the 68th and 69th blocks of flats, a collector, and a water tower had been approved, the latter lagged considerably, so that the blocks were frequently flooded with raw sewage including feces and water did not reach the fourth and fifth floors.[81]

Urbanization means migration and settlement. It means young people coming to the city in search of opportunity, meeting other young people, falling in love, starting families, and having children. School crowding was a huge problem, only in part because of the Khrushchev-era shift over to universal middle education in the Sixth Five-Year Plan. The Severodvinsk schools were critically overcrowded, and they would be even more crowded in the 1960s and 1970s as fertility rates increased. Roughly 6,600 children attended schools in two to three shifts, and another 1,200 children would enter in the next years. Construction had begun on two new facilities, but without additional funding neither would be complete by September 1955, and the Ministry of Shipbuilding refused to provide additional funds. By 1960 local governments forecast 15,000 students versus only 6,772 at present. This meant the need to build seven more schools not considering the two currently being built, plus a nursery school and kindergarten.[82]

Molotovsk and Iagry grew into a city, and the city into a large city of a quarter million people. But the authorities could not feed them properly. Gorkom secretary Glukhov wrote Latunov on January 29, 1952, about the lag in construction of the Bread Factory with sixty ton/day capacity. Glukhov had blame to go around—for Russian Republic's ministry of the food industry and in particular the Trust Rosglavkhleb (yes, the Russian Administration for Bread) and its Arkhangelsk division, which had failed to provide either the right equipment or assemble it properly. But Glukhov's admonitions fell on deaf stomachs.[83]

The problems extended to the cemetery, maternity hospital, pioneers' club, and train station. The clinic and hospital were in temporary wooden buildings. They lacked a pediatric facility. They needed an orphanage for 120 children. Town fathers reiterated the request for 3,500,000 rubles for a new maternity hospital, three million rubles for a new House of Pioneers for the thousands of children of school age, and two million rubles for roads and garages. We can only imagine what it means that the municipal cemetery was short 120,000 m^3 of sand and one million rubles.[84] Nor would the authorities consider a proposal to build a train station to accommodate 100 passengers for Severodvinsk.[85] Later the budget included the station but not the track.[86]

Loginov and Molikov tirelessly pursued various ministries to secure urban renewal in Severodvinsk. They wrote the minister of shipbuilding,

I. I. Nosenko, in late September 1955 about the fact that a whole series of construction projects languished, that architectural firms failed to produce blueprints, and that existing plans had long been superseded by the demands of a rapidly growing city. They outlined a series of demands: an increase in the square meters of housing from 85,000 m^2 to 129,000 m^2; rapid completion of the new hospital with surgery and seventy-five beds at two million rubles, and a TB wing with fifty beds at 1.5 million rubles; a maternity ward increased from the plan of fifty to 100 beds at 2.5 million rubles; replacement of the wooden dam built by prisoners on the Solze River in 1938 with a concrete one to secure the municipality's reservoir at six million rubles; cardinal reconstruction of the central heating network in eight blocks of the city, much of which was built in 1937–45 above ground because of swampy conditions in wooden facilities in the neighborhoods of the city consisting of wooden buildings, at five million rubles; replacement of thirteen kilometers of sewerage at a cost of seven million rubles; ten million more rubles to finish construction of 189 wooden apartment houses; and eight million rubles for music at school, or a total of forty-two million rubles.[87] We do not know the precise response of Ivan Nosenko, a true Soviet citizen, the son of a worker who rose up through training in a shipbuilding institute to become director of the Baltic Shipbuilding Factor in 1938, rear admiral, minister of shipbuilding industry and simultaneously during the war deputy minister of the tank industry, from 1947 minister of transportation machine building, from 1952 minister of shipbuilding again. He was briefly removed by the Beria clique, then from 1954 served as the minister of shipbuilding again, and for his devotion to the seas and the motherland was entombed in the Kremlin Wall.[88] But we know that over the next thirty years the city fathers of Severodvinsk repeated their entreaties to many other ministers to ease constant growing pains.

At the End of the Earth: An Outpost of Civilization

Amderma truly was the end of the earth. According to one story, a Nenets boater on the shore of the Kara Sea found a stone with interesting colors and flecks and sent it to Moscow. In 1932 a prospecting group of the Vaigach Expeditionary Force of the OGPU appeared on the Iugorsk Peninsula (see

Chapter 3). In the fall of 1932 the force identified a fluorite deposit in the hills at the Amderminka River and on a sandstone spit of land without any shelter set up camp. In the spring of 1933 from Vaigach they sent another group under Evgenii Pletnev to establish a mine. In July 1933 a settlement was established under mining engineer Evgenii Livanov to begin extracting the fluorite; a year later, ahead of the gulag administration's schedule, they mined the first fluorite. In 1934 they produced 5,711 tons of fluorite, in 1935 8,890 tons, and in 1936 15,195 tons of fluorite. This permitted the USSR to end imports of the ore. The miners and prisoners lived in tight barracks two to a plank and worked in terrible conditions where the average temperature in the mine was –6°C.[89] But why would they build a mine so far from need and demand?

Amderma means "walrus rookery" in Nenets. On the shore of the Kara Sea, to the east from the Iugorskii Strait and 420 kilometers from Naryan-Mar, 1,260 kilometers from Arkhangelsk, and 270 kilometers from the nearest train station in Vorkuta, Amderma's sense of isolation was reinforced by a polar night from November 27 through January 16, if mitigated slightly by a polar day from May 20 through July 30. The Kara Sea here is called an "ice vault" because for more than eight months of the year it is iced in, in many years constant northeast winds push ice onto the shore, and the sea is free of ice only in September. Nine hills mark the city, rising to 155 meters above the sea. Getting to Amderma required an airplane (from the 1930s Amderma had a "runway" of sorts and could serve as a stop on the way to Rudolf Island and the North Pole) or steamer. By ship, captains and crews passed through the Iugorskii Strait between Vaigach Island and Eurasia and between the Barents and Kara seas, itself often filled with ice. The Englishman Arthur Pitt first entered the strait in 1580. A polar station opened in 1911 and was closed in 1993.

In spite of the harsh conditions and low population densities, in the 1930s the authorities promoted the development of Amderma and an "Iugorsk Industrial region." Relying on slave labor to develop resources, it would have consisted of coal and other ore mining operations, settlements to support the mines, and a railroad, built across the tundra, from Vorkuta to the coast, to the Iugorksii Strait and the town of Khabarovo, located fifty kilometers from Amderma, with a new port capable of handling freighters to carry coal and fluorite, and who knows what else, along the Northern Sea Route. The Vorkuta coal was seen as the richest in the entire Pechora River basin—and a key to developing the Northern Sea Route and to extracting Vaigach Island and Pechora basin resources. On the surface, the roughly 400-kilometer rail was irrational, given the fact that Khabarovo Port might be open in all only 110 days per year (August–November). But Stalin's rapid industrialization included distant outposts where even the faint promise of industry on a distant, icy shore led to resolutions to urbanize at Stalinist tempo.

In 1932 the government ordered a railroad to be built from Vorkuta to the Iugorskii Strait and ordered engineering firms to finish the project within a year. The proximity of Vaigach mining gave officials the confidence that such a project would be successful, even if the Vaigach mines operated poorly. Officials transferred one geologist, one topologist, two medical workers, and scores of prisoners from Vaigach to Khabarovo, where they were joined by workers from the Commissariat of Ways of Communication, with supplies to last several months. An expedition to chart the railroad commenced work in June. The engineers discovered not surprisingly that permafrost, the nature of the soil, the low population densities (nomads appeared only twice yearly), and the absence of local materials for construction would make this a very complex endeavor. The head engineer estimated the need for as many as 16,000 laborers in a five-year project. Gulag prisoners, "voluntary laborers," 35,000 tons of supplies, 500 horses, cattle, and thirty-five tractors would arrive by steamer at the coast. And hence, the Vorkuta-Khabarovo railroad, an official component of the Second Five-Year Plan to develop the coal of the Vorkuta River basin, fell to "technical problems," that is, tremendous human hardships. Only on the eve of World War II, only after the Nazi invasion did trucks, horses, and other machines arrive to accelerate the construction. In fact, of all great plans for the region, only the Amderma fluorite mine operated, and in 1942 all other work was put on mothballs. When Vorkuta was incorporated into the Komi Republic, work on the creation of a Iugorsk Industrial Region ceased, and future industrialization focused on resources located closer to the shore—in essence only on Amderma.[90]

The general point of Amderma was to develop fluorspar for use in the cement industry. But surely there were more easily accessible and friendly places to extract fluorspar than a peninsula sticking out into the wild

Kara Sea and a mine burrowed through permafrost about 400 to 500 meters from the surface. From 1933 to 1936 the mine was run by the gulag for no one in his right mind would move to Amderma unless a prisoner, from 1936 to October 1939 as part of the Mining-Geological Administration of Glavsevmorput, still with prisoners, and from October 1939 to September 1951 under Glavruda (the Main Administration of Ore of the Ministry of Non-Ferrous Metallurgy)—with prisoners. Amderma reserves were perhaps 1,272,000 tons with CaF_2 47 percent, $CaC0_3$ 34 percent, SeA_2 at 16 percent, and some residual iron and zinc compounds. Initially the hope was annual production of 6,000 tons by 1954 and 25,000 tons by 1956. But Amderma lacked a roadway (a treacherous 1.5-kilometer road from the mine to the town was the only hope), not to mention dump trucks, backhoes, and cranes.[91]

Perhaps Amderma received impetus from the plans of Glavsevmorput to expand aviation. Amderma was fortuitously located at the crossroads of the Northern Sea Route and the Arctic air route. Airplanes used a small spit of sandstone between the sea and laguna as a runway. Otto Shmidt's well-known 1937 expedition to the North Pole used the runway on the return so that the airplanes could replace skis with wheels. In as much as most of the snow had melted at that time in June, the inhabitants of Amderma were mobilized to broaden and lengthen the runway with snow they brought on sleds and in wheelbarrows. Among those receiving medals for the Shmidt expedition were I. D. Khromov, the director of the mine,[92] no doubt for rousing prisoners to haul and pack snow and ice

Amderma consists of three districts: Primormsk, the town itself on the shore of the sea, Rudnichnyi (Mining), and Poliarnyi (Poliarka or Polar) in the hills 1.5 kilometers from town. The main streets of Amderma are Central, Lenin, Revutski, and Dubrovina. Was there any Soviet city without Lenin Square or Lenin Boulevard? From the start, Amderma had only one enterprise, the mine, which defined the entire life of the inhabitants. The mine had drilling, mining, and enrichment divisions; operated a chemical laboratory, an electrical power station, garage, repair shop, water works, radio station; and had capital construction and housing departments and a club. A weather station, post office, middle school, hospital, and newspapers (*Za Bol'shuiu Amdermu* [To Big Amderma], *Poliarnyi Shakhter* [Polar Miner] later *Poliarnaia Zvezda* [Polar Star]) followed.[93] During World War II new sources of fluorite were found and the mine was closed, although Amderma continued to grow as a base for Arctic assimilation.

The mining operations were synonymous with "fiasco" from the start. In June 1952 the secretary of the Nenets regional party committee, A. Gudyrev, reported to the Arkhangelsk obkom that the Amderma mines were being removed from operation because they filled with water. Engineers who designed the mines failed to consider the impact of the Kara Sea on the mines. But "the plan" required continued operation even in this miserable environment. "Having determined that the government will not sanction the removal of the mine from an operational state, and considering the flooding of the mine to be wrecking," Gudyrev wrote, "I ask you to raise the question in the government about how to require Glavsevmorput to keep the mine in working condition . . ."[94] In July, Gudyrev repeated his concerns about the fact that the mine was already in "wet mothballs" and it was pointless to start up operations anew in as much as nothing was being done—What could be done? Lenin might ask—to remove the water.[95]

Kuznetsov, the director of Glavsevmorput, responded to Obkom secretary Latunov. He noted that given the "significant time, money and manpower" to evaluate the situation, they would wait for a report from "specialist permafrost experts (spetsialisti merzlotniki)" before deciding how to fix the situation in the mines.[96]

Not only mines were in grave health. According to regional officials and economic managers whose bread was buttered with mineral deposits, prospecting and surveying began to lag in the Russian northwest in the early 1950s, which put strain on economic and political organizations of the region that relied on natural and mineral resources to feed their factories. On November 15, 1952, Latunov wrote Minister Anastas Mikoian in Moscow about the fact that "the condition of geological service in the north European part of the USSR is in a most unsatisfactory condition, and in the last years the volume of geological work being carried out in the north has exhibited a clear trend toward decreasing." This "abnormal situation" did not meet the demands of local industry for new sources of ore. The plan for 1953 approved by the Leningrad Geological Administration for the Northern Complex Expedition "not only does not create potential" to exploit new regions, but prospecting in areas of known reserves were not even included in plans. These included the Volongskoe coal reserves, the central part of Bolshezemelskaia tundra, the polar Ural region, and Amderma.[97]

Specialists needed to evaluate the lead-zinc deposits in the region had not been assigned, nor in many places had drilling and other equipment, tractors, tents, cots, and winter clothing been provided; workers were also short 200 pairs of boots. The Northern Geological Expedition had been forced to cut its work sharply—but not because of a boot shortage. Regional officials painfully claimed this had direct and immediate consequences on the assimilation of natural resources in the entire region. This in turn served as a brake on other areas of the economy: for example, coal was needed to support the steamship authority, the canning industry, electrification, and heating.[98]

Through 1952 letters flowed more insistently from party officials in Arkhangelsk to Moscow. One indicated that more than two million square kilometers of terrain remained "poorly evaluated in geological surveys," which meant ignorance of coal, fluorite, lead, zinc, and other ore deposits of industrial significance. The number of geological surveys conducted by the Northern Geological Expedition from 1949 to 1952 dropped from thirty-five to thirteen, and in 1953 only three were planned.[99]

Officials in Moscow disputed this picture. The deputy chairman of Gosplan, the state planning administration, argued that the volume of geological survey work undertaken by the Ministry of Geology in the north (in Arkhangelsk province and the western Komi Republic) was, in fact, growing. Because of their strategic interest to the state, such resources as minerals and fossil fuels remained in the center of attention of the Ministry of Internal Affairs. It sponsored exploration in the Komi Republic and the Bolshezemelskaia tundra to the tune of nearly 195 million rubles in 1952 and more than 250 million rubles in 1953. Moscow explained the decline in funding for the Northern Expedition in part through the increase in support to the Ministries of Internal Affairs, Geology (MinGeo), and Oil Industry. MinGeo work in the northern European part of the USSR more than doubled from 1951 to 1952 and was planned to double again in 1953, with drilling to grow from 8,480 to 63,200 meters.[100]

It seems that government officials had determined that basic research was more costly than it was worth and that resource development could be achieved more efficiently through the research and development activities of economic ministries and the camps of the secret police. The government had also apparently decided that economic enterprises—mines, forestry operations, construction brigades, and so on—had enough information about the location, extent, and accessibility of resources to move ahead with large-scale development programs without significant future expenditures on basis research. This emphasis reflected the continuing debate among party officials, managers, scientists, and engineers over the proper levels of funding for basic and applied research and technology, with the state generally pushing applied research and accountability to economic programs and with scientists striving to protect basic research and a modicum of accountability in determining the direction of research programs. Yet under pressure from regional officials and specialists in the Northern Expedition, Moscow planned to keep carrying out work on Northern Onega bauxites, non-ferrous ores in the Troitsko-Pecherskii region of Komi, and nickel ores elsewhere in the north, with a decision to be made on how to proceed after evaluation of data from the Northern Expedition.[101]

The mine at Amderma, an outpost at the end of time and space, operated miserably from the start, was costly to keep open, was poorly designed and dug, and should never have been built in the first place except to use prisoners as fodder and extract a bit of fluorite from the earth. At best, employees who were not prisoners were paid with great delay, on occasion reaching three months. This led many workers to stop showing up for work in the winter of 1951–52 and in subsequent years.[102] As the Barents Sea winter bore down on the outpost, the Northern State Marine Steamship Authority reported that in 1951 that coal deliveries to Amderma at roughly 3,500 tons were half what had been ordered. In 1952 they delivered 5,275 tons thankfully, although still some 600 short of the plan, and then a steamship turned back to Arkhangelsk after unloading only 169 tons of its 2,025-ton load because of worsening weather conditions.[103]

Symbolically, the Amderma cemetery serves unknown hundreds in unknown disposition. It sits 1,500 meters from the town on the shore of the Kara Sea. The earliest dated burial was 1948, although it is likely that many occurred earlier, but the authorities never put up gravestones. Earlier several graves were in town, but in connection with barracks construction after the stationing of a military detachment in Amderma they moved a number of them to the town cemetery in 1970. According to some residents, from 1930 to 1950 there was a cemetery in the region of the mine, only for prisoners, and because of the continued assimilation of the land into Amderma, information about that cemetery and who is buried there has been lost.

After the temporary closing of the Amderma Mine Administration in 1951, a port and freight organization, Torgmortrans, were created on the infrastructure left behind. The military significance of the city grew during the Cold War with an aviation wing established in 1956; the revitalized Amderma airport could handle virtually all types of planes. Telephone service for the isolated Amdermanians would have to wait until 1966.[104] During the 1960s extensive construction took place including docks and breakwaters in the port, housing, schools, and kindergarten, all connected with Amderma's transformation into a military base. Amderma was always an important radio station for navigation. Yet the effort of the Northern Sea Route Steamship Authority in 1964 to organize a passenger line from Arkhangelsk to Amderma on the *Bukhovina* ceased after one trip because of nonexistent demand.[105] Two- and three-story wooden and brick apartment buildings went up in the 1960s and 1970s that have central heating, hot and cold water, and sewers and with fresh water piped in from the tundra some five kilometers away and later from Tointo Lake eleven kilometers south of the city on the peninsula. The town's electricity comes from a diesel-powered generator. In the 1980s in Amderma "a new page in the history of construction was opened: the equipping of a building with all conveniences from aluminum construction." Diksonstroi began this experiment, having manufactured the Complex Permafrost Laboratory; the Amdermastroi Trust "Akrtikstroi" put up another ten buildings of this sort.[106]

In the 1980s the mine briefly reopened, only to close just as quickly, and in 1993 the military base in Amderma also shut down, followed by the shutdown of the Complex Permafrost Laboratory and Torgmortrans in 1995; in 2000 the construction company Amdermastroi and in 2002 the Amderma territorial administration of hydrometeorology and environmental control were subsumed in the Northern Territorial Administration of the Hydrometeorological Service, with only a skeleton crew remaining. A dead city, like many cities of the Russian north in the twenty-first century, Amderma was created for a narrow mineralogical reason in the Stalin era and built and operated by prisoners. Without orders from an authoritarian regime, how could it have remained open, let alone opened in the first place? It remains to be seen if ongoing exploration of oil and gas of the northern Timano-Pechora site will lead to Amderma's reawakening.[107] Its population has dropped from a maximum of 5,495 in 1989 to 556 people who drift aimlessly about. Those who succeeded in fleeing have left the Soviet industrial legacy behind. Russian's Ministry of Natural Resources and Environment ordered that up to 18,000 tons of scrap metal be shipped from Amderma harbor, to be followed by the cleanup of Franz Josef Land and Wrangel Island in 2011. The scrap is spread haphazardly through eighty-two square kilometers of the municipality.[108] The remaining residents make do by scavenging copper and considering the short summer.

Arctic Urbanization Reconsidered

Villages, towns, cities, and single-resource company settlements dotted the Soviet Arctic Circle. Many of them were established by engineers and party officials, administered by gulag administrators, and settled by innocent prisoners, peasants, and volunteers. They reflected the uneven and costly desiderata of Arctic urbanization. The same patterns of hope for resources and hardships for inhabitants, budgets for industry and shortfalls for housing, and heroic but irrational plans persisted across the tundra and taiga.

A few other cities that have been carefully studied bear mention. No less than ore, gold, coal, and Siberian timber also fell to prison labor. One gulag camp, Igarka, 400 miles inland on the Enisei River, focused on these resources. Maxim Gorky, who had glorified the White Sea–Baltic Canal in *Belomor* (1933), encouraged schoolchildren in Igarka to celebrate Igarka's glories—and hence to ignore its inhumanity and inefficiency. He addressed the Young Pioneers as "future doctors, engineers, tank drivers, poets, pilots, teachers, artists, inventors, and geologists!" In a few years, he continued, when their upbringing by the harsh climate was complete, as "iron Komsmoltsy (Communist Youth Leaguers)" they would go to work and study and would see the vast expanses of the nation, "its gigantic factories, colossal electrical power stations, cotton plantations of Central Asia . . ." He closed his letter urging the pioneers to write a book, which appeared in 1938 as *My iz Igarki* (We Are From Igarka).[109]

By the early 1930s Igarka grew rapidly to 14,000 people including 4,000 kulaks exiled in 1930 and 1931 during collectivization. A western observer was convinced that the prisoners were well dressed against the climate and well fed for their backbreaking labor and that they received "normal wages for their work, are free to move about the town, and their children study side by side with the children of free workers. . . . Of course they cannot leave Igarka." He also apparently approved of the impact of Soviet practices on the "natives" living in the Arctic and subarctic, members of tribes living at an "extremely primitive level." The presence of tribal councils enabled "red missionaries" rapidly to create indigenous Soviets. The red missionaries organized Red Teepees to gather the newly red reindeer hunters whom they considered to be deeply backward and teach them to read, write, and socialize their herds at purported economic advantage, with fewer men tending and herding and more in town working alongside their children to study great Russian achievements.[110]

Igarka's town father, Boris Lavrov, was determined to bring a bit of the urban to the Arctic Circle. He ordered a brick factory and four-story buildings to be erected on permafrost. For Lavrov this was not a utopian thought; he believed that Igarka, as a contemporary polar city, must have stone houses. Igarka, on the shores of the Igarskaia tributary of the Enisei River, 163 kilometers north of the Arctic Circle, was perhaps the first city built on permafrost—the average annual temperature of Igarka is −10.5°C. In 2009 6,700 inhabitants remained in Igarka, a drop from its height of 23,300 in 1939 and 18,800 in 1979 that resulted from abysmal living conditions in the Arctic after the breakup of the USSR, a problem exacerbated by the end of government subsidies. In the 1930s until 1950 Igarka was a prison destination with its port, shipbuilding, fish, and lumber operations, with the latter reaching one million cubic meters of exports annually in the 1970s and 1980s, and scores of ships using its ports. Its air wing supported icebreakers and freighters including the first icebreaker to the North Pole, the *Arktika* in 1977.[111]

The Arctic city Vorkuta, founded in 1932 as the site of a large Soviet forced-labor camp to develop Pechora basin coal, grew rapidly in the postwar years because of its contributions to defense industries. Before the war only one mine had opened. During the war ten more mines opened, a power station came into operation, and the North Pechora railway connected Vorkuta with the nation's rail network. Pechora coal was crucial to supply blockaded Leningrad, the northwest and the Baltic and Northern naval fleets during the war since the Don basin had been lost to the Nazis. In early times, high cost prevented the development of Pechora basin resources, although specialists knew of their extent from the mid-nineteenth century. Without a railroad, the ore and coal had to be shipped by boat with five or six transfers and finally to rail. Often freight was unloaded at waterside to wait for up to six months to reload onto smaller river-going vessels. Even with the gulag system and with the creation of Vorkutstroi in 1938, the dispatch of supplies to Vorkuta and raw materials out moved glacially—and costs were outrageously high, but the authorities justified them through gulag labor. Delivery over a sixty-four-kilometer-long narrow gauge railroad from Vorkuta to Rudnik took up to ten hours even when the railroad operated at full capacity. Engineers, managers, and prisoners had no idea how to build roadways, structures, or mines in the permafrost. Visitors noted that power-generating equipment was in complete collapse, that buildings needed repair the moment they were finished, that bridges were untrustworthy, and that housing, stores, schools, and hospitals lagged in construction.[112] Camp prisoner uprisings occurred in the 1950s, and there is evidence that some of the camps in the region remained open until the 1980s.[113] A bleak, concrete town of 200,000 individuals in 2005, in Vorkuta today some five of thirteen mines remain open. At its peak in 1967 some 205,000 non-prisoners inhabited Vorkuta.

Or one could take the city of Norilsk on the Taimyr Peninsula that grew out of the Norgulag and from 1935 to 1956 employed tens of thousands of prisoners to extract ore from the largest nickel-copper-palladium deposits in the world underneath the permafrost. Thousands died at forced labor especially during the war and immediately after because of short supplies of food. By 1953 the combine was producing 35 percent of the Soviet Union's total nickel output, 12 percent of its copper, 30 percent of its cobalt, and 90 percent of its platinum group metals. In the twenty-first century Norilsk Nickel, having forgotten somewhat its gulag roots, is the world's leading producer of nickel and palladium and a top producer of platinum and copper.

Norilsk, like Monchegorsk, Apatity, Medvezhegorsk, Murmansk, Severodvinsk, Vorkhuta, Naryan-Mar, and many other polar cities, was built, at first, with picks, shovels, hammers, wheelbarrows, pack horses, and an occasional tractor or bulldozer by an assortment of devoted and unenthusiastic, inspired and deadened, hardworking and bored officials, specialists, laborers, and prisoners. They shared the upward movement of production indices and the downward drop of temperatures into the –30s and –40s. Many of them engendered a distant attitude toward indigenous people or displayed the outright conviction that the backward reindeer herders would never contribute to socialism, and thus they required the firm grip of the modern culture of copper and nickel. Of course, the millions of prisoners who helped transform the Arctic into a site of industry and infrastructure experienced at best disinterest, ennui, and fleeting hope, and tens of thousands of them perished. When Stalin died, most of them were quickly released from prison camps, but rarely were permitted to travel directly back to their former homes, and they became, like many of their neighbors, former prisoners who trudged about silently.

To save time and money, planners designed cities on top of ore and industry. Workers and their families lived near the din of machines and the smoke of smelters. Worker safety also lagged behind the needs of the working class—although the worker was the privileged citizen in Soviet society judging by newspapers, journals, and proclamations. For the state Arctic urbanization was costly, and for the worker it was dangerous and even unhappy.

When the USSR broke apart in 1991 and subsidies disappeared for cities, industries, and their workers, a rapid outmigration commenced in search of a better livelihood and gentler climate. Many smaller cities and towns whose *raison d'être* dated to the Stalin era suddenly lacked financial support, let alone economic rationale. Their infrastructure rapidly decayed, the costs of daily life skyrocketed, and they lost contact with the modern Russia of Moscow, Petersburg, and a few other cities. But several of them have experienced rebirth with the discovery of oil and gas; the growing demand for nickel, platinum, and copper; the rejuvenation of the Northern Sea Route for commercial reasons; and the transformation of harbors into transit points; as well as because of the growing ease of shipping in Arctic waters that has resulted from global warming. The administrations of President Dmitri Medvedev and Vladimir Putin embraced the sale of Russia's vast gas, timber, and other resources and the redeployment of the Northern Sea Route as a panacea for the nation's economic recovery and rediscovery of its status as a superpower. The cost of living continues to rise as shift workers who man oil and gas fields have replaced the hardworking Soviet proletariat. International trade has grown. Soviet cities that once dotted the Arctic Circle may be transformed into Russian cities with environmental risks and human costs that must be addressed in a market, not a Soviet, system, by a government still tempted to limit any protest among those who disagree with a future pegged to rapacious exploitation of resources.

6

Transformation of Taiga and Tundra

The boondocks, drunken cops,
And at the endless miles of fields
The dry eyes of the elements stare,
As at white empty sheets.
. . . .
And the little steamboat has not yet sailed
Along the muddy river to us from far away,
On the freezing watery surface, with a godforsaken whistle
Which brings both joy and sorrow—sorrow.
. . . .
On the Onega River everything is in white snow,
But only one light is looming.
Like a talisman on the far shore
Your lamp shade in the window, like a tangerine.

—OLEG MITIAEV, "The Boondocks"

The heroic Soviet effort to industrialize the Arctic, modernize small fisheries, establish large-scale forestry operations, overlay taiga and tundra with centralizing technologies of production, communication, and transport, and collectivize reindeer herding—simultaneously transforming the local and indigenous people—so that the vast region operated according to plans commenced under Joseph Stalin. While in many cases the industrialization and collectivization efforts that spread across northern latitudes resembled those in other regions of the empire, the uniqueness of the Arctic climate, the need to settle vast regions, and the challenges of building transport and communications infrastructure ensured a difficult, distinct, and costly effort. We can better understand the challenges that party officials, planners, and workers faced in transforming the Arctic landscape to meet the needs of the state through examination of a series of technologies, rather than strictly a chronological approach: transport, reclamation, electrification, fisheries and forestry, and communications.

Soviet economic policies to develop these technologies underestimated the human and environmental costs of industrialization and collectivization. Economic plans were anything but objective and rational as their framers asserted, but reflected the political and ideological priorities of the leadership. Of course, in capitalist nations similar human and environmental costs accompanied industrialization, and indigenous people in Alaska and Canada felt the burden of securing the Arctic during the Cold War. Yet most evidence indicates that these costs were significantly larger in the USSR. In spite of being a workers' paradise, in the Soviet Union party leaders and economic planners worried more about production than worker safety in terms of the pace of work and the nature of the technologies of production involved. They assumed that such resources as water, land, and ore were free or nearly free; they paid for them by throwing wave after wave of poorly equipped laborers at them. They were convinced that collectivist attitudes would protect nature from the degradation endemic in predatory capitalism.

Regional party officials, planners, and managers competed for investment funds. Not only in the industrial heartland, but in the union republics, the Urals, Western Siberia, and the North and Far North these individuals advanced ambitious programs for resource development and settlement that made capital even tighter. In Arctic regions they focused on extraction of raw materials—lumber and pulp, oil, gas and coal, iron and non-ferrous ores, and so on—because of the fact that they were relatively plentiful. Inadequate transportation infrastructure handicapped efforts and is only one of the reasons that they erected as quickly as they could, right on the spot, rudimentary factories to mine, refine, enrich, pulverize, and process these materials for shipment to the central industrial region and also for export. Given the relative poverty of the USSR, officials of the central government decided to find investment by extracting it from the agricultural sector and by underfunding social programs (medicine, housing, consumer goods, stores, schools). The combination of unrelenting pressure to tame nature for its contribution to the national economy, the primacy of the five-year plan for increasing production, and the shortage of capital to pursue more modern, efficient production processes contributed to the significant environmental costs of Arctic development, as an analysis of industries and approaches to resource development in the Russian northwest—Arkhangelsk, Karelia, and Murmansk—reveals.[1]

The Soviet effort to assimilate Arctic resources relied upon the creation of extensive production, electrification, communications, and transport infrastructure. Like the construction of the railroad across the American West in the 1880s, the interstate highway system, or the ongoing Brazilian engineering program to create a trans-Amazon highway system, planners expected Arctic infrastructure magically to support the expansion of settlement, commerce, and military security. Although Soviet road, railroad, and communication builders succeeded in overlaying the entire northwest with this technology, it was insufficient to the tasks at hand in total length, quality, and equity. While increasing in reach manyfold from 1920 to 1970, roads remained by any measure—kilometers per capita, kilometers per square kilometer, passenger and freight kilometers, and so on—significantly lower than for the central regions of the USSR. On top of this, roadbeds were poorly prepared and quickly deteriorated. Potholes appeared like dandelions. In spring, lumber roads were sooner swaths of afterthoughts through the taiga than gravel roads, and even tractors and bulldozers found it hard going through the mud. The telephone and telegraph, so vital to ensuring contact between economic and political organizations in the countryside and central authorities in town, city, and capital party organizations, spread through the Arctic at a cold snail's pace and in small numbers; postal service was more reliable but extremely slow even by Soviet standards. How could Soviet planners and officials ensure the development of Arctic resources according to plans if crucial infrastructure lagged?

Of course, much of the reason for the poor state of transportation and communications concerned climate. It is no easy matter to build in and over frozen ground, nor to run telephone lines over bogs, nor to move material and equipment to building sites in temperatures of –30° to –40°, especially if initially your labor force consists largely of poorly dressed slaves working with hand tools. Over the years engineers gained experience and expertise in pushing roads and grids across the taiga and tundra. Local party officials, collective farm managers, and mill operators slowly convinced ministries to fund telephone service to what must have seemed to those sitting in comfortable Moscow offices as the end of the earth. They gradually hooked the remotest ends of the Kola Peninsula to "civilization."

Unfortunately, officials concentrated resources on central industrial regions of the country, and the Arctic was relatively starved for investment capital not only for such things as factories, mines, power plants, roads, dams, and melioration projects, but housing, stores, hospitals, and schools. It made sense to focus investment in central regions with higher population densities and resources that were easier to extract, especially because in the overly centralized Soviet system investment generally served the privileged members of society in Moscow, Leningrad, and republican capitals. Urban residents, with their opera and ballet, their subways, their well-furnished apartments, libraries, and clothing stores looked at the remote regions of the USSR (called in Russian *glukhie mesta,* literally "deaf places," and what we would call the "boondocks") with disdain, even if they were militarily and economically vital to the nation. Policy for transportation, communications, and other infrastructure therefore reflected one of the broader inequities of Soviet society—technological inequity—as this tour of taiga, tundra, swamps, wetlands, and other ecosystems will reveal.

Transformation of the Taiga

In so many ways, the Republic of Karelia resembles Maine or Minnesota with its multitudinous rivers, lakes, bogs, and swamps, with its thick boreal and Acadian forests, with its seasons of harsh winter freezes, mud, and warm summer swarms of mosquitoes. Soviet planners were determined to turn this difficult environment, terrain, and climate into a machine-like entity. In addition to the massive prison labor construction project of the White Sea–Baltic Canal, they had in mind turning lakes and rivers into a transportation network. They ran into the problems both of unique climatic, topographic, and geological conditions and of an extremely rudimentary level of machine technology. They had no tractors, graders or bulldozers, or motor boats. Shovels and rowboats had to suffice. Significant losses in the Winter War with Finland further delayed the flowering of the planned economy, so party officials and engineers returned to nature transformation with vigor and vision in the late 1940s and 1950s. Reclamation of swampland expanded and construction of hydroelectric power stations accelerated, and yet projects lagged far behind targets.

Many individuals and societies have pursued reclamation of land—draining of swamps, building on flood plains, pushing back forest, using fill to create new lands (for example, the Dutch who have retaken land from the sea, the Bostonians who built Back Bay, La Guardia Airport in New York City, and so on and on)—and regulation of river flow toward the ends of transport, power generation, and flood control. The history of the American West may largely be viewed through water and reclamation projects—and the pushing aside of Native Americans.[2] Yet the Soviet effort in northwest Russia was unique in its scope and vision—and its failings.

As part of the Stalinist Plan to Transform Nature (1948), Soviet visionaries turned to Karelia. The ministries of state farms, agriculture, reclamation, and animal husbandry of the USSR and Karelia were engaged in this effort, as were design and construction institutes, and local party, economic, and other institutions. The reclamation projects involved building open and closed drainage ditches and canals that relied largely on gravity rather than pumps to remove water. Pine trees on the site would be milled for pipes, culverts, and drainage systems. This reclamation work, so crucial to the rebuilding of Karelian agriculture, lagged into the early 1950s. At one meeting after another the authorities complained about shortages of labor, materials, and equipment owing to excessive targets, poorly functioning centralized distribution of resources of capital and labor, and rudimentary understandings of how to reclaim land.

Postwar Karelian agricultural officials faced constant pressures from Moscow to create an agricultural wonderland; they shared the vision of transformation of the sodden, green landscape into a powerhouse of food, crops, paper, pulp, and wood production. This led to an extensive reclamation program to increase the amount of sown land. In November 1948 the total area of sown land and meadowland in state farms was 19,500 hectares that were fed with reclamation canals and mechanisms of roughly 18,000 kilometers in length, fully 90 percent of which required repair or reconstruction. On top of this, Karelia had lost roughly 20,000 hectares of reclaimed land to war damage and the peace treaty with Finland. The goal of the first postwar five-year plan was to return agriculture to its earlier modest production levels by rebuilding irrigation systems, transfer canals, and ditches and to put fields and roads into good shape using hundred of prisoners from the Karelian gulag.[3] But in the absence of appropriate machinery or workers, reclamation barely moved forward. In 1946–48 the ministry of state farms of Karelia acquired only one grader, and the tractor promised in the spring of 1947 still had not been delivered. The minister therefore requested in short order: three backhoes, two graders, two S-80 tractors, two bushwackers, five tractor attachments for clearing swamp and meadow, and twenty-three disk harrows.[4]

Because of the highly inefficient central supply system, not only heavy equipment was missing. In winter of 1949 local managers turned to the head of the Moscow-based State Farm Bureau of Reclamation (Sovkozmeliovodstroi) with a request for 110 pairs of rubber boots and ten pairs of leather boots. The bogs would be soggy, but without boots laborers had cold, soaked feet and legs.[5] Poorly equipped laborers were only part of the problem. Sometimes managers could not assemble a brigade, and when they succeeded in sending work brigades into the bogs, they were often poorly organized. Entire groups of men stood around, waiting for instructions or equipment.[6] Reorganization of workers into labor brigades was ineffective without more than axes and skidders and other equipment. Often the workers had no horses, and when a horse, no saddle.[7] It would be a challenge to reclaim 70,000 hectares (260 square miles) as planned.

Reclamation lagged everywhere, leaving frustrated officials to complain. At a 1952 meeting of leading Moscow and Karelian officials, attendees lamented the fact that in five months of work in central regions only 8,400 m^3 (two-thirds) of planned 12,500 m^3 had been excavated. The reasons were clear: the one steam shovel operated only one shift in the winter dark without lights. Yet the officials blamed local authorities for incompetence and replaced local bosses. They ordered the new head of Agromekhanitsiia, the Karelian bureaucracy responsible to modernize agriculture, to assume full control, secure ten days' reserve of fuel oil, and erect a bulletin board to exhort the workers. Of course, none of this had much to do with mechanization. They needed to train another excavator operator or two and get their hands on dump trucks and tractors.[8] The same stories were repeated throughout the republic.[9]

Electrification and Hydroelectricity

Hydroelectric power stations were part of the reclamation effort to transform river basins, provide electricity to pump water, and enable transport and irrigation. From the point of view of power generation, the twentieth century is the century of hydroelectricity as much as of nuclear power. Engineers and construction trust managers have learned how to alter the flow of rivers, build dams, establish regimes for reservoirs, employ larger and larger turbines and generators, and even build "cascades" of stepped dams to turn entire river basins into machines, with the largest ones being 1,000 megawatts (MW), 4,000 MW, even 12,000 MW.[10] Permafrost, rock, and subzero temperatures were hardly obstacles to Soviet engineers sitting in design firms in Leningrad and Moscow.

Promoters of hydroelectricity tout cheap electricity generation, flood control, irrigation, improved transport, and an end to the frustrating seasonality of rivers. Wet and dry seasons and high and low flow periods disappear under the direction of engineers who have straightened, dredged, and captured the river. Yet the construction and operation of stations always reveals significant and often underestimated long-term costs. Dams trigger economic growth in some sectors of the economy while interrupting other activities. For example, fish habitat is disrupted as water velocity, temperature, and chemistry all change under the new regime of operation, and the dam serves as a barrier to migrating (anadromous) fish. Where forestry operations depend on the spring float to bring lumber downstream to mills, a dam impedes smooth operation of the float. Inundation of flood plains, often the most fertile farmland available, destroys entire communities where people have lived and toiled for generations. Dams transform 100-year flood events into fifty-year, even-more-catastrophic events; siltation and evaporation cut engineers' predictions of energy production. The greatest impact may be on "oustees," those individuals whose homes, lifestyles, ways of life, and memories also are inundated. For major dams—Three Gorges in China on the Yangtze River, the Sardar Sarovar on the Narmada River in India, the Sao Francisco in Brazil—hundreds of thousands, even millions of people are displaced, and the promised new settlements, with promised schools, housing, stores, and a promised better way of life, never materialize.[11]

In the USSR this story was repeated from north to south and from the European USSR to Siberia. Electrification was a crucial and early feature of Soviet efforts to produce more kilowatts than the United States, illuminate agriculture and urban environments, and lessen the burden of labor in the Arctic as well. Yet goals for electrification exceeded designers' capabilities, and Stalin intentionally starved the countryside and the peasant for electricity until well after World War II. As a result, rural electrification barely met the needs of people in the countryside and in the Far North. Whether measured in kilowatts of capacity, generation, or consumption or some other standard, and whether in towns or villages, residents of northern latitudes lived in great deprivation compared to those citizens of the USSR in central cities. Many villages and collective farms were literally left in the dark. If lucky, collective farms and forestry operations made do with small generators or electrical supply secured from a local industrial concern. The further from a city, the more likely that tasks had to be performed by workers without electrical labor-saving machines, tools, or equipment and in the dark through long winters.

With the exception of industry, when electrical service was supplied, it came in small amounts, directed to places of business. In Arctic regions, power lines were strung along concrete or wooden poles that had to be fixed into swampland, forest, or tundra. None of this construction was easy to accomplish and none of it built well enough to last a lifetime against the elements. (Because of this, Russia's northern rural power infrastructure today requires billions of rubles to upgrade and repair; tens of thousands of poles have fallen or decayed.[12]) Beyond the difficulties in construction, great distances across the taiga and tundra, and low per capita demand per ruble of investment, another reason for the lag in bringing electricity to the countryside was a concomitant aspect of collectivization: Stalin and his colleagues had determined to extract capital from the countryside for the industrialization effort, and until the Khrushchev era the rural resident, the peasant, the collective farmer, the lumberjack, and others were lucky to see a lightbulb, let alone a generator.

After the war Stalin finally approved the expansion of rural electrification using smaller, rudimentary hydropower stations on scores of smaller rivers complemented by small engine generators. In the Russian northwest, especially in Arkhangelsk province, rural electrification involved attempts to tame seemingly every potential river site. Hydroelectric power stations were designed for rapid construction and low cost. They included earthen dams and usually one turbogenerator of standard manufacture. Work crews had limited access to bulldozers, tractors, or backhoes; construction brigades attempted to economize on concrete, steel, and other material costs. Similarly, the distribution systems were rudimentary, as were the motors and illumination systems that in many villages or collective farms involved no more than a dozen 25-watt light bulbs. Yet for the peasants and lumberjacks who toiled in the dark for much of the year these simple comforts were welcome.[13]

In Karelia and sparsely populated Murmansk region, while the stations were smaller scale than those along the Volga, Dniepr, and a variety of Siberian rivers, they reflected engineering hubris to electrify the Arctic. They set out in Karelia and Murmansk to dam the Suna, Vyg, Kem, Niva, Tuloma, and other rivers. Even though Soviet citizens faced great pressures to conform to specific narratives about state economic development and technologies crucial to that development, some engineers, writers, and policy makers worried about the seemingly unstoppable way that hydroelectricity design and construction firms attacked Soviet rivers and reclaimed Soviet land. The engineer Peter Palchinsky paid with his life for suggesting a more balanced approach to the development of the Dniepr basin in the late 1920s. A leading Soviet engineer of the postwar era, Petr Neporozhnii, who rose to become minister of electric power development and electrification of the USSR, worried about the breakneck approach to Siberian hydroelectricity, especially under Khrushchev, without sufficient consideration of lag in demand. Widespread opposition to the industrial despoliation of Lake Baikal broke out in the 1960s. A great dispute arose in the late Gorbachev era about the outrageous plans to divert the flow of Siberian rivers to Soviet Central Asia through a series of transfer canals, some 1,500 kilometers in length and excavated in part through "peaceful nuclear explosions." Through literature, such representatives of the genre of Village Prose as Valentin Rasputin addressed the destruction of mythic Russian nature by the Soviet development model.[14]

But the most important narrative was embodied in the slogan that "Communism equals Soviet power plus electrification of the entire country." The effort to bring electricity to the countryside through a kind of Soviet Rural

Electrification Administration began in the late 1940s and early 1950s, with impetus often coming from below.[15] This was a reversal of Stalin's initial unwillingness to invest in electrification in the countryside during the first five-year plans.

The first child of Karelia electrification, the Kondopolzhskaia hydropower station, dates to 1923. This station was part of the State Plan to Electrify Russia (GOELRO) signed by Lenin in 1918 with great fanfare and was connected with the construction of the Kondopolzhskaia Paper Factory, which at the time was one of the largest in the world. They built the station primarily with wheelbarrows and shovels, with laborers moving a total of 626,000 cubic meters of earth. The first unit came on line in 1929 at the very modest 5,100 kilowatts (kW). From 1937 through 1941 additional work in the Suna River basin expanded power to 27.5 MW. This supplied the factory until World War II when the equipment was evacuated to Uzbekistan to keep critical technologies out of Finnish and German hands. In the fall of 1944 after liberation, reconstruction began, and by 1947 two turbogenerators with a total power of 23 MW began spinning. In March 1951 a third turbine was added, and later the power was raised to 31.4 MW, in part by replacing all wooden pressure tanks and so on with metallic parts and equipment.[16]

Impoundments destroy natural phenomena. Hydroelectric engineers try to select sites for dams where high flow and a large volume exist; one such area is a waterfall. In the 1940s and 1950s, encouraged by engineers from the Tennessee Valley Authority, Brazilian engineers flooded the falls for the Angiquinho Hydroelectric Power Station on the Sao Francisco River, as the poet Castro Alves lamented:

> The waterfall! Paulo Afonso! The abyss!
> A colossal fight of the elements!
> Centaur claws in paroxysm
> Shaving the flanks of bloody parts.
> Reluctant in the pain of the cataclysm
> The giant's sweaty arms
> Supporting a grinding (wonderment! amazement!)
> The whole river which falls upon her shoulder![17]

While engineers preserved the Kivach Water Falls on the Suna River, they inundated the Girvas Falls another 32 kilometers upstream with the Paleozerskaia station (started in 1950, completed in 1954), a 25-megawatt station with two turbines. I am unaware of a calculation of how many waterfalls ten meters or higher have been inundated for hydroelectricity, but it must be very large—considering the United States, Brazilian, Chinese, German, and Soviet experiences.

To meet the growing demand of industry, construction on the Vygskii and Kemskii cascades commenced in the 1950s. On the Vyg River, five stations with a total capacity of 240 MW were built between 1956 and 1967 spread out along 100 kilometers to the White Sea from Vygozero. These stations gave great impetus to the economy of the region including the Nadvoitskii Aluminum Factory (which opened in 1948 and as part of Russian Aluminum in 2012 had almost 1,500 employees) and the Segezhskii Pulp and Paper Combine (1939, and in 2012 employed roughly 5,300 in a still ramshackle operation to produce Sack paper, SP kraft liner, paper sacks, and wood chemical products: tall rosin, distilled tall oil, fatty acids, and turpentine).[18] The Kemskii cascade with a total capacity of 330 MW includes the Iushkozerskaia, Krivoporozhnaia, Poduzhemskaia, and Putkinskaia stations and the as-yet-unfinished and now mothballed Beloporozhskaia.[19] These dams average up to several thousand meters across and the reservoirs behind them average twenty or so square kilometers of surface area.

In Karelia, chapter-and-verse description of the construction of the Putkinskaia hydropower station on the Kem River provides some indication of the cultural significance of "inundation." People, buildings, cemeteries, and other objects must be removed or properly sealed before inundation. The zone behind the Putkinskaia station required the special attention of military and other organizations: a military settlement, plus six houses, two water towers and pump stations, electrical power lines, substations, roads, and a local cemetery would be flooded. Engineers and builders got into a dispute with military officials because the latter dragged their feet over dismantling a small base in the area to be flooded. The engineers also apparently did not plan properly for the rerouting of the Leningrad–Murmansk Highway—a major oversight. Both fisheries and forestry operations were interrupted. One of the last activities was the removal of the remains of 2,230 individuals from the local cemetery, soon also to be flooded.[20]

Like at the Aswan Dam years later and hundreds of other hydropower stations throughout the world, construction of the Kem River dam led to destruction of irreplaceable archeological sites.[21] Specialists at the Institute of Literature, Language and History in Petrozavodsk, the capital of Karelia, complained to the main engineer for capital construction in 1965 that funds obligated to pay for archeological excavation ahead of inundation of the Putkinsk reservoir had not been delivered. More to the point, the head of the archeological survey protested that twelve ancient sites had already been destroyed. He begged that other sites uncovered during the research higher along the Kem River be investigated fully before it was too late.[22]

Having struggled mightily to develop standard practices and approaches beginning with the Volkhovskaia hydropower station in the mid-1920s, engineers pushed onward into the taiga and tundra to build new stations to power burgeoning industry. On the Kola Peninsula the major early station was the Lower Tulomskaia hydropower station on the Tuloma River. Falling more than fifty meters in the seventy-six-kilometer stretch from Notozero to the Kola Fjord at the Barents Sea, engineers calculated it was an ideal site for two units on the Upper Tuloma at 100,000 kW. They would also build a 450-meter-long fish ladder. The engineers proudly noted that the Lower Tulomskaia would be the northernmost in the world and built in just two years in spite of polar conditions. A Leningrad design firm, Gidroelektroproekt, began site investigation in 1932. By 1933, a commission of the Supreme Economic Council under electrical engineer A. V. Vinter had decided upon the two stations with the lower to be built first. Gidroelektroproekt completed the design specifications in April 1934; the main engineer of Belbaltkombinat of the NKVD immediately approved it. The authorities just as immediately transferred a large number of gulag prisoners and construction experience from the Belomor Canal to the site.[23]

The engineers chose a rather simple design, one with a double layer of bitumen covering a sand and gravel dam. Problems arose with the low quality of equipment provided by ElektroAparat (gear drives, pinions, miter gears, and so on) and centrifuges from the Perm Factory. But an investigative commission was happy with the design of the fish ladder, the organization of work, the productivity of labor, and Stakhanovites. Indeed, labor rather than

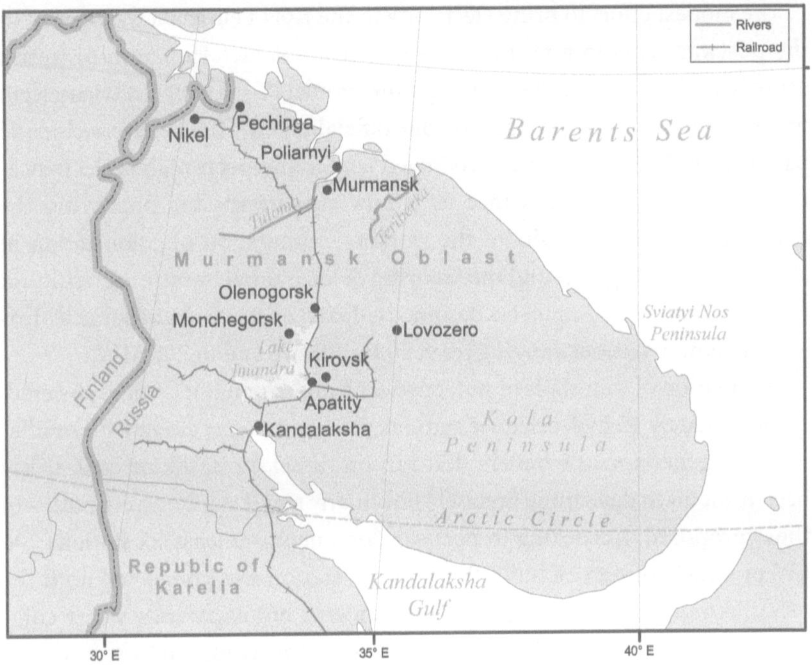

Kola Peninsula. (Map by Manny Gimond)

technology was the key, for the dam took "only" three years to build, with 90 percent of the workers Stakhanovites who fulfilled new norms at no less than 150 percent. They needed these devoted workers because they had but one 375-liter (100-gallon) capacity cement mixer, two dump trucks, a five-ton derrick crane, one compressor, and two excavators for the entire project.[24]

In spite of published reports to the contrary, the Lower Tulomskaia station never operated at design capacity, in part because of the natural aging of the plant, siltification, and inadequate funds to undertake repairs. Accidents and waste in work, as well as tardiness and absenteeism, also had a negative impact on operation. The station operated at one-quarter power in 1937, half power in 1938, and production of electricity in kilowatt/hours reached only 60 percent of targets before World War II. While in 1950 four turbogenerators each at 12,500 kW were spinning in the Arctic chill, the station continued to run at well under capacity. But the plant had a stable if inefficiently large labor force of 210 employees.[25]

The modest effort to bring electricity to the Kola Peninsula was reborn in the postwar years to power collective farms, fish factories, administrative centers, and other facilities, although the reindeer's share of electricity went to metallurgy and the military. As one official noted, "In the severe climatic patterns of Murmansk with a prolonged winter and polar nights, electrification has a special significance as a powerful means for improving the cultural-daily life conduits of the workers, stimulation of colonization of sparsely settled regions, and the securing of local inhabitants in agricultural inhabited points."[26] Engineers designed a dozen or so small and large hydroelectric power stations ranging from 50 to 120 kW and to 200 MW.[27]

Inundation of waterfalls or not, postwar Karelia remained underpowered. Approximately 15 hydroelectric stations provided power along with smaller stations, generators, or boilers that ran on diesel, peat, oil, or coal. It has been difficult to determine precisely how many small stations were built—of those proposed. According to postwar documents, at least six stations 200 kW or larger were to be built. But no large station was completed until the 1950s. Among those referred to in documents but apparently never completed were the Pudozskaia, Olonentskaia, Petrozavodskaia, Ukhtinskaia, Spaskaia Guba, Sheltozerskaia, Velikogubskaia, Padanskaia, Sortoval'skaia, and Solomenskaia stations.[28] Planners were not bashful about their hopes for Arctic electrification.

Sel'elektro, the organization responsible for distributing a modicum of electricity to the countryside, struggled with restarting, rebuilding, and building anew after the war. The Lovozersk collective farm hydropower station at 100 kW at Pal'go Lake (the lake itself is now the site of a nature preserve)[29] had been mothballed at the beginning of the war. Murmansk officials implored the central authorities to secure funding in March 1946 to finish construction and to provide one box of glass, 200 kilograms of nails, three tons of oakum, and ten tons of cement.[30] On June 9, 1946, Murmansk authorities asked the Council of Ministers to assign 150,000 rubles toward expansion of electricity production for collective farms, fish processing facilities, and district centers. They had to apply to the very top to get things moving in the far-off Arctic. Lytkin, chairman of the executive committee, and Kutyrev, secretary of the Murmansk obkom, wrote directly to Anastas Mikoian, chairman of the council of ministers, noting the approval to build small hydropower stations on mountain rivers for operation of fisheries collective farms—but they needed funds for capital investment.[31]

One of the ways to promote electrification of the countryside was the construction of power lines from existing power stations. Even this easy path was no guarantee. Power lines of ten to twenty kilometers were added to illuminate and power the tundra, although these often failed to reach very far in the absence of poles, even along such a crucial stretch from Apatity to Khibiny.[32] For example, the directors of the "Red Pulozero" collective farm on the Kola Peninsula gained funding to electrify because their laborers toiled in constant darkness; they had secured 500 poles of required size and quality, and they had labor reserves to erect them, but they could not get agreement from Murmanselelektro to install the poles for some reason.[33] The poles themselves were no promise of delivery of electrons.

In spite of general support and an abundance of ideas, in fact plans for electrification were rather modest. For 1946 for Murmansk region power stations and grid, including district centers, officials noted the construction of two power stations at a total power of 120 kW, and in 1947 three stations with total power of 120 kW ("Krasnyi Sever" at Kolvitsa, "Red Star" at Belokamensk, and at Povir "Red Pulozero"). This would serve only four collective farms, but no state farms or machine tractor stations.[34] Other plans were equally as modest.[35]

Shortages of basic supplies waylaid Arctic projects. Korsak, director of Murmanselelektro, reported that plans for a power line from Apatity to Khibiny demanded both documentation and an answer to the question of where 1,000 tall and 300 short poles might come from.[36] A generator that served the "Tundra" farm and district organizations, a "Pobeda (Victory)-25," produced only 14 to 15 kW although it was victoriously rated at 25 kW. But Korsak refused to provide a 50-hp motor because the Ministry of Electrification refused to supply it to him.[37] Under Petrishev, its chief engineer, Murmanselelectro engaged in an orgy of hydropower station planning, seeking to locate stations wherever rivers permitted and providing estimates on the power demands of nature preserves (!), collective farms, regional centers, agricultural councils, and so on. In its planning documents Murmanselelectro provided details down to numbers of lightbulbs and poles.[38] But planning did not lead to poles or electricity. Continued problems led Korsak to write the heads of the Murmansk Obkom, Oblispolkom and Glavselelektro, six months later for help in building supply roads and getting construction materials to the work sites.[39]

The same story of great plans to bring electricity to the taiga and the same slow progress occurred in the forests. A multitude of organizations with jaw-breaking names and acronyms that were often branches of Moscow and Leningrad establishments were involved. In Arkhangelsk province, such construction trusts as Zapselelektroset'stroi (the Western Agricultural Electrical Network Construction Trust) responded to lumber enterprise managers who had repeatedly requested provision of line electricity to transform mills that ran on generators, small motors, and a lot of human power to guide the lumber through dangerous gang saws into more modern facilities. Engineers were brought together in such design institutes as the Northern Institute of Electrification of the Forestry Industry (SevNIIELP) to wrestle with the technical problems of creation of these networks.

The Main Rural Electrification Administration (Glavselelektro) provided funding for rapid expansion of generation and distribution systems in its many provincial branches. The Arkhangelsk branch worked with engineers at the regional design administration of the Northwestern Electrical Assembly Trust (Sevzapelektromontazh) in this work. The engineers calculated illumination targets based on square meters of the facilities (barns, hog pens, milking facilities, warehouses, clubs, and so on). They based their designs upon detailed correspondence with managers of collective farms, forestry enterprise, and small factories. Distribution systems consisted of power lines of three, five, and up to twenty-five kilometers in length at 10 to 35 kilovolts (kV).[40] From the Soviet archive materials we know precisely the number of supports, poles, standards, cross pieces, insulators, and so on; the poles were often built from concrete that decayed rapidly or fell over in the taiga or swap, also often with pine that quickly rotted. We know the kind and number of transformers, and we know that in many transformers the level of oil was not at full level, resistance oscillated wildly, the wiring lacked insulation, and other such problems plagued normal operation.[41]

Unfortunately, construction on both stations and distribution systems lagged years behind schedule in virtually every instance, cost overruns and labor shortfalls plagued most projects, and accidents abounded at construction sites. In the Emetsk region in southern Arkhangelsk province, a

Raispolkom (district Party committee) commission investigated a disaster at the Chachevskaia hydroelectric power station on the Chacha River in the Kholmogorsk region. This station was intended to supply several collective farms. Construction began in the late 1940s and stretched endlessly into the 1950s, with almost unfathomable delays given the small size of the facility: roughly ten kilometers of power lines, twelve kilometers of wiring, 2,017 bulbs, and an earthen dam of 342 cubic meters.[42] Even before the leaks threatened station operation, a poorly designed spillway and penstock and malfunctioning turbogenerators caused problems. Repairs and restarts were fraught with disaster—both because of poor construction and failure to take into consideration local hydrological conditions.[43]

Murmansk officials hoped to use electricity get things moving in fishing and other activities on the northern Kola Peninsula shores. The Leningrad division of Gidroenergoproekt designed a 1,500-kW station at Teriberka where four hard-to-operate diesel generators used 350 tons of fuel annually that was always in short supply and hard to deliver.[44] In the summer of 1946 site selection for the station began along the Teriberka River that flows 190 kilometers and drains a basin of 2,227 square kilometers. Reflecting both the challenges of building in the Arctic and of building in the Arctic with the Soviet planning system, engineers succeeded in opening the Upper Teriberskaia hydropower station, twelve kilometers from the delta of the Teriberka River, only forty years later in 1984, and in 1987 the Lower Teriberskaia hydropower station opened. The stations provided electricity to fifty villages and towns, fish state and collective farms, and MTS and government centers. Until that time, only half of them had power, and the poorly operating diesel generators often broke down.[45]

Teriberka is an administrative center on an isolated peninsula jutting into the Barents Sea about 100 kilometers east of the Murmansk fjord and a site of some mineral wealth including diamonds. At its peak in 1959 almost 4,800 people lived in Teriberka; Teriberka lost importance in the 1960s when ship tonnage increased and fishing fleets were able to travel more easily in the open ocean, coastal fishing shrank, and a newly constructed fish processing complex in Murmansk put smaller processing facilities in Teriberka out of business. At the same time, the collective farm "Murmanets" was abolished along with its mink farm, the reindeer herd was transferred to Lovozero, and the fish processing plant was shut down because large fishing vessels were not able to enter the river from the ocean. On top of this salmon disappeared with the construction of hydropower stations on the Terberka River.

By 2012 there were likely under 2,000 individuals in Teriberka, but with the opening of the Stockmann oil and gas fields in the Barents Sea, the village is growing again, and officials have revised dreams of a 12-MW wave-power station, in the works since the 1980s, but mothballed upon the breakup of the USSR, a few dozen kilometers away in the Dolgaia Bay. But it will take a great deal to make Teriberka anything more than the ends of the earth. In the summer of 1996 a young couple traveled there from Murmansk by bus over dirt roads, a bit more than 100 kilometers, a trip that took five hours. Patches of snow here and there sat on the permafrost.[46]

Inland Fisheries and Transport

With proper reclamation and dredging, the lakes—and the rivers and streams that connect them—would be turned into a seamlessly operating passenger transport, freight, and fisheries system. The White Sea–Baltic Canal took advantage precisely of this hydraulic situation. Accordingly, other lakes might be tied together for freight and passenger travel and for development of the fish industry. To take just one example: Leksozero, located in the Muezerskii Region in the western part of Karelia and part of the Lenderka River basin, has a surface area of 164 square kilometers. Most of the shoreline is rocky and covered with forest. Leksozero joins the Rebol'skoe and Lenderskoe lakes with bodies of water in Finland, with the general flow through the Vuoksa River into the Ladozhkoe Lake. The lake has very complex relief with a large number of discharges, confluences, and pools. Ice disappears from the lake in mid-May and freezes only in the middle of November, but this still limits transport. Fourteen kinds of fish inhabit the lake including trout, a number of salmon species, grayling, pike, perch, and other fish that are considered economically valuable.

After the Bolshevik revolution party officials worked quickly to organize small-scale fisheries on these lakes and along the White Sea shoreline into larger cooperatives, artels, and collective farm units. They viewed individual fisherman both as inefficient and potentially dangerous because the wealthier ones among them must be bourgeoisie and therefore hostile to Soviet power.

Because of the sailors' and fishermen's distrust of the Bolsheviks, the collectivization of inland fisheries preceded in fits and starts with the hope that the newly organized state fisheries would produce significant catches to feed burgeoning industrial centers.[47] Industrial and municipal pollution threatened inland fisheries and destroyed ecosystems. Reclamation projects ruined spawning areas. Overfishing nearly destroyed what was left. On top of this, deep sea fishing expanded rapidly after World War II and especially from the 1960s onward. But the demands of trade to generate foreign currency and of feeding burgeoning cities meant that the effort to fish both inland waterways and the ocean grew in any event.

A dozen or so Karelian processing operations existed, each connected with a region, a lake, or a river to receive, store, process, or dry the catch. The Karelian fisheries were rich but poorly exploited owing to underpowered and creaky old motorboats. Because of technological backwardness, the authorities turned to tried-and-true methods of exhortation of workers to increase production. No matter the geographic region, the area of industry, the nature of the personnel, or the time of year, the authorities employed Stakhanovism to establish control over the labor process, overcome backwardness in the factory or field, and ignore difficulties in motivating workers owing to low salaries, modest benefits, and few consumer goods.[48] They were rarely able to secure new boats or repair old ones.

Without a fleet, how could freight, passenger, or fishing operations move forward? The Kiurshevskii Fish Station owned two rowboats, the Goslova brigade and Andomogorskii station each had one, while the steamboat station, as befitting its name, operated four motorboats between six and ten tons and with 20- to 25-hp engines, all of prewar manufacture.[49] According to a plan for assimilation of small rivers and lakes for the first half of 1948, planners and inspectors discovered that in Kestenge the hulls of three wooden boats were in no condition for further operation. Wooden cutters in Kaleval, a village of several hundred people on the northern shore of Topozero Lake, with capacity of five tons of freight and ten passengers, also demanded immediate repair. On top of this, neither the Pudozhsk nor Kondopozhsk region had a boat, and orders for new boats destined for the Vodlozero, Sandal, and Topozero lakes were taking forever to be fulfilled.[50] On nine major lakes the inland fleet of the Republic of Karelia in the 1950s had access to only ten boats with engines ranging in horsepower from 45 to

90, in capacity from one to nine tons, and crew size of twelve to twenty individuals.[51]

Other fisheries in the Russian northwest suffered similar problems. Inland water transport was critical for the USSR for shipping wood, food, coal, building materials, machinery and equipment, agricultural goods, and so on. The USSR was first in the world in total length of rivers, so what easier approach to expanding transportation networks could there be? During the Stalinist five-year plans, "rapid assimilation of sea and river transport that will tie the far north . . . to the cultural life of our great nation." This work "demands from sailors and river workers of north clear-cut Bolshevik work." Yet as Stalin declared in his speech at the 18th Party Congress water transport remained one of the lagging branches of the economy, for example, regarding ship repair, with the *Ia. Sverdlov* sitting at the Laisk Docks for the winter with no repairs or materials, the *Rusanov* at Krasnaia Kuznitsa since 1939, and so on.[52] The same story of no ships, dilapidated boats, no sailors, and dilapidated sailors held throughout the Russian northwest, on top of which the ports were so underpowered and understaffed that they could not pull ships from the harbor for wintering on shore so that eight ships burned coal to keep steam engines warms, while the ships blocked tug, freighter, and other traffic.[53]

When ships arrived at docks there was no guarantee of their seaworthiness. In January 1948 the Arkhangelsk-based fishing trust "Sevryba" received thirteen trawlers from "Murmanryba," eight of which could fish, four which were under repair, and one of which sank in the Northern Dvina River delta. It appears that two more ships were pulled from service for repairs shortly thereafter. It mattered little because they could not staff the ships fully or pay them on time, and their crews were poorly qualified: The captains for these ships had only middle technical education, although three were communists.[54] By July Sevryba could fill only 577 of 681 positions (489 sailors, twenty-three administrators, and sixty-four support staff), and only forty-seven were party members (and only four of the captains). On top of this, seventy sailors had already quit.[55] The administrators of the Northern River Steamship Authority penned missives to the ministry of the river fleet on the crisis situation regarding staff and crew. They needed nearly 7,000 workers but had barely 4,701, and the demand for stokers, sailors, grease monkeys, and navigation officers was quite high. Even if Arkhangelsk province got first pick of future graduates of various naval schools, they would still be 1,300 employees short.[56] Yet who could blame the workers as few of them wanted hard work at low pay far from home?

An aging inadequate fleet compromised operations everywhere. As Arkhangelsk Obkom Secretary Savelii Loginov wrote to the minister of the merchant marine fleet in 1952, transport of freight from Naryan-Mar to shore points on the Barents and Kara seas (Khabarovo, Karskaia Guba, Korataina, Varendei, Alekssevka, and Sinkin Nos) had been to this time handled by two small boats each at twenty-five tons with 65 hp. They were replaced by a ship from the "Red Profintern" factory at 150 tons capacity and 160 hp that itself was held out of operation in 1951 because it was not seaworthy. The port had to rely on the goodwill of ships used for hydrographic and military expeditions.[57]

The further from the center, the greater the technological lag. In addition to a very short season because the Pechora River and its tributaries were iced over much of the year and old and because of underpowered boats whose seaworthiness was a question each spring, the absence of modern facilities in the Nenets region prevented increased production in the fish industry. The Pechora Motorized Fisheries Station sought the construction of an auxiliary ship repair facility, but Sevryba and the provincial administration of fisheries stations opposed the project as redundant to other facilities. Only one thing was clear: the spring catch of herring on the western shores of the Cheshsk Bay with motorboats breaking the ice was not very efficient, so Sevryba promised to send a 300-ton vessel and a Finnish boat with a 20 hp motor.[58] The secretary of the Nenets party committee, A. Gudyrev, called for the expansion and mechanization of fish operations and the construction of housing for workers as ways to increase the processing of the harvest. The Cheshsk, Indigsk, and Pechorsk fish factories of Sevryba had limited means to receive catch of navaga, smelts, and herring, let alone process the catch at docks in dilapidated buildings, and the factories had few cranes or other machinery to offload and handle the fish. On top of this, as usual, few workers had housing or access to cultural organizations or schools for their children.[59] But the head of Sevryba rejected these worries. He argued that it was superfluous to build new facilities near Naryan-Mar when Perepuski was only twenty kilometers away. He acknowledged the need, however, for a Fisheries Technical School to train a permanent labor force of indigenous people and agreed to ask the Ministry of the Fishing Industry to provide factories and trucks to facilitate handling of fish and to acquire refrigerated seiners from the Petrozavodsk Shipbuilding Factory.[60] Given the constant struggle over limited processing facilities for inland and ocean fisheries, it is understandable why one of the most powerful fishing nations in the world could never get good, fresh, canned, filleted, or other fish to the Soviet table.

While catch of fish and marine mammals in the Nenets region had grown from 342,000 centners (one centner = 100 kg) in 1940 to 1.2 million centners in 1954 and included such valuable fish as various cod, salmon, sturgeon, and herring and while the fleet had grown in size and reach, improvements in processing capacity on shore had not followed. Most of the buildings and equipment dated to the 1920s and 1930s, the decrepit wooden structures had begun to collapse, and it was impossible to produce a high-quality product given poor canning and smoking facilities, the absence of refrigerators and freezers, and no processing equipment. Things were worst of all in Arkhangelsk where the Fish Combine faced deliveries from thirty trawlers in 1954, up from only thirteen just four years earlier but with no increase in handling capacity.[61]

In the mid-1950s the fish industry began to make progress with the expansion and partial mechanization of fish processing facilities and of ship repair factories. Canning had new production lines. In Naryan-Mar the Pechora Fish Factory opened such a line, as did the Kaninsk Fish Factory, with new facilities scheduled to open in Soiansk, Onega, and Pinaga.[62] Yet the closer one moved to shore, the more the uneven development of the industry came into view. Most ships were poorly equipped, underpowered, and of small size, housing and cultural "stock" lagged, and workers were most dissatisfied with the quality of life and work.[63]

In addition to shortages of modern equipment, another problem concerned workers who felt poorly motivated and often lacked adequate training. The result was accidents. In 1946 a fire on the *Sedov* killed a large number of personnel and required a mass funeral service, with a commission established to locate the relatives of the dead, fly them to Arkhangelsk, and bring them by train to Molotovsk to organize viewing, service, and reception for employees and relatives and arrange the pensions of the sailors to be paid out without delay.[64] The authorities more often than not blamed personnel for accidents. No matter the size of the economic unit—the factory, ship, railway station, and so on—they counted scores of cases of "violation of labor discipline" that ranged from tardiness to truancy, leaving a work station unattended, drunkenness, failure to follow orders—that led to reprimands and firings.

One has the sense that high accident rates resulted from a combination of worn-out equipment, constant pressure to meet targets, poorly motivated workers, and a low level of education. It is not possible to determine if accident rates were higher in Arctic regions. Accidents and other persistent safety problems convinced members of the Northern River Steam Ship Authority to focus more effort on training of the merchant marine during the inter-navigational period. Baev, an instructor in the department of water transport of the Obkom, argued that the high accident rate indicated insufficient preparedness among the fleet commanders and inadequate training. Only 22 percent of leading personnel had general middle education, 23 percent middle technical education, and the remainder only elementary school education. Seventy-one captains, forty mechanics, thirty-six motorists, 127 wheelmen, and 360 stokers lagged in training. The authority asked for the creation of a marine technical school.[65] According to one official, "Fifty percent of our mechanics do not correspond to their position," and eighty percent of accidents resulted from this.[66] This was the administration's fault as much as anyone else's, for as a union official pointed out, "Our personnel don't know where they will winter, which does not create the proper conditions for a settled lifestyle of the merchant marine."[67]

A high level of alcoholism plagued the development of the Northern Sea Route, including the problem that a man fired for drunkenness on one ship might easily get employment on another, so rapid was labor turnover. All the captains seemed to be aware of this.[68] In other cases, sailors somehow provided papers to show they were fit for service when they were not. In spite of the cost, one captain called for a medical exam before each expedition. And, because the older sailors who had served more than twenty years at sea more frequently fell ill, he also called for a better educational system to ensure the flow of younger cadres from, say, the Arkhangelsk Preparatory Marine School.[69]

Technology in the Forest

If in Karelia planners hoped to use modern melioration and reclamation techniques to drain swamps and wetlands and across the tundra of the Kola Peninsula sought hydroelectricity to power mining and metallurgy, then in Arkhangelsk province—and other regions of the USSR—they intended modern technology to permit efficient harvesting of lumber. The taiga was certainly rich with lumber—and the Russian republic of the USSR had roughly one-half of the timber in the world. But owing to undercapitalized forestry operations, clear-cutting near to existing road and railroads, and workers who were underpaid and poorly rewarded, nothing close to modern selective cutting took place. Indeed the absence of modern technology made the forest a bleak place to live and work during Soviet power. A comparison of Arkhangelsk forests with the Northern Forest of New England gives a sense of the technological lag in the USSR and the grotesque waste of resources from forest to float to mill.[70] Lag in technology in the forest and for workers at rest made forestry operations inefficient and dangerous.

Kotlasles (the Koltas Region Forestry Trust in central Arkhangelsk province) consisted of twelve forestry collective farms and fifty-eight logging points, of which only thirty were mechanized. The trust maintained only 117 kilometers of narrow gauge rail, fourteen gravel and dirt roads at 140 kilometers, and twelve tractor roads also at 140 kilometers to connect all of these operations. It acquired such equipment as tractors, winches, and generators over time, but remained underpowered for this equipment and for locomotives, automobiles, bulldozers, and steam-powered devices.[71] Its Konetsgorsk forest collective farm failed to fulfill plans because both tractors and winches were in repair. Of sixty-two KT-12 tractors, only fourteen to sixteen were on the job every day, while of two winches, only one operated—and only for ten days in June 1954.[72] For every year and every forestry operation I examined in the archives, the lag in technology meant perennial underfulfillment of plans for felling and yarding, while tractors and winches and other equipment remained unavailable, in repair, and underutilized.[73] These shortages continued throughout the decades.[74]

Glavleszapchast (the Main Administration for Forestry Parts) could not provide the staff (it was more than 200 employees short) or the parts to repair ten S-80 and fifty K-12 tractors, five bulldozers, fifty ZIS-21–150 trucks, cranes, generators, saws, and so on and thus was well behind on all repairs for 1953 and through the first nine months of 1954. The administration suffered from labor turnover and lack of machine tools.[75] One Khalchev, senior engineer of Arkhangelskles (the Arkhangelsk Forestry Trust), reported that repairs for 1953 on tractors were only 28 percent compete, no trucks had been repaired, parts supply remained a problem, the garage was too small, and they had fifty-five of the 147 mechanics they needed who were not very qualified and quickly moved on to easier work.[76]

The taiga never willingly gave up its lumber and pulp to workers, no matter how hard the workers tried or how actively party officials exhorted those workers. For the 1949–50 season with only days left, the plan was only three-quarters fulfilled, and only in the Upper Toemsk and Cherevkovsk[77] forests did the situation look promising, while at other yarding operations in the middle of the forest points the situation was bleak and soon mills would be screaming for logs.[78] The productivity of laborers at most enterprises in the region measured at cubic meters per day was 50 to 65 percent of plan, in part owing to shortages of equipment and horses. At the Leisk, Krasnodrask, Shenkursk, and Kolmogorsk forests even laborers on the "mechanized" roads could hardly move 1,000 cubic meters per day, and they were was short hundreds of lumberjacks and horses.[79]

Nature, not bold Soviet plans, determined the cutting season. Preparatory work on roads had to be finished before the fall freezes and winter blizzards. But for Severoles (the Northern Forestry Trust) lumbering remained seasonal, with 70 percent of the plan fulfilled in seven months from June to the end of the year. Latunov wrote Khrushchev on February 20, 1954, reporting on the massive forestry products industry in Arkhangelsk of nineteen mills with 133 saws. Yet Severoles did not work year-round because concentration of cutting in the summer and fall that deprived mills of logs for five months out of the year. Because of this seasonality, they let workers go, and as soon as deliveries arrived they had to call them back to the mills at great cost and effort. In 1953 they hired 14,200 workers and let 11,600 go. Latunov begged the busy Khrushchev to order the Ministry of the Paper Industry to get on board to operate the mills year-round.[80]

Arctic industries generally were unable to attract skilled employees. Many employees in the taiga had incomplete education, some of them were barely literate, and many turned to alcohol for company during their long, hard labors. Dvinosplav, the organization responsible for moving lumber from forest along river to mill had but eighteen engineers and forty-five technical specialists in all of its enterprises, many of whom were inexperienced. Of ten bosses of the float, only one had higher education and one had unfinished middle education. Of 100 directors at holding grounds (river storage booms), nine had middle and ninety-one elementary education. Dvinosplav was so short of qualified individuals that it kept people on who were even incompetent or filled the wrong position based on their (low) qualifications. One Comrade Chakhotin on the huge river drive "Dvinovazhskoi" was allegedly systematically drunk and had mismanaged 12,000 m^3 of lumber on a narrow tributary, yet was permitted to work another seven months.[81] Perhaps it was better to have a drunkard than no worker. In 1948 in the eight major forestry trusts (Dvinoles, Kotlasles, Onegoles, Mezen'les, Sevtransles, Konoshtranles, Dvinosplav, and Severoles), they were short 444 engineers and 767 technical specialists—two-thirds of the requirement.[82] Not surprisingly, across the Dvina, Kotlas, Onega, and Northern forestry trusts, only 55 percent of workers fulfilled norms, and for several of the regional operations fewer than 25 percent fulfilled norms.[83] On top of this, a large number of workers who completed forestry school programs were sent to work in other regions, which left Arkhangelsk forests short of lumber camp managers and tractor and truck drivers. Even sending loyal communists into the woods did not do the trick. In 1947 party officials sent 1,069 communists out to fell trees, 455 from Arkhangelsk, 614 from other cities, and 121 simply left, fifty deserted, twenty-eight retired because of illness, and twelve took leave for family reasons.[84]

It took a long time for the authorities to realize that they contributed a great deal to the anarchy in the forest by punishing, transferring, or firing employees at a moment's notice so that there was little continuity of leading personnel and experience. In this way they often treated people like the capricious Arctic climate they believed they could ignore.

The Slow Pace of Road and Railroad Construction

The Dwight D. Eisenhower National System of Interstate and Defense Highways (the "interstate," for which congress found $500 billion in part by adding "Defense" to its name and to the title of the bill) contributed to new patterns of economic growth and settlement in postwar America, may have singlehandedly destroyed passenger rail traffic, and encouraged a second boom in the opening of the American West and Southwest. The 45,000-mile-long interstate highways simultaneously triggered the transformation of rural America. In some cases they speeded the decline of small towns located on the previous national highway systems whose residents were suddenly bypassed by the multilane, limited-access roads. Pushing across the West and Southwest, the highways led millions of people to leave the rust belt. The highways also drained valuable federal support from municipal transport systems. Without Amtrak, the poorly funded but crucial passenger rail, passenger traffic would simply have disappeared. Finally, the interstate gave impetus to large-scale transformation of the environment, not only through the construction of thoroughfares with thirty- to seventy-meter-wide rights-of-way, but through the spread of franchises of fast-food restaurants and stores that homogenized the landscape.

In the former Soviet Union, roads and railroads also had economic, social, and environmental impact, although in comparison with the United States on a much smaller scale, especially because the roads and railroads were not as extensive, and into the twenty-first century do not meet needs of length, regional reach, quality, and safety. Russian highways remain among the worst in the world in terms of quality and safety according to the World Economic Forum, for reasons ranging from outmoded construction practices to corruption. The poor quality of roads limited economic growth "by slowing the movement of freight and increasing its cost but also resulting in the second highest rate of highway fatalities in the world."[85] In 2012 Russia had 47,300 kilometers of federal highways and an annual budget of $17 billion versus perhaps $60 billion in the United States, which has ten times as many paved roads; only 5 percent of Russian roads are of good quality. The distance between Arkhangelsk and Petersburg is less than 800 kilometers as the crow flies, but 1,300 kilometers by two-lane roads taking seventeen hours at best to drive, and the "modern" train takes twenty-five hours. The pace, scale, and history of transport infrastructure have been uniquely "Russian." Regarding the fragility of Arctic ecosystems, the impacts of roads and railroads have been extensive and irreversible.

During World War I the Tsarist government finally actively supported the construction of a rail line to Murmansk using prison labor, in part to ensure that military supplies from allies could be delivered and also in part in answer to businessmen who had long railed for the line. Soviet planners accelerated the construction of roads and railroads, especially with the promulgation of the five-year plans and the expansion of the network of the gulag system across the northern latitudes. Still, the transportation system remained inadequate to the needs of the state and economic managers until the end of the Soviet era, especially in Arctic regions, in terms of total length, freight capacity, quality, safety, and other indices. Narrow two-lane roads, many of them unpaved and with nary a guardrail in site, slowed traffic. Few trunk lines facilitated travel between axials radiating from Moscow, requiring freight to move greater distances and through a series of transfers. The further from Moscow and Leningrad, the lower the population densities, the harsher the climate, the less local officials found it possible to impress upon their ministerial and party superiors the dangerously critical state of transport, communication, and other infrastructure.

The poor situation with roads and railroads interfered constantly with the operations of the dozens of forestry, transport, construction, and other organizations spreading inexorably across the Arctic and subarctic. Even when harvest targets were raised, road-building budgets remained inadequate. Dvinoles, the Dvina Forestry Trust, completed very few roads and had few tractors to build or maintain them. In one year, it ordered 205 tractors, but factories delivered only eighty-six of them. Dvinoles planners indicated 275 kilometers of roads of various kinds, but only 140 kilometers had been built. Comrade Medvedev, the main engineer and deputy director of Dvinoles, concluded, "I personally consider it a great mistake that we don't have any technology whatsoever in the forest. It is a basic law on the foundation of which we should carry out all activities, to which everything should be subordinated, on which the brains of our scientific-technical personnel ought to work. We do not have technology in the forest . . . Is it possible that [officials of the Commissariat of Forestry have] have not been in the forest? But it turns out that the Commissar is also an enemy. If he were a man who was devoted to Soviet power, he would look at what goes on in the forest." Indeed, Medvedev asserted, they had "retreated" from modern technology, which was the same thing, in his mind, as wrecking.[86]

Considering the size of the territory of the Russian northwest and the complexity of communication and transport networks, it was no small feat to create transport infrastructure and rebuild it after the war. Arkhangelsk province had five train departments with a total of 1,148 kilometers of track, dozens of train stations and depots, plus the addition of the Konosho–Kotlas line in 1942 at 350 kilometers (Konosha, a forestry town of about 10,000, located in the southwest corner of Arkhangelsk province), with a second Konosho–Arkhangelsk line to be finished by September at 423 kilometers. In addition the authorities had installed nearly 51,000 radio public address systems, 10,000 of which filled the countryside with joyous reports of enlightened party leadership, and 32,998 telephone and telegraph connectors, 1,932 kilometers of roads of all-union significance, and 12,588 roads of regional and local significance (which meant they were dirt and gravel) to bring each nook and cranny, peasant, horse, and nanny into the family of Soviet brotherhood.[87]

In mid-1947 Sevlestranstroi had spent only 5.4 percent of its annual budget on road construction, and a series of its lumber operations simply had no roads. Sometimes construction crews had equipment (but often not), and sometimes they had met employee targets (but often not). In Edemsk the 120 workers including thirty women and girls sat around idly, rarely showing up for work in the face of no equipment, no bosses, and no monthly pay. Construction crews lacked every kind of tool: chisels, gouges, pries, planes, joiners, sledge hammers, awls, levels, steel measuring tapes, and so on.[88] In the summer of 1947, among its various operations Sevlestranstroi had only 1,176 of its planned 2,126 workers and fifty-seven of 141 horses. Another construction organization, Sevlesstroi, had acquired sixteen ZIS-5 and thirteen Gaz-AA vehicles (designs based respectively on American Auto Dispatch and Ford AA trucks), but only ten of twenty-six trailers required to move the lumber to yards or river banks, of twelve tractors only four were in working order, only three of twelve winches, and only two of nine bulldozers. The upshot was that the construction of forestry roads remained a labor-intensive job at every step: earth moving, stump removal, grading, and so on.[89]

Arctic conditions, low population densities, and equipment shortfalls frustrated all plans and activities. One Comrade Ippolitov, the deputy director of the Main Administration of the Northern Forest Operations, pondered how to speed up lumber operations if they had only five narrow gauge railroads. More to the point, many roads operated only seasonally, not year-round, especially the northern trusts of Onegoles and Dvinoles, not only because of the poor condition of roads, but because of the absence of workers and of machines. Ippolitov in part blamed managers and engineers in forestry operations, especially within Dvinoles, for the failure to mechanize the industry and for a lack of understanding about how best to organize available machinery, equipment, and lumberjacks into efficient operations. More "technical discipline" was required, by which he meant more efficient use of existing machinery and equipment. Obkom secretary Latunov lamented the fact that year after year "we declare and declare for one and the same reason—that we use technology poorly and workers have very low productivity."[90]

In the railroad industry, train engineers faced the unenviable task of trying to keep locomotives running at higher speeds and unloading freight quickly, yet running on rails prone to failure. The Pechora department of the Arkhangelsk-Moscow railroad line discovered that at least 32,000 rails, or more than 200 kilometers of poor quality, lighter rails, were subject to failure. Crashes and accidents had resulted. Engineers had to slow their trains in some cases to 15 km/hour. A heavier rail was being manufactured by 1952, but only eighty-five kilometers of rail had been replaced. On top of this, 1,148 railway bridges built during World War II of low-grade, untreated pine had begun to rot from truss to weight-bearing beams, and only a few hundred of them had been replaced.[91] If in 1948 there were four crashes and 264 cases of "substandard work," then in 1949 there were ten crashes and 377 cases of substandard work. The authorities reported a "huge accident" with a passenger train at the beginning of 1950, although not how many injuries or fatalities. But they determined the cause was a "low level of labor and government discipline among workers connected with train traffic." In the previous fourteen months they had issued more than 1,000 administrative punishments including 212 involving legal proceeding, 673 for "violations of technical norms," and 197 for truancy or skipping work.[92] In other words, the worker, never the technology, was the source of failure. Arctic roads and railroads had begun to fray around the edges as soon as they had been laid.

In spite of regulations, unsafe conditions prevailed. Latunov alerted the minister of the ways of communication, B. P. Beshchev, in May 1954 that the Niandomsk rails were in miserable condition, the majority of the system consisting of rail laid as part of the push to Murmansk in World War I had worn to nine millimeters, and scores of rails failed each day. In 1951 10,316 highly defective rails were replaced, in 1952 another 9,277, in 1953 13,200, and in the first half of 1954 5,482 more. But the number of defective rails continued to grow. Four hundred kilometers of rail required immediate repairs; crashes were epidemic. At least the central authorities provided the Arkhangelsk–Moscow trains Nos. 71 and 72 with "new" passenger cars—which citizens found more comfortable—that had seen duty in the Siberian rail system.[93]

The two Ts—employee turnover and backward technology—created problems along the track everywhere. The Lavelsk forestry farm was in a critical state. In 1954 Lavelsk had 639 employees and in that year had hired 513 persons, yet released 402 for various reasons. Not surprisingly, the plan was only 85 percent fulfilled. The major reason was the workers in the lower yard did not have access to loading platforms, and in the upper yard there were no empty railway wagons. The narrow gauge rail had been hurriedly built, "as a result of which frequent accidents occur." On October 15, 1954, seventeen wagons and locomotives derailed, and it took the workers forever using pries to get them back on track. Truancy led to the loss of an average of 11.3 laborers daily, and in October up to seventeen per day. Not surprisingly, if we can judge from archival sources, the inadequacy of lectures and discussions of foreign affairs hardly made inroads in fighting crime, hooliganism, and drunkenness. Perhaps the poor attitude of workers had something to do with cramped housing with no furnishings, no water, no dryers, and no curtains or tablecloths.[94]

Everywhere one looked—the Upper Toemsk, Velsk, Rovdinsk, Shenkursk, Ust'ianksk, and Cherevkovsk regions—road and railroad technology lagged. The operations were short narrow roads that were classified as gauge rail, truck, tractor, ice-horse, rails-horse, and snow, as well as generators, tractors, winches, saws, trucks, cranes, and sleds. Not surprisingly they also lagged in *sotskul'tbyt*—housing, clubs, red corners, schools, first aid stations, and banyas, as well as supplies to stores.[95]

Passenger rail was an afterthought. The Kotlas–South Train Station was planned at 350 square meters and capacity of 250 people, but each day 850 intercity passengers and 900 suburban passengers passed through. The design failed to include a place for baggage operations, kept baggage, or mother and babies.[96] Not only train stations failed to meet up to already low Soviet standards. Intercity trains—the Arkhangelsk–Isakororka, -Molotovsk, and -Obozerskaia lines—lacked even simple comforts "from the moment of the purchase of the ticket until arrival at the destination point."

On top of this, transport from Arkhangelsk proper to the train station on the other side of the Northern Dvina River was by ferry with no passenger, automobile, or rail bridge. The station's platforms were too low and too short, and the station was filled with garbage.[97] Secretary Latunov wrote Stalin directly concerning construction of a dedicated steel bridge across the Northern Dvina to replace the ferry that operated no more than 225 days each year, of which forty days required two powerful icebreaking tugs to accompany it.[98] During World War II the authorities ordered a temporary bridge built cross river to the Bakrits and Isakorgka passenger railroad stations.[99] This decision apparently reflected the need to transport wounded soldiers arriving at the Arkhangelsk Pier Station by boat or rail to hospitals on the other side. In summer time, special steamships equipped with ambulatory facilities carried the wounded across who were then carted by ambulance to hospitals. The most seriously wounded would have to be taken to hospitals in Molotovsk and Isakorogka, while in the winter an ice bridge was built.[100] Until 1964, Arkhangelsk residents had to take a ferry—or walk across ice—to get to the Arkhangelsk train station. Further improvements in daily life waited for the late Khrushchev era.

Qualitatively these lags persisted into the late Soviet era. Train station buildings arose with delay and then melted away with the spring thaws. They rarely had bathrooms or running water. Motor pools and supply depots came into existence as an afterthought. To this day the only way into Naryan-Mar in the Nenets Autonomous Okrug is by airplane or helicopter, the railroad serving Norilsk having passed the Nentsy by and a road slowly centimetering its way from the Ural Mountains to Naryan-Mar that may arrive in 2020.

Talking with Vanya at the Collective Farm

Spotty service of all sorts was particularly pronounced in regions far off the beaten path. The very limited availability of telegraph, telephone, and postal service in the Russian Arctic reflected more than high capital costs, great distances, and low population densities. While regional officials, managers, medical personnel, and others recognized the role that telephones played in promoting economic growth, political control, and public health (for example, emergency medical service), precisely low population densities and great distances convinced planners that limited investment funds might better be used in other sectors of the economy, notably in heavy industry. As a result, central planners and party officials determined not to fund infrastructural technologies that were crucial to forestry, farm, and other economic operations. Officials instead supported the growth of telephone service in larger cities and made phone numbers available largely to well-placed individuals. Yet even being designated a priority was no guarantee of success. Officials of the Arkhangelsk Automatic Telephone Station pressured the "Sevkabel" factory to produce wire for telephone lines after years of unfulfilled production promises. Sevkabel production of telephone line reached twenty-four kilometers in September 1935, thirty-nine kilometers in October, and thirty-five kilometers in November. Soon they would have 3,000 customers in this important city of lumber, fish, a trade port, and the home of the region's party apparatus—but only if the Red Dawn factory manufactured phones.[101]

In 1917 the nation had about 240,000 telephones versus 700,000 for Britain. While in Europe and the U.S. telephone service expanded as a matter of course in the early twentieth century, in the USSR it actually shrank in the 1920s both for reasons of technology and reasons of conscious decision in spite of the fact that the Bolsheviks and Lenin in particular understood the importance of employing telegraphy, the telephone, and the postal service toward the ends of political control.[102] The civil war and growing shortages of parts and wire led the Bolsheviks to order the volume of telephone communications cut and the remaining capacity to serve primarily the government and economic instances. This led to communalization of phone service for citizens, a decline in their access to it, and five-year and longer wait periods for lucky folk to get a phone line.

During the NEP the authorities likely deliberately suppressed service to prevent unmonitored horizontal communications, at the same time learning to monitor conversation. Another reason for the shrinking of phone service was that the Commissariat of Heavy Industry dumped unprofitable telephone exchanges on local governments that had even fewer resources; this stifled development for the long-term. The suppression of numbers of subscribers was surprising given both Lenin and Trotsky's belief that communication and transport technologies would help create the all-important *smychka,* or the bond between the city and the countryside.[103]

Communications lagged throughout the Arctic. In 1938 the Commissariat of Communications learned that wiring in Murmansk region was in a "catastrophic condition" with a series of cables out of service and proper tools for linemen unavailable. Makovetskii, the head of the regional communications office, attributed part of the problem to the fact that "the building season in Murmansk is extremely short" and that staffing levels remained a problem. Local authorities asked for a half million rubles to solve these issues.[104]

Telecommunications infrastructure remained inadequate through the war.[105] The Murmansk Administration of Communications freely admitted in 1944 that the existing system of city-to-city communications in no way answered the great demands which inhabitants and enterprises placed on it. Rapidly unfolding construction along the entire peninsula, the mining of nickel ores in Monchegorsk region, the development of Kirovsk region, and other industrial development and reclamation projects demanded "stable and strong connections." Yet the regional centers (Kirovsk, Monchegorsk, Pulozero, Apatity, Kandalaksha, and so on) were able to carry out telephone conversations along iron wiring, which made for nearly inaudible connections. In most places, bosses could not just pick up the phone, but had to order calls ahead of time over low-grade steel wire; copper connections were always overburdened by talks with Moscow, and this in the region that produced the copper. Most regional centers had at best telegraph, but lags in repair and capital construction made the situation hardly operative. In 1940 in all of Murmansk, only twenty-seven enterprises had telegraph, and only forty-four could receive and send telegrams by telephone. There were three intercity phone centers, and only 1,624 kilometers of lines with 6,171 kilometers of wire but only 832 kilometers of which was copper.[106]

In the absence of telephone, radio and telegraph had to suffice. Yet technological lag hindered the operation of these technologies, too, in the Arctic. Of the regional centers of Kirovsk, Ionanga, Kandalaksha, Umba,

Monchegorsk, Lovozero, Ura Buba, Teriberka, Kola, and Poliarnye Zory, only the first two only had radio telegraph connections. In Kandalaksha region, important for its lumber industry, only five of sixty-four rural councils had telephone connections. They operated poorly at best, with inordinate waiting times before connections were made, and suffered from a large number of dropped calls. The system simply could not handle the volume, insulating capacity was limited, and few repairs were carried out.[107] There was but one phone per 1,000 residents in Murmansk in 1944, and four in all for 16,000 Monchegorskians. To put it bleakly, modern telephone service came to the Kola Peninsula only in the twenty-first century with mobile phones.[108]

In Arkhangelsk province things were no better. Local party officials implored the Arkhangelsk party organization to provide resources for improved telephone and telegraph service. In the winter of 1952 the Pinezhsk raikom (regional party committee) reported that the Kuloi village council was forty kilometers from regional communications with letters delivered at best three times per week, but often with no delivery at all, with the Ezhugsk forestry operations 157 kilometers from the nearest post office, while the Plesetsk Party committee complained that telephone service was nonexistent in the countryside; at least Plesetsk village itself had 100 numbers with 132 clients. In many places they were lucky to have hand-cranked radio connection. For example, Safokovo, Soina, Moseevo, and Kargopol village had radios but no electrical connections. The hand-cranked generators had burned out; the generators were sent for repairs in June 1951 but by February 1952 still had not been returned.

When it existed, telephone service was miserable. The telephone lines of the Plesetsk district—that would soon become home to a major cosmodrome—were overwhelmed 50 to 60 percent of the time by telegrams. This meant that six collective farms, eleven forestry organizations, twelve communications groups, and a series of other social, cultural, and industrial enterprises had access to phones only an hour or two per day. A 1954 plan to extend phone services was only partially fulfilled. In Mekhrengsk region, with five villages, eight collective farms, four forestry groups, and so on, there was but one telephone line stretching 101 kilometers that served eleven telephones.[109] In 1954 the Uftiug railroad had no telephone, nor did the regional political center; telegrams came by train over two to three days "losing their significance."[110] Postage and telegrams, already moving slowly, slowed to a crawl.[111]

The provincial authorities planned to improve telephone connections markedly during the fifth five-year plan (1951–55). If in the province on January 1, 1951, there were 10,230 phone connections, in 1954 there were 11,240, and by the end of the plan they hoped for 17,700 in total, with the addition of phone lines in the cities of Niandom (360), Kotlas (2,000), Molotovsk (3,000), Arkhangelsk (6,000), Plesetsk (240), Velsk (720), Kargopol (240), and Semenovsk (240).[112]

In the 1950s Secretary Kozhevnikov of the Belomorsk regional party committee complained to the central authorities about the need to improve the local telephone system; his letters to the provincial phone authorities had gone unanswered. Only four of fourteen agricultural councils and four of sixteen collective farms had telephone service. This required officials themselves to go to the closest post office to arrange calls at great expense of time and money. How could local enterprises conduct business properly if they could not be in contact with regional planning bodies? Kozhevnikov requested a modest sixteen additional telephones for the entire region. Through the intervention of the Obkom, the phone company agreed with the request for telephones, although refused to supply up to ten phone operators for switching stations.[113] The fact that Obkom intervention was required for such seemingly minor items as telephone service was a special feature of the Soviet system of centralized planning and distribution of goods and services in the absence of market demand.[114] On the eve of the collapse of the USSR in 1991 most people still waited years to have a landline installed.

Petty disputes among local people—that reflected the absurdities of bureaucracy—also interfered with regular communication. While steamboats carried both passengers and letters and parcels, the boats did not stop long enough to ensure that mail was loaded and unloaded. On June 3, when the *Ivan Kalaiev* bound from Arkhangelsk to Kotlas stopped at the Topsa dock, the postal clerk ran out to the river to find the boat had already pulled anchor and was steaming away.[115] Of course, things were no better with air delivery; even in the summer the 72nd Aviation Detachment could not manage to serve Dvina regions regularly—and often not at all.[116] The absence of modern communications made the cold Arctic expanses seem more cold and isolated.

Drafty Dormitories and Dingy Dining Halls

The technological inequity in the Soviet Arctic was reflected above all else in the lag in bringing technologies of daily life to the people who filled Arctic spaces. In the twenty-five years that I have worked in Soviet and Russian archives and libraries, I have never encountered an industrial or agricultural setting in which *sotskul'tbyt*—housing, stores, hospitals, schools, roads and municipal transport, sewerage and water—was completed in a timely fashion to provide decent daily life for workers. In the Arctic regions, my sense is that sotkul'tbyt lagged even further behind than in other regions of the country. This may be because budgets were tighter for enterprises that, like elsewhere in the USSR, were responsible for providing housing for their employees and because of the challenges of building in such difficult meteorological and geological conditions and with such low population densities. This occurred in taiga, in tundra, at industrial concerns, in burgeoning cities.

The Kola Energy Company, Kolenergo, for example, could not guarantee to feed or house its workers because of more pressing concerns: building dams. Kolenergo's directors complained to the central authorities that construction of apartments lagged. More crucial was the poorly functioning bakery—given the wistfully remembered slogan of the Russian Revolution, "Land, Peace, Bread!" Where was the Arctic bread? The town at the Tuloma power station grew rapidly in postwar years, while the number of ill-provisioned stores and so on remained the same, and a terrible shortage of food persisted. What could they expect of workers who often put in eighteen-hour days and had no regular work schedule or vacations, lived on top of one another in miserable apartments, and couldn't even count on potatoes? One worker wrote that "lunch in our dining hall constituted a great strain on the nerves" because you must wait and wait and wait then "quickly gulp down your food which often consists of an inedible meal." On top of this drunkards lay about in the cafeteria, and there were often fights.[117]

Another aspect of technological lag, as noted in Chapter 4, concerned worker safety. Not only did workers suffer at home, they might never return home from accidents at work. They did not have hard hats, steel-toed shoes, special clothing, safety goggles, or ear protection, and the machines they operated might bite. On the railroad, safety lagged in conditions of rotting bridges and ties, unhappy workers, and overburdened facilities. If the train traffic on the Kuloisk section of the Pechorsk Railroad was indicative, then accidents were epidemic.[118] Unfortunately, investigators merely accumulated statistics and did not search thoroughly to discover the reasons for accidents because they were too often willing to blame the worker.

Legion problems faced every branch of transport, communications, and construction. The Isakorgorka branch of the railroad had woefully inadequate medical facilities. It operated one hospital with thirty-five beds, ten for the maternity department, that served 12,000 workers and their families, and the clinic was in a small room of a wooden building dating to 1913.[119]

Everywhere he looked, the Arctic worker saw his frozen compatriots and their families without basic needs, crowded into more and more congested housing. Town fathers and mothers penned missives on a monthly basis about the rapid increase in production under their lead, but their missives on the need for schools, better transport, public baths, dining halls, and housing went nowhere. To take one example, because of the growing importance of the Pechorsk railway, Solvychegodsk had grown to 6,000 inhabitants and soon would top 9,000 people. Already 200 families of the employees lived in the Kotlas train station and other stations along the route.[120]

There was something behind the drumbeat of blame assigned to the worker: High labor turnover and limited experience among employees plagued all sectors of the Arctic economy. But poor *sotskul'tbyt* led to high turnover and low job satisfaction. In the Arctic fleet constantly new faces with insufficient training and experience led to an increasing number of serious accidents—in 1946 there were eight cases, in 1947 fifteen cases, in the first half of 1948 five cases. The main reason was "insufficient technical literacy, violations of technical norms . . . and violation of discipline." Officials must have known that this resulted in part from the absence of housing: in 1946 fifty-four people left the fleet, in 1947 113 left, and in the first half of 1948 there were already seventy-six departures, with the rate doubling in two years. Five percent of the sailors had no apartment or lived temporarily with others, 40 percent lived with big families in rooms from 100 to 160 square feet, and they were all poorly paid.[121]

The rivers and seas were a point of real concern. Obkom secretary Latunov wrote Molotov, deputy chair of the council of ministers, in 1952 to repeat his concern over the growing shortage of personnel for the Northern Steamship

Authority yet rapidly growing tonnage of freight, especially lumber, and increasing targets year by year. They could not attract workers without housing—279 families lived in kitchens and cubicles unsuitable for inhabitance. Young specialists as a rule lived six to eight people per dorm room. Even many captains lived together—with five to six people in 100 to 150 square feet, not to mention the absence of a workers' club or any entertainment. The Ministry of River Fleet responded with the offer of a 10 percent salary bonus for each six months of work, plus a one-month salary bonus for workers signing a one-year contract, with subsidized groceries, toward the end of hiring 500 individuals. Housing would come only at a later date, but the ministry promised it, too.[122]

Party archives were filled with pieces of paper, reports, indications, evaluations, and other "-ations" of employees in the forest, on ship, and on shore, on their personalities, qualities, ability to work with others, their levels of political consciousness and engagement, whether "technically literate," whether they had initiative and ought to be considered for special courses to raise qualifications. But also, judging by the archival materials, decisions about whether to free a promising worker to attend an institute to raise qualifications would be decided first of all by whether he was needed to meet production targets.[123] The archives reveal terrible problems in spite of the tireless care of the party. Beyond high rates of turnover, they included the youth and inexperience of sailors; a hands-off approach on the part of captains; drunkenness; workers unhappy with shore leave so short that they could not wash their clothing; tardy return to ships because their apartments were so far away; miserable living conditions on board; inedible food; rats who ate the sailors' boots; filthy quarters; and so on.[124]

It was no better on land, although one might think in the forestry industry it would be easy to build a wooden house. Yet workers were always miserable—when they woke in the morning in cold dormitories in cots without linens and pillows and at night when they returned to the drafty domiciles and dingy dining halls. Throughout Arkhangelsk forests the story was repeated. Housing lagged significantly for Dvinoles, Onegoles, and Kotlasles; as of February 6, 1950, only 521 Dvinoles workers had housing versus 1,352 by plan, and Onegoles was little better.[125] Woe be it to new employees who were unlucky to get a spot in a cabin without a stove, bed, or any other comforts. Dining halls were no better.[126]

The new engineering department at the Arkhangelsk Forestry Institute that had been organized in 1951 to improve the qualifications of sector employees could not provide good housing for potential students. The dormitories had "no housekeepers, the sanitation is miserable, and they are poorly equipped . . . All of this negatively affects the study of students."[127]

The regional union representative added his voice in a complaint to the Obkom's forest industry department in October 1954. He noted the work of brigades, workshops, and forest collective farms was a mess. Workers stood idly by, machine and equipment was of low quality and in ill repair, and workers had turned to booze rather than work. They left early. They despised the dining hall with its tasteless noodle soup followed by cold noodles. They could not get soap, salt, potatoes, vegetables, or bread. Their living conditions left much to be desired. They were bored in their spare time by the absence of radio and film, while the red corners and workers' clubs were best for ennui. Knit goods were a fantasy. Things were particularly bad in Voliuksk, Andelsk, and Makoveevsk. In the Makoveevsk store, the salesman was drunk every day and the workers helped themselves to the goods.[128]

The Velsk forest collective farm reported to the secretary of the Central Committee of the Moldavan Communist Party, Gladkin, on the work of Moldselles (the Moldavian Forestry Trust). Moldavan and Bulgarian "gastarbeiter" often toiled in the taiga of Arkhangelsk and the Komi Republic. The report indicated low productivity of labor, the absence of machinery and equipment, and backbreaking manual labor. In the second half of 1953 more than 17,000 man-days lost were lost to truancy, tardiness, fighting, and "manifestations of amorality." The Moldavans preferred to stay home, sell goods from their houses at speculative prices, and drink.[129] Why would Russian Arctic lumberjacks be any different?

Toiling in the Tundra

Soviet authors addressed the isolation and technological inequity of the collective farm and the countryside. The writer Nikolai Kuz'mich Zhernakov (1914–1988), from Kholmorgory, a village southeast of Arkhangelsk on the Northern Dvina River, the home of eighteenth-century native-scientist Mikhail Lomonosov, at one-time a religious center, and still known for its bone-carving handicrafts, had a difficult early life. He lived in an orphanage, had to drop out of school to work as a lathe operator, baker, fisherman, and longshoreman, and finally served in the Red Army on the Finnish front, where he was seriously wounded in war. Unable to do his accustomed heavy work, he bounced around for a while before finding his calling in 1955 on the editorial board of a regional newspaper, *Kholmogorskii Kolkhoznik* (Kholmogor Collective Farmer). He began to write poetry and prose about life in the countryside, collectivization, and problems of rural life. He transferred to *Pravda Severa.* He wrote historical novels, in 1959 becoming a member of the Union of Writers. Zhernakov was a devoted communist and member of a number of editorial boards, and he won a series of awards and medals.

In *Krutye Suvoi* (Deep Drifts, 1975) Zhernakov addressed the problem of the hard life on the collective farm. In one scene, because of a hard, early frost, farm director Kostin goes to the regional council to talk all of the farms into getting their workers out at night to pull up the potatoes illuminated if need be by "Bat" flashlights.[130] In another scene, Communist Youth League members introduce a new way to fell lumber largely by organizing lumberjacks differently,[131] not through acquisition of powerful machines. Life was hard, but devotion, innovation, and communist labor would overcome technological lag.

By the late Brezhnev period, authors dealt with the difficulties of life in the Arctic in a more direct fashion, largely abandoning the formulaic approaches of inevitable victory over nature and enemies in the genre of socialist realism. Evgenii Bogdanov (1923–1999) was born in Oshevskii village of Arkhangelsk Province. A volunteer who served at the front during World War II, he turned to prose after the war and published with Arkhangelsk and central publishing houses. He enjoyed drawing and painting with oils.[132] Bogdanov addressed aging, growing isolation, and inefficient productivity in the Soviet countryside.

In *Leap Year,* which takes place at the end of the 1970s, the young director of the "Borok" State Farm, Stepan Lisitsyn, encounters a series of problems that require all of his efforts. Among the problems is the aging of the rural population as young people leave for cities: "Go to the city, become some one important, you will have a different fate," say mothers to their children."[133]

Lisitsyn believes that education and technology will help solve problems of underproduction and outmigration, but the obstacles were great, and the population of country residents aged in the late Soviet period.

Another challenge to the collective farm was the struggle between the old, rooted peasant way of life and the new way, which was not fully embraced even decades after collectivization. Even though collectivization was to end the three-field system of rotation and erosion, many peasants still used horses, not tractors, on small plots of land.[134] State farms were money-losing endeavors that required state support. Peasants tended their own private plots in an ideal way, but at the farm itself no one paid attention to the dirt, garbage, and broken glass. As Bogdanov pointed out, you go into any street in any village and you find "filth, pot holes, and ruts." People considered communal property "not theirs." And the state didn't even give a ruble for repairs.[135]

Modern farming machines had not found a place in "Borok." The tractor remained a stranger to many people. A wife yelled at her husband-mechanic for returning home in dirty, greasy clothes. She tells him she works with people. He responds, "For me the machine is also a living essence. Sort of."[136] One worker remembered the penetration of the tractor into the state farm, first a "Fordson," then a KhTZ (Kharkov Tractor), then the power "Stalinets," a ChTZ (Cheliabinsk Tractor, the manufacturer earlier of the "Stalinets"), and now a modern diesel. For the character Chikin the loud sound was wonderful, calming, like music, for it meant "everything is ok, there will be bread!"[137] Yet machines were few and far between, even if they were crucial for life in the taiga. The winters required tractors with snowplows to operate day and night, for otherwise how could you get food to the animals or bring in wood? Even with the tractor the "Borok" farm had been cut off from the outside world, become a kind of "autonomous geographical point," that lived on what the peasants managed to store in the autumn. They might have to bring in flour and other supplies by plane or helicopter.[138] In *Windblown Songs*, Bogdanov wrote about remote villages where almost nothing happened, although they still sent lecturers from the "Knowledge" (Znanie) society to meet with the old ladies who stayed behind. The store was closed because there was no one and nothing to buy. They brought in bread, tea, sugar, and something else once a week by car or horse. The school had closed, too; the children were long gone.[139]

Simply put, technological transformation of the Arctic environment did not extend to the workers themselves whether in literature or in reality. They were left to toil in dangerous, dirty, and crowded conditions, and they lived from meal to meal and moment to moment. Arctic workers and managers were starved for modern infrastructure and investment capital. No matter the ecosystem, they had to work with machinery and equipment inappropriate for Arctic conditions. More frequently than workers elsewhere in the USSR, they operated outdated, underpowered, and poorly functioning machinery and equipment. They were forced to use hand tools and human and animal power in unbearably challenging weather. Laborers were forced to fashion wheelbarrows out of scraps and sweat, a practice that some observers celebrate for showing ingenuity rather than confirming the inhuman conditions in which they toiled. When they had tractors, they were fewer in number than ordered and frequently broke down without spare parts. Their horses were sooner nags with cute nicknames than draught animals. In a word, brute force labor in brutal conditions prevailed above the Arctic Circle. This was true for all workers, but especially for the hundreds of thousands of prison laborers who transformed Murmansk, Arkhangelsk, and Karelia into Stalinist economic units.

Second, industrial safety and public health lagged significantly. Whether they were behind notoriously low standards for the rest of the country is unclear. But archival materials reveal that on top of legion pollution and despoliation of the environment, factories lacked even modest fire brigades, emergency plans, protective clothing, and other technologies that would reflect basic concern for the laborer.

Third, the extent and quality of power production, transport, communications, and other technologies lagged significantly behind the rest of the USSR. When you look from a train window across the tundra or swamp, you see telephone, telegraph, and power lines out of kilter. These poles, usually single reinforced concrete poles that taper from the base to the top and that bleed and crack as soon as they "cure," rickety-looking metal standards, and A-frames never stood with proper Soviet posture like well-bred Soviet poles. It was hard to drive them into the rock and frozen earth in the first place or to affix them in the swampy soil, and many fell, or turned to the side, or rusted, or melted, and those that managed to carry electrons according to plan eventually gave in to changes in the season and extreme temperatures.

But the poles marched forward further and further into the Arctic plains and thereby extended the reach of political and economic organizations.

Workers in the tundra and taiga lived a remote existence. They were intermittently linked by radio or post to economic and political organizations, but rarely to faraway family and friends. The Russian highway system is nearly 1,000,000 kilometers in length (although only one-third of those roads are paved), but on the Kola Peninsula roads extend only 4,400 kilometers, and the eastern part of the peninsula is sparsely inhabited and has neither roads or rail; people can be reached only by helicopter and sometimes boat. The entire Kola railway system measures 891 kilometers in length and consists essentially of the St. Petersburg–Murmansk highway. On top of this, the polar night adds to isolation, running eight weeks from late November to mid-January. In Karelia there is one federal road, the 780-kilometer-long stretch of the St. Petersburg–Murmansk, and because of poor and hurried construction practices the other roads seem always under repair. Such goods and services as food could be transported with delay at great cost. Similar lack of service surrounded the telegraph, telephone, and postal systems and created profound isolation.

One did not have to journey to the countryside, to the collective farm, to the forest, to smelting operations in the tundra, or to a small town built to support a hydroelectric station to find lags in modern technology. Even historic Arkhangelsk, a crucial port on the White Sea, where Peter the Great ordered a shipbuilding yard constructed in 1693, the site of many of the great lumber mills of Revolutionary Russia, the seat of the northwest region's political and economic power, the disembarkation point for Arctic-bound icebreakers, research vessels, trawlers, and freighters loaded with supplies, lumber, and coal, with a harbor that received lend-lease supplies in World War II, even a city of Arkhangelsk's importance and size was starved for electricity, infrastructure, and worker's quarters into the 1960s.[140] The provincial communist party secretaries wrote letter upon letter to the Central Committee to request more investment; their secretaries for departments of transport and communications, electrification, and industry appended reports and analyses indicated how this city of 250,000 in 1939 and 285,000 in 1959 required modern technology to serve the state.

Yet Arkhangelsk, built on delta mud and swamp largely out of wood structures, remained nineteenth century in terms of public health and transport until the late twentieth century, with sewage often flowing into basements untreated. In a letter to the council of ministers of the RSFSR (the Russian Republic), Obkom secretary Latunov tried to play up Arkhangelsk's importance as an international port where more than 200 foreign ships arrived annually to take advantage of various wood, lumber, paper, pulp, and shipbuilding facilities. In spite of the economic importance of the city, Latunov worried that construction of housing, cultural, and other facilities in fact "far lag[ged] behind the demands of the workers." The city had 10,000 wooden, old, and worn apartment, barrack-style, and other houses. The local Soviets managed 1,317 communal homes; one-tenth of them were in emergency situation in which 568 families with 1,740 members lived under the constant threat of falling into the muck. Further, war interrupted 1940 targets to build 13,500 m^2 of living space; the city had fewer than 5,000 m^2 and required tens of thousands of square meters more by the mid-1950s The Obkom had received 1,287 complaints from citizens who lived in kitchens, corridors, and service areas. The obkom therefore requested budgets for 1955–57 to build 47,000 m^2 of housing, drain swamp and wetlands, and reconstruct plumbing, sewage, electricity, and city lights. The city also requested funds to build an asphalt factory, install eleven kilometers of tram lines, and add twenty buses, ten taxis, six microbuses, and ten tram cars to the municipal transport network.[141] Arkhangelsk party and government leaders pressed the central authorities to provide three diesel cutters for the Northern Dvina and Pechora deltas; workers in summertime could not get from their homes—often on islands or shores—to collective farms or fisheries.[142]

Indicating a distinction of an authoritarian political system—the crucial position of the leader in greasing the wheels of decision making—Nikita Khrushchev himself triggered the rebuilding and modernization of Arkhangelsk housing stock with visits to the city where he saw firsthand the filth and backwardness. After his visits the first large-panel construction commenced in December 1960.[143] Khrushchev was a fan of industrial construction techniques, rejected the "lowly brick," and worshipped mass-produced panels of concrete that could be used interchangeably in building-block apartments. Soon thereafter, the authorities finally approved the construction of a railway and automobile bridge across the Northern Dvina. In the twenty-first century, in a period of frontier capitalism, with

Moscow controlling resources and the provinces struggling to meet the needs of modern society, Arkhangelsk's housing, transport, and public health have begun to age visibly again. Here and there costly business centers rise, while there and here streets and older apartment buildings continue their slide into the muck.

Arctic transformation never met the designs of visionary planners and determined Party officials. They assumed they might overlay the tundra and taiga with modern infrastructure and that nature would then follow the dictates of the plan. While they succeeded to some degree in making Arctic regions more habitable, they could only hope their investments had been worthwhile.

7

Rediscovering the Arctic

Pevek—a sea city-port in Chukotka—became one of the symbols of the thousand-year advance of Russian into the polar countries.

Ледокол, ледолом, ледорез, ледодав, ледобой, ледогрыз . . . И это все—о нем!

Soviet joke: What were main feelings of the Chukchi people before the Revolution? Cold and hunger! What were their main feelings after the revolution? Cold, hunger and eternal gratitude!

In August 2007, Russian parliamentarian and explorer Artur Chilingarov, long connected with AARI, engaged in what some observers called a publicity stunt by planting a Russian flag on the bottom of the Arctic Ocean at the North Pole. The government supported the expensive expedition as part of the Russian contribution to the third International Polar Year. All of the components of Russia's quest for strategic advantage, economic growth, and superpower symbolism were present. A nuclear-powered icebreaker, *Rossiia*, cleared the way for a research ship, *Akademik Fedorov*, staffed by approximately 130 scientists, to get into position for Chilingarov's descent. President Putin welcomed Chilingarov's flag-planting expedition as confirmation of Russia's claim of the Lomonosov Ridge, which would enable Russia to extend its exclusive economic zone toward the North Pole and include several vast oil and mineral deposits. In 2009 Chilingarov announced another expedition to launch a floating research station in the Arctic. The station—together with an icebreaker and a research ship already in position—would gather fresh scientific evidence to bolster Russia's claims to regions of the Arctic as an "exclusive economic zone." Putin endorsed this research program, noting that Russia's history was closely linked to Arctic exploration. Tying these Russian efforts to the great power status of the USSR, he referred to Soviet efforts to build major facilities and cities in circumpolar regions and to the

Artur Chilingarov. (Courtesy of the Russian State Museum of the Arctic and Antarctic)

Northern Sea Route, established in the 1930s with great fanfare.[1] Russia's vast Arctic spaces were again the stage for big science and big technology, nuclear energy, and state power to come together.

Although they knew little about him in 2000, many people embraced Vladimir Putin as Russia's savior. In contrast with the impetuous and unpredictable Boris Yeltsin, whose government had become increasingly stained by the air of corruption—and a drunken Boris, Putin would put the economy in order, bring stability to the political system, and restore Russia's pride and superpower status. In his two terms as president, Putin accomplished precisely this, yet also chipped away significantly at several democratic reforms and the fragile rights of nascent civil society. He established control by fiat over the regional governors, who had become, in his mind, too independent. He gained the power to appoint them to avoid the creation of fiefdoms in the periphery. About 90 percent of all media outlets are now state controlled. In the name of rapid development of Russia's great natural resources, Putin disbanded the federal environmental protection agency and gave the provincial governments responsibility for enforcement of weakened laws but without adequate staff or personnel, and the FSB closely monitors—some would say interferes with—the operations of nongovernment organizations, especially those with political agendas or strong foreign contacts.[2]

Putin served as prime minister under President Dmitrii Medvedev and returned as president in 2012, continuing big science programs that he and Medvedev have determined will secure Russia's position in the world arena. As part of the effort to rebuild Russia's sense of itself as a superpower, Putin supports increased expenditures on space exploration, nuclear power, and the military. Yet the infrastructure of provincial cities has collapsed from lack of funding. Many provincial municipal governments have often been unable to provide basic services, let alone snow removal, on the basis of extremely modest local tax revenues. Since 2006 the federal government has recognized the need to make significantly greater resources available in the provinces and municipalities, although Moscow continues to be a kind of "black hole" of power and money. The Arctic Circle, with its vast resources and strategic significance, has benefitted from the Kremlin's attention, especially because of oil and gas. But the government ignored vast and growing income differentials, decay of infrastructure except in Moscow and St. Petersburg, and a growing demographic crisis especially in the Arctic region. Several Arctic cities have lost 20 percent of their population since 1991 largely due to outmigration, while the nation as a whole has seen its population shrink as birth rates plummeted.

In the second decade of the twenty-first century, Russia's Arctic policy reflected all of the paradoxes of a government laying claim to economic and political power, fearing alternative voices, and displaying infatuation with big technology to support its agendas. Once again, as with the Bolsheviks in the 1920s, scientists have had to reach an uncomfortable accommodation with the central government. They seek funding without excessive strings, arguing that they know best which directions of research to pursue. Yet the Putin administration has determined to cut funding to all but a few major institutions and universities,[3] to focus educational and research priorities on applied results, and to buy such superpower baubles as reactors, rockets, and icebreakers. Administration spokesmen referred without irony to the importance of the enforcement of environmental laws, but the government enacted policies to gut national environmental protection, weaken NGOs, and give up on the Kyoto Protocol as economically unsound and failing to prevent climate change—which many Russians actually welcome for facilitating Arctic development. The costs of assimilation of Arctic resources will be always be high because of harsh climate; will environmental safety and cleanup have a sufficient role in northern latitudes in the 2010s, 2020s, and beyond? The ordinary citizen—the oil rig shift worker, the power station operator, the teacher and doctor, the Nenets herder and Saami fisherman—may be left in the cold again. The nation has become an oligarchy of wealthy oil magnates, state-run businesses for the exploitation of oil, gas, and minerals, a vigilant state security police, and a distant and powerful presidency. This political and economic constellation of forces resembles that of Tsarist and Soviet political culture in many ways, with risks and benefits for people in the Arctic. A coterie of middle-aged men served as Putin's shadow cabinet. Wealthy individuals with access to the Kremlin established control over natural resources and staked Russia's future to their exploitation, similar to the situation under Soviet power.

Collapse of the Arctic in the Yeltsin Years

People left in droves. They could not take the loss of income, the harsh climate, the uncertain future. When the USSR collapsed, so did the subsidies in salaries, construction, and snow removal in Arctic cities and towns. Able-bodied young people sought hope in Moscow and St. Petersburg. The urban population of northern regions dropped from roughly 2.7 million to 2.0 million inhabitants. Left behind were ghost towns like Amderma and Belomorsk. Across the Arctic high mortality owing to cardiovascular disease, cancer, and environment far exceeded natural increase. Out-of-marriage births and levels of abortions were one-third higher than in the rest of Russia.[4]

In summer 2007 I visited Belomorsk, the terminus on the White Sea of Stalin's great gulag canal. In the nineteenth century, this was the vibrant village Soroki with lumber mills, and by the 1920s it gained importance as a railroad terminal. During the construction of the canal it became Belomorsk and served as the capital of the Karelo-Finnish Soviet Republic from 1941 to 1945. Yet its population dropped from a high of 19,000 in 1989 to 11,000 people in 2010. Few residents had regular work. It was unclear what they did; they moved without purpose or direction, in the summer in a drunken stupor, in the winter in small steps to a cold hallway or apartment to crash. Streets and sidewalks melted, cracked, and filled with hillocks. Public transport and private cars were smoke-belching old models "imported" from more prosperous regions of Russia. It's a direct train to Petrozavodsk and Petersburg to find a future, but inertia, fear, and hopelessness kept 11,000 people in modern ruin. De-industrialization has hit other cities and regions of the former Soviet Union.

During the last years of Soviet power, pronatalist policies and an anti-alcohol campaign produced a natural increase in the population of nearly one million annually. Since 1987, however, the number of deaths has increased by 720,000 annually, while the number of births has declined by close to 1.2 million. By January 1, 2002, Russia's population was only 144 million, down 4.3 million people from its peak at the beginning of 1992, and the pace of natural decrease (the surplus of deaths over births) intensified after 1997. Until 1998, migration into Russia from other former Soviet republics gave the appearance of slowing the natural decrease. According to Tim Heleniak, a geographer with the University of Maryland, "While the high increases in deaths among middle-aged men from cardiovascular disease and such external causes as murder, suicide, accidents, and poisoning has received the greatest attention, it is actually the decline in the birth rate that has had the greatest impact on population size." He continued, "Of the decline in the number of births from 1989 to 1999, about 9 percent can be attributed to decreases in the number of women of prime childbearing age and the remainder to real declines in childbearing."[5] Russia's birthrate dropped to 1.17 in 2004 from 2.08 babies per woman in 1990, below the 2.4 children required to maintain the population. Abortion rates were very high and male mortality exceedingly high. The result was a population that shrank by an average of 700,000 people each year. Russian leaders noticed the demographic crisis but did little until 2003 or 2004, when they provided modest grants and subsidies to young families as part of a pro-natalist policy, although the high cost of living tempered the impact of grants. These demographic trends had significantly greater impact on Arctic regions because of outmigration of hundreds of thousands of people.

Soviet policies contributed to the rapid development of the Arctic and Siberia from the 1930s until 1991, when the population tripled, from ten million to thirty-two million inhabitants, while in the rest of the country, the population grew by 64 percent. Interestingly, if in the rest of country there were 88 males for 100 females, in the north there were 101 males per 100 females. This was a younger population than on average in the remainder of the empire—and a highly Russian population (71 percent). It was also more urban than the rest of the country (79 percent for the north versus 73 percent for the country as a whole). In 1991 when the nation's population peaked, there were eleven cities with greater than 250,000 residents in the north, with Murmansk the largest at 472,900 people. The disintegration of the USSR set in motion processes that weakened the state's ability to support the Arctic and hence the population of the Arctic. In the absence of agriculture, the government and its enterprises had to ship food in. In Soviet times foods were subsidized, in the market system not, so prices went up substantially. In addition, wage increments, bonuses, and vacations disappeared. When the USSR broke up, there were ninety-five people per 100 km^2 in this region, which made it 2.5 times denser than Alaska and fifty times more than the Canadian Arctic. Of the sixteen regions that comprised Heleniak's Russian "north" and Siberia, all but one experienced outmigration from 1989 to 1999, and ten of them lost more than 10 percent of their population, with migration far outpacing a higher fertility rate than in the rest of the country and a 223,000 natural increase.[6]

Eventually President Medvedev addressed one underlying reason for the outmigration in September 2008. He said, "I stress that these huge investment and economic opportunities are still only potential for now. Development of the Arctic region is hampered by poorly developed economic infrastructure and unresolved social problems, including a lack of affordable passenger transport and a severe housing shortage. A look at the figures says it all: there are 650,000 people on the waiting list for housing subsidies in this region. As a result, people continue to leave the far northern regions and there is a serious depopulation problem there."[7]

If Arctic cities and families suffered through economic uncertainty, then the scientific establishment, including Arctic science, nearly collapsed. This crisis was connected with a sharp decline in the level of financing that called forth the involvement of foreign governments and foundations to salvage the enterprise and occasionally provoked xenophobic concerns that foreigners were stealing the scientific legacy. Another problem was the "brain drain" of young people to other professions at home and of more established scholars to research centers abroad. The latter brain drain was a shock and, to some people, an embarrassment, but the failure of young people to enter scientific careers, largely to pursue business or law, has had a longer-term negative impact in the "aging" of science. Officials in the executive branch and the Parliament focused exclusively on the political and economic controversies of the day, including hyperinflation and Russia's decline as an international power, not on education, environment, public health, or science. The result was the near destruction of fundamental research. The latter was saved only by the intervention of a variety of foreign governmental and private endeavors, with the most publicized of the private endeavors underwritten by the financier George Soros.

Finally, conflict arose between members of the prestigious Russian Academy of Sciences, the locus of basic research, and a federal government determined to rein in its independence and perhaps to gain control of its vast resources. There was no question that the enterprise begged for reform. "Rather than being the best science in the world," Graham and Dezhina write, "science in the late Soviet Union was crying out for reform. It was a system that emphasized quantity over quality, seniority over creativity, military security over domestic welfare, and orthodoxy over freedom." At its nadir, late in 1993, financing for science was perhaps one-twentieth of its Soviet level. The USSR had been a leader in many fields, but now several major scientific schools were on the edge of collapse. The Yeltsin administration stumbled, not knowing how to deal with all of the crises it faced, let alone for science.[8] Since that time, the Putin administration tried to gain control of Academy resources, especially its real estate and other assets, and to put state-appointed bureaucrats in charge of administering the enterprise; in the summer of 2013 the government moved to take over the Academy, a move with unclear outcome that provoked a storm of dissent among academics.[9]

President Yeltsin, whose administration took power with great hopes of building a democratic future on the foundation of a market economy, did nothing to support fundamental science in a time of crisis as great as that just after the Russian Revolution in 1917 and 1918. Even at that time, scientists managed to expand the scientific enterprise, not watch its collapse. In this environment, Arctic research came to a near standstill. During perestroika and especially after the breakup of the USSR, difficult economic conditions led AARI directors to cut staff, the scientific fleet, and the number of research bases at both poles; they closed the Murmansk Branch of AARI in 1995. Researchers could afford neither drift stations nor summer high-latitude expeditions. In July 1991 researchers at the SP-31 left the ice; for the first time since 1952, there was no drift station in the Arctic. Northern aerial expeditions ceased in May 1993. Yet, as with other institutes, AARI succeeded in preserving basic cadres and schools. Researchers focused in the 1990s on examining all of the data gathered in the entire period using IT. Work with foreign partners led to a series of electronic climate atlases for the Arctic and Antarctic with a large response at home and abroad. Research also turned toward ecology in connection with the industrial assimilation of the shelf. In 1992 AARI organized a laboratory "Arctic Shelf" under G. N. Zubakin. Projects with Norwegians, Americans, Germans, and Japanese proliferated.[10]

Researchers still managed to carry out a few expeditions in the face of catastrophic financial circumstances. In 1992 AARI scientists visited and restocked a small research team on Zhokhov Island in the northwest part of the East Siberian Sea discovered initially by the Vilkitskii expedition in 1914. In order to proceed, their flights had to take on freight from private concerns. The airplane delivered diesel fuel and linens and carried away trash, including a year of soiled linens. In the autumn an aerial expedition secured supplies for two people for zimovka on the island. Several other research teams carried out hydrological and hydrochemical research in the Kara, Laptev, East Siberian, and Chukhotka seas. In 1993 the last expedition until 2001 involved aerial oceanographic work, while a base on Zhokhov Island was put into mothballs. The reason may have been a terrible accident. On May 15, 1993, between Cape Shmidt and Uelen, an Mi-8 helicopter carrying a Russian–French expedition crashed, killing eight people, after which oceanographic photography was forbidden in this region. But helicopters continued to carry researchers for hydrological measurements in the seas of the Siberian shelf closer to shore. Then the funding for this research program dried up, too. There was one great success in the Yeltsin era: In October–November 1998 researchers from AARI and the Institute of Oceanography of the Academy carried out an expedition north of Franz Josef Island and Spitsbergen under I. E. Frolov and Captain V. A. Viktorov on the *Akademik Fedorov*.[11] As the authors of one book noted, "Many scholars, polar explorers, pilots, sailors, and many millions of other people remained without work and were required somehow to survive, not being occupied with their favorite work, but to do something to feed their families."[12]

At the same time, the United States, Germany, Sweden, and Canada were actively engaged in research at northern latitudes, and Japan and China also increased their presence. This rekindled both scientific and governmental interest in Russia. In 2001 AARI turned to representatives of the Russian Foundation "Polar Expeditions" and a French company to explore the possibility of resuming research on the drift ice of the central Arctic basin with a tourist component to the program called "Arktika-2001." The hope was to create month-long scientific expeditions over the next three years, which in 2001 and 2002 were realized, while in 2003–4 drift station research commenced again: SP-32 station operated for nearly one year, followed by SP-33. In August 2004 the *Akademik Fedorov* quit St. Petersburg, left the Baltic Sea, sailed around Scandinavia to the Barents, Kara, and Laptev seas, and headed north to the 85th parallel where, after three days of search, it found SP-33. A helicopter facilitated the transfer of crews. Eleven people manned the station. They studied ice regime, dynamics, mechanics, and aerosol-optical, ozonometric, and actinometric observations. The camp was well equipped with prefabricated structures, tents, radios, generators, tractors, and snow removal and transport vehicles. SP-33 drifted 2,225 kilometers over 360 days. SP-34 led up to the IPY. Begun with the twenty-third trip of the *Akademik Fedorov*, it commenced on September 17, 2005, and lasted 252 days with ten to twelve personnel. On April 16, 2006, the head of Rosgidromet[13] (the inheritor of the Soviet Gidrometsluzhba), A. I. Berditskii; the minister of communications, L. D. Reiman; and deputy chair of the Duma Artur Chilingarov visited the station, taking an Mi-8 helicopter through a private airline. The men lived in nine comfortable prefabricated boxes with work areas that included computing facilities. Scientists produced new data on the morphometry, structure, and local dynamics of ice cover; the characteristics of water masses; the chemical and gas makeup of sea ice and snow cover; meteorological and radiational processes in the atmosphere, snow cover, sea ice, and upper levels of water; and changes in background characteristics of ecosystems.[14]

State Power and the Arctic

All of this was a prelude and background to the intense interest of the government in securing the Arctic militarily and economically as a symbol of the nation's recovery from the embarrassment of the Yeltsin years. The Arctic has become transformed from a site of Cold War competition to one of fierce economic competition.[15] Unlike Antarctica where collaboration prevailed by treaty, the Arctic remained a contested place between Russia, Canada, Denmark, Norway, and the United States. While policy makers and specialists tempered competition with cooperation in areas of scientific research, notably on climate change and other environmental studies, the nations of the circumpolar world rushed to chart and develop natural resources—notably oil and gas, but also nickel and copper, titanium, iron and coal, and fisheries and forestry. The Russian government pursued these resources with renewed vigor because of its security and economic interests.

Russia's industrial policy of pushing economic growth on the basis of extraction of raw materials was a central aspect of Vladimir Putin's worldview. He has long believed the development of natural resources was the key to Russia's economic future and rediscovery of its status as a superpower. He addressed precisely these issues in his candidate thesis that he defended at the St. Petersburg Mining Institute in 1997. In the thesis he urged the country's oligarchs to focus on national interests and contended that "the state has the right to regulate the process of their development and use." Putin acknowledged that technological lag would slow Russia's assent, especially if it could not become a competitive manufacturer. He encouraged western investment, but with the proviso that Russia would retain control over all operations. He argued that Russia's great natural resources were the key to remaking the country into a great economic power with a high standard of living based on the "fatherland's processing industry based on the extractive complex."[16] "Putinism" is a kind of economic, not ideological, program, although strongly decorated with superpower designs and patriotic slogans and furnished with big state development programs such as the rejuvenated Northern Sea Route.[17]

The upkeep of secure navigation along the Northern Sea Route has necessitated new laws to secure Russia's claims and enable it to control the increasingly active Northern Sea Route. A 2001 bill that established Russian

Maritime Policy through 2020 according to which Russia reasonably asserted "sovereign rights in the exclusive economic zone for exploration, development and conservation of natural resources, both living and non-living at the bottom, in the subsoil and the superjacent waters, the management of these resources, energy production through the use of water, currents and wind, the creation and use of artificial islands, installations and structures, marine scientific research and conservation of the marine environment," including rights over "the continental shelf of the Russian Federation." The policy referred to "the increasing importance of the Northern Sea Route for sustainable development of the Russian Federation." Maritime policy established such long-term objectives as "research and development of the Arctic to the development of export-oriented economic sectors, priority social problems," and "the creation of ice-class vessels for shipping, specialized vessels for fishing, research and other specialized fleets," all toward the ends of state defense and resource development, and with reference to the Arctic, "establishment of sovereignty, sovereign and international rights of the Russian Federation."[18] In this light, Chilingarov's audacious flag-planting claim becomes understandable.

According to *The Basics of State Policy of the Russian Federation in the Arctic Region*, the Arctic accounted for 90 percent of the recoverable hydrocarbon reserves found on the entire Russian continental shelf, including the 70 percent of reserves that are located in the Barents and Kara seas, in addition to facilities producing natural gas, apatite concentrate, and many strategically significant non-ferrous and precious metals (nickel, copper, cobalt, and so on) Russian state and state–private ventures, including several with multinational corporations, have continued to develop these resources.[19] The "basic objectives" of state policy according to the document were: to secure solutions to social and economic problems facing the country through the expansion of the resource base of the Arctic zone; to ensure military security, defense, and protection of the state border of the Russian Federation lying in the Arctic zone, including through its Northern Fleet; to preserve and maintain unique ecosystems; to remediate the ecological consequences of economic activities in the conditions of increasing economic activity and global climate change; and to rebuild the Northern Sea Route, a fixture of superpower status in the Soviet era, in part through the modernization of Russia's nuclear icebreaker fleet. Toward these ends, government officials declared the intention to support basic and applied science and technology at a level sufficient to create a foundation to manage the economy and security of the north in the "natural climatic conditions of the Arctic." Russia's leaders insisted that they will accomplish these goals in the spirit of international cooperation through the maintenance of mutually advantageous bilateral and multilateral agreements and treaties and through the sharing of information about the Arctic zone.[20]

The economic importance of the Arctic explains why the Russian government filed a claim with the UN Commission on the Limits of the Continental Shelf (CLCS) in December 2001 with the hope of securing the rights to areas lying 150 miles (241 kilometers) beyond its 200-mile zone with territory exceeding 1.2 million square kilometers (463,000 square miles)—in the Barents Sea, the Sea of Okhotsk, the Bering Strait, and the ice-free waters of the Arctic Ocean—which Russia views as its possessions. This claim rested on geological research according to which the Earth's crust at the Mendeleev Elevation in the Arctic Ocean proves, according to Russian scientists and politicians, the continental nature of many sections of the oceanic floor. This claim is in keeping with the Convention on the Law of the Sea passed by the UN in 1982 that enables littoral countries to expand their sovereign rights beyond the 200-mile exclusive economic zone over those sections of the seabed where the continental origins have been proved conclusively. According to the convention, if such a government can show that the depths of the continental shelf are a natural continuation of the mainland, then it may make claims up to 350 nautical miles (648 kilometers). To prove this Russia has drilled in the region of the Lomonosov Ridge. But the CLCS rejected the claim on the basis that Russia had not presented geologic and geophysical evidence that the Mendeleev and Lomonosov submerged ridges were extensions of Russia's continental shelf.

Russian bureaucrats have tried a way around the CLCS—simply using paper. The Ministry of Economic Development (Minekonraz) proposed to change the manner of calculating the territory of the shelf and exclusive economic zone (200 miles). The wonder of such changes consisted in the fact that they don't require the approval of other governments even as the boundaries could move by hundreds of miles. Ministry officials asked Russia's Maritime Collegium to examine all boundary lines as part of the "Basic Government Policy in the Arctic for the Period to 2020 and Long-Term

Prospects."[21] This was an effort to update situations that date to the Soviet period from which many laws and conventions prevail. To establish coastal and exclusive economic zones, governments use a combination of the territorial water line (the contiguous zone) and coastal baseline (mean low water mark). Minekonraz proposed to use the coastal baseline except when impossible and then to use the line of the contiguous zone, which moved the limit of the exclusive economic zone several miles closer to the North Pole.[22]

While Russia played in the snow, the United States sat on its hands. Because, against the advice of former presidents, secretaries of state, admirals, business people, and others, Republicans in the U.S. Senate have refused to ratify the 1982 Convention or even consider it, the United States may neither assert any additional rights beyond 200 miles from Alaska nor serve on any commission to consider others' claims. This has not prevented other countries, for example Denmark, from laying claim. In pursuit of exclusive rights to a series of oil and gas deposits in the Arctic, the Ministry of Foreign Affairs of Denmark intends to present by the end of 2014 documents to the UN that support its exclusive economic rights to parts of the Arctic continental shelf. Danish scientists assert that the Lomonosov ridge under the North Pole is a continuation of the North American and Greenland tectonic plates and hence (through Greenland) indicates Danish ownership. Of course, Russia drew parallel conclusions supporting its own claims. Aleksei Fenenko, of the Institute of Problems of International Security, observed that Denmark's actions will contribute to sharp disagreements about the status of the Arctic, in addition to which Canada, which has remained neutral to this point, may also begin to act. Denmark's claim raises a series of questions concerning the status of the Northern Sea Route. Russia claims that the Lomonosov Ridge runs into the Novosibirsk Islands. All of this means that a transit zone and an exclusive economic zone may soon overlap. Currently, Russia has a monopoly on control of the Northern Sea Route—corresponding to international norms that say where ice cover is more than six months, there the coastal state regulates—as Aleksandr Danilov, director of AARI, explained.[23]

Chilingarov set out determinedly to secure Russia's claims. He enjoyed the limelight. He relished big splashes—in the Arctic, in Lake Baikal, perhaps to the bottom of the Mariana Trench. With several decades' experience as an explorer and from a position as deputy chairman of the Duma, he pushed the Russian Arctic agenda. Never one to self-aggrandize, Chilingarov wrote, "Polar explorers are a special breed of people: a lazy-bones at high latitudes won't survive long, and of course a rascal can hard live among pure-hearted and selfless people. Otherwise there are no limitations for becoming a member of the cohort of arctic researchers. As an old, experienced polar explorer, I'd like to tell you: do not be afraid to cross over the invisible line which is called the Arctic Circle, take that step, risk it! You will never regret the fact that you spent time in the harsh regions of the Arctic. And if it turns out they take you to the North Pole, I assure you: something very special will happen in your consciousness. That invisible frontier will divide your life into 'before' and 'after' the Pole. And you will have time to think about this wonderful minute of revelation of nature."[24]

While his activities heightened concern outside of Russia about growing competition for Arctic resources, he downplayed the notion that the Arctic had become a "hot spot of the twenty-first century." Chilingarov participated in a roundtable discussion of that very subject held by the RIA-News Agency in April 2007, months before he planted the Arctic flag on the ocean floor. The news agency presented the results of a recent survey that the newspaper *Komsomolskaia Pravda* had conducted indicating that 65.8 percent of respondents believed the government ought to make the north more attractive to Russian businesses and inhabitants and fully 18 percent more supported large-scale government projects such as those of the Soviet era to achieve those ends, while only 9 percent rejected Arctic development as a pressing issue.[25]

Chilingarov said he did not consider the Arctic a "hot spot," for the exclusive economic zone in the Arctic had been determined and the Duma was in the act of "preparing still more laws." One law concerned the Northern Sea Route to make it international, but protect Russian interests. Fortunately, the government had begun to pursue a policy to protect Russian interests at high latitudes with new bases, new icebreakers, and so on. But Chilingarov rejected this as provocative. First, sixty countries cooperated in the IPY that had *no* relationship to the Arctic. Already a French–Russian expedition had been on the ice for five months. Even if the major presence was Russian, other nations contributed. He added, "There is worry, there is interest, but if we ourselves are not occupied with the Arctic then, of course, someone else will be." This comment may have referred to disagreement with Norway over access to the

Barents Sea; in September 2010, Norway and Russia settled their forty-year-long dispute over a 175,000-square-kilometer area in the Barents Sea essentially north of both countries. The agreement, signed by President Medvedev and Norway's Prime Minister Jens Stoltenberg, apparently will foster oil and gas exploration and put an end to unregulated fishing.[26]

Also at the roundtable, a staffer of the Ministry of Foreign Affairs, Dmitrii Chumakov, who defended his candidate of science degree in economics on "Energy Resources of the Russian Arctic in the World Economy" in April 2009, pointed to the Russian determination to get the UN CLCS to approve Russia's third effort to lay claim to a zone stretching 350 nautical miles from shore. Chumakov acknowledged that the words *struggle* and *fight* for resources exist in the Russian lexicon and likely come from the Soviet mindset to the present.[27] But at the same time as Chumakov and Chilingarov downplayed the notion of a conflict, the Russian government pursued policies that revealed that officials found the Arctic to be of direct strategic, military concern and that global warming will help Russia in meeting these concerns.

Russian commentators often made bellicose comments about the military designs of other Arctic powers and NATO, but downplayed Russia's military efforts to protect its interests through the expenditure of billions of rubles on an increased military presence in the Arctic. According to one report in *Kommersant,* Canada determined to conduct more extensive military exercises in the Arctic called "Nanook" with at least 1,000 soldiers, not as an annual event, but as a sharp response to Russian pretenses to control more of the Arctic through its claims on the Lomonosov Ridge. The United States and Denmark, both NATO nations, participated to a small degree in Nanook, and NATO has openly declared its strategic interests in the Arctic. *Kommersant* concluded these actions were directed against Russia.[28] But the "Nanook" exercises involved "a scenario involving Canadian Forces intercepting an ecotourism boat carrying migrants attempting to enter Canada illegally"; "Nanook" is nothing new, but an annual exercise that gives the Canadian Forces a chance to practice their skills in responding to emergency and security situations.[29] In fact, Canada has embarked on a $70 million project to map the sea bed on its side of the Lomonosov Ridge, and the government has begun to fund expansion of the Canadian Arctic fleet and perhaps construction of new deep water ports.[30]

In September 2012 President Putin called for a "great leap forward" ala Stalin's industrialization campaign of the 1930s to rejuvenate Russian military industry. While he did not mention Stalin, it was clear to members of his Security Council that he had Stalin in mind.[31] Reforms in the Russian army to handle new challenges in the Arctic were already in place. A 2008 government decision on the basic directions of Russian Arctic policy through 2020 included a call for the creation of special units of Arctic troops; in 2011 the first such units appeared. A special motorized brigade prepared for Arctic conditions would be stationed at the Kola Peninsula 200th motorized regiment base. The Ministry of Defense had taken into account the experience of Norway and Finland in creating these brigades and would provide special equipment and clothing to permit soldiers to fight even in conditions of deep freezes. According to a journalist for *Komsomolskaia Pravda,* "The international arms race in the region of the North Pole, as is well known, was begun by the neighbors of Russia long ago. Since that very time at the depths of the Arctic Ocean have been discovered great reserves of fossil fuels. Now dozens of these 'gentle' countries are striving to divide these territories, of course, for their own benefit. To this date by international-juridical means. But the 'army' argument is being prepared just in case."[32] It was as if Russia only protected its interests, while other nations were the aggressors.

Icebreaker Envy

Russia will assimilate Arctic resources through big science and technology—airplanes, satellites, drift stations, research institutes, and crucially icebreakers, including a third generation of nuclear icebreakers. Russian shipbuilders, administrators, and officials exhibited great nostalgia for the Soviet Union, which created the world's greatest icebreaker fleet. Russia remains the only country to operate civilian nuclear-powered icebreakers, although the icebreaker fleet has aged considerably and a number of them have reached the end of their service lives.

Selective nostalgia for Soviet achievements—but never its many technological failures—fueled the rejuvenation of Arctic conquest in the twenty-first century. Russians of the Soviet generation reminisce about the *Lenin* icebreaker, which was launched in December 1957 and sailed on its first mission in September 1959. In the celebratory journalism about the glorious Soviet heritage not once will you encounter the complete story of the dangers involved in the rapid, and perhaps premature, embrace of nuclear icebreakers. When referring to competition for Arctic resources instead journalists pointed out that Russians are a full quarter century ahead of the other nations. At the end of the 1950s "we left the Americans behind and first built a nuclear icebreaker," the chief engineer of the *Lenin* atomic icebreaker recalled."[33] Yet the *Lenin* had two serious accidents in 1965 and 1967, both of which released a significant amount of radioactivity and required illegal dumping of wastes and reactors at sea. One of the accidents in 1965 required cutting out the reactors and reactor vessel and submerging them near Novaia Zemlia in the Tsivol'ka Bay after permitting a loss of coolant accident; there was almost a meltdown and more than half of the fuel rods were damaged. In the 1967 accident, workers fortunately noticed a leak in the coolant system after reloading fuel. But they used pneumatic hammers to repair the system, in the process of which they destroyed the reactor. They cut this one entirely out of the ship and also dumped it and 3,500 tons of radioactive waste near the previous site.[34]

Russian leaders have been increasingly vocal in celebrating Soviet achievements: Stalin's joyous establishment of the nation as a military power in the 1930s, Sputnik, the fiftieth anniversary of Iurii Gagarin, the first cosmonaut, in 2011, and of course various nuclear achievements—with the exception of Chernobyl. On August 17, 2012, the nation observed the twenty-fifth anniversary of the sailing of the *Arktika* to the North Pole, the world's first surface vessel to do this. The 1977 feat to take the North Pole celebrated atomic transport, subjugation of the Arctic, and expansion of the polar economy. It also celebrated the career of Minister of the Navy (and member of the Central Committee of the Communist Party) Timofei Guzhenko, whose entire career outside of bureaucracy was connected with Arctic marine activities. As minister, Guzhenko saw the Navy became a major economic actor in the USSR. The order indicating Guzhenko's removal as minister for reasons of health in 1986 did not mention the 1986 sinking of the *Admiral Nakhimov* and the loss of life of 423 people. But Guzhenko who died in 2008 was recognized by the naming after him of an Arctic tanker with deadweight of 70,000 tons in 2009.[35]

The *Arktika,* a second-generation nuclear icebreaker, was retired on October 3, 2008, after thirty-six years of service. The *Arktika* was built at the Baltic Shipbuilding Yards along with the *Siberia, Rossiia, Soviet Union,* and *Yamal* nuclear icebreakers. In 2012 Russia maintained six atomic icebreakers—but five were built in the Soviet period and had reached the end of their service. Construction on the most recent addition to the fleet, the *Fiftieth Anniversary of Victory,* commenced in the Soviet era, but the ship was put to sea only in 2007 owing to extensive construction problems including a serious fire. The difficulties facing the Russian government in its desire to expand the fleet should have been clear from the fact that it took twenty years to build this ship.[36]

Russian officials worried about a pause between the second and third generation of icebreakers because experienced sailors, mechanics, engineers, and others might be lost to other employment opportunities. The average age of fleet personnel in 2010 was forty-three to forty-five years; young people did not enter the industry. The crews preferred foreign fleets or Russian oil and gas companies that paid much more. Rosatom doubled the salary of icebreaker personnel on average to 55,000 rubles per month, still under $1,900 monthly but well paying in Russia.[37]

In October 2010 before the Maritime Collegium in Murmansk deputy prime minister Sergei Ivanov called for development of the Northern Sea Route through new icebreakers, satellites oriented to produce accurate data, and container shipping. He pointed out in 2010 that the route had carried a bit more than four million tons of freight, while plans called for increases in tonnage and number of trips. Ivanov asserted that the future would see winter shipping as well. But all of this depended on icebreakers and, reminiscent of Cold War competition, he pointed out the construction of a new icebreaker in Canada in partnership with the United States as a partial justification for a third generation of icebreakers.[38]

On August 6, 2011, in Naryan-Mar, the secretary of the Russian Security Council, Nikolai Patrushev, convened a meeting with the leaders of Arctic regions on development of mineral resources and possible sources of financing. He noted that all of the ambitious plans of Russia for assimilation of the Arctic were connected with the Northern Sea Route. To support the route, he suggested a special administration would be created—perhaps a new Glavsevmorput? The main problem, Patrushev noted, was to build

Artkika nuclear icebreaker. (Courtesy of the Russian State Museum of the Arctic and Antarctic)

icebreakers. While ten of them were still in service, only two of them had not reached the end of their planned livelihood (*Yamal* and *Fifty Years of Victory*.) According to the Ministry of Transportation, by 2020 the Northern Sea Route would carry sixty-four million tons of freight, by 2030 eighty-five million tons, while in 2010 in all only 1.8 million tons and in 1987 6.6 million. Oil and gas and nickel and other valuable commodities across the Arctic—all of these resource extraction projects required icebreakers and modern infrastructure along the entire Northern Sea Route.[39]

Plans for Russia to spend thirty-seven billion rubles on a new atomic icebreaker *Arktika* were revealed in a contract between the Baltic Shipbuilding Factory and Rosatomflot, a subdivision of Rosatom. Rosatomflot officials worried that until 2018 there would be insufficient icebreaking power to solve all of the transport and other problems of Russia in the Arctic. According to one observer, the new *Arktika* "will differ from earlier atomic icebreakers by the fact that it is capable of working both in estuarial conditions which demand a small draft, and in deep waters which demand a large draft." This will permit it to escort tankers up to 70,000 tons in displacement. Another observer was fairly certain there will be cost overruns,[40] and as of 2013 funding already lagged.

Andrei Smirnov, the deputy director of Rosatomflot, argued that icebreakers will enable development of difficult-to-exploit fossil fuel reserves and contribute to a five- or sixfold increase in shipping. He appealed to economic, strategic, and nationalist (geopolitical) reasons for supporting the construction of a new icebreaker fleet. Finally, Smirnov pointed out that from Kamchatka to Murmansk takes but seven days, whereas through the Suez Canal would take twenty or twenty-five days. And while the northern latitudes had ice, the southern had something more dangerous and long lasting: pirates. And pirates "cannot exist in the Arctic in principle: they will freeze."[41]

Playing with Oil

The most important activity of icebreakers was to facilitate oil and gas field development. According to several estimates the Arctic may have 30 percent of undiscovered natural gas fields and 13 percent of undiscovered oil. While accounting for only 1 percent of the nation's population, albeit a relatively "urban" population, the Russian Arctic in 2012 contributed 11 percent of the country's gross domestic product and 22 percent of its export earnings. Overall, 30 percent of the nation's reserves of gas and 13 percent of reserves of oil were concentrated in the Arctic.

In April 2000, addressing a special conference on the Northern Sea Route and Russian shipbuilding on board the *Arktika* in Murmansk, President Putin endorsed breakneck Arctic development. Like others, he claimed that the volume of cargo shipments in the Arctic might reach more than ten million tons a year in the not-so-distant future (while the actual volume barely exceeded one million tons). In terms, rhetoric, and thinly veiled nostalgia for the heydays of Arctic discovery in the 1930s, Putin pushed the state to take the lead in its rebirth. Putin referred to the great riches of the north and called the Artic the nation's "strategic reserve." He declared, "Never forget about the primary significance of the Northern Sea Route—its transport potential. Today it is used primarily for domestic shipping and for export. In the future it may become a corridor for international commercial shipping."[42] In July 2012 Putin signed a law establishing a new Glavsevmorput of ships and airplanes and economic visions.[43] At a press conference in Tiksi in December 2012, on the day the law took effect, Putin reiterated the fact that the economic future of the Arctic depended upon the development of the Northern Sea Route. It would grow in importance given climate change and the growing number of days open to navigation. Putin underlined the concrete tasks the government had taken to develop the sea route. He claimed it was economically more valuable "than the Suez canal."[44] Putin called the formation of contemporary infrastructure for the Northern Sea Route "the most important socio-economic problem . . . that will guarantee the dynamic development of the northern part of the country, and the creation of new industry and jobs." He called for expanding existing ports and building new ones, for example Varanda on the Iugorsk Strait and Sabetta on the Yamal Peninsula, and modernizing river, railway, and northern airports in support of polar aviation.[45] Presidents Medvedev and Putin repeatedly visited Severodvinsk to praise nuclear shipbuilders at Sevmash and Zvezdochka as part of their pep talks for the Northern Sea Route.

Big state development projects have also returned in full force and symbolic display. At first building on technologies imported from abroad and now increasingly built in Severodvinsk and other closed military and industrial cities, Russia embarked on a federal program for the "Creation of High-Tech Drilling Units, Machinery and Equipment for the Deepwater Production of Oil and Gas and the Development of Hydrocarbon Resources on the Arctic Continental Shelf from 2003–2012." Under this program, Russian industry will manufacture extractive technology to develop the Arctic hydrocarbon wealth. After the collapse of the Soviet Union the nuclear fleet advanced the idea of using nuclear submarines as underwater ships to transport cargo along the Northern Sea Route, an idea abandoned as costly and risky. With the discovery of oil and gas fields on the shelf of the Barents and Kara seas, designers of nuclear submarines embraced the idea of building nuclear-powered underwater drill ships, as well as using nuclear icebreakers and floating nuclear power plants to power exploitation of Arctic Ocean oil and gas, as a way of keeping shipyards opened after the United States and Russia had agreed to decommission submarines.[46] Russia has been somewhat successful in transforming shipbuilding technology into oil drilling technology. The government plans to see forty ice-resistant oil platforms and fourteen off-shore gas platforms built by 2030 in the effort to tap Arctic oil and gas, plus build fifty-five ice-resistant tankers and storage tankers and twenty gas carriers, all by establishing new public–private government holding companies because of the inadequacy of civilian shipyards to the task.[47] Rosatom (the Ministry of Nuclear Power) has also embarked on a program to build dozens of nuclear power stations, a number of them in Arctic regions, by 2030.

Yet joint projects with other nations were required because of the high cost of exploitation of Arctic deposits and because Russian technology lagged far behind that in the west from the point of view of efficiency and environmental and worker safety. The experience with the Shtokman gas fields indicates this problem in high relief.

Launched like the *Stalin, Lenin,* and *Arktika* icebreakers with anticipation of great economic benefits, but like the *Sedov* caught in the ice, the "Shtokman" fields have not opened in spite of the hundreds of millions of dollars thrown their way. Initially owned by Gazprom (51 percent), Total S. A. (France, 25 percent), and Statoil ASA (Norway, 24 percent), the Shtokman Development AG had a budget of more than $800 million for 2008–9,[48] based on a promise of access to reserves estimated at more than four trillion cubic meters; an agreement signed in 2008 saw production beginning in 2013–14. But just four years later, the development of the "enormous field of gas condensate" under the Barents Sea was on hold. According to the Bellona Foundation, "The announcement from the Russian state gas monopoly indicates that even the Russian government cannot, for the time being, see its signature gas project yielding a healthy financial return." This was a disappointment for President Putin and his economic strategy, for Russian leaders saw the production of liquefied natural gas for export as key to the country's financial growth. Both Total and Statoil gave up their shares of the field. Representatives of the Bellona Foundation see this result as confirmation of their position that it is simply too challenging—and too expensive—to drill in the center of the Arctic Ocean. It does not help that Russian laws and tax incentives make it difficult for western companies to join any Russian project.[49] Finally, the European partners have grown weary of the potential of a disastrous oil spill—even with the best technology and first-rate scientific research. If Rosneft's performance was any indication (it was responsible for 75 percent of oil spills in the Khanty-Mansi Autonomous Region in 2011),[50] then fear about the environmental challenges of extracting gas condensate in the Barents Sea with Gazprom may be well placed.

After two years of waiting to announce the obvious critical delays, Gazprom recognized that in its current form the Shtokman project could not go forward. Since Statoil and Total pulled back, Gazprom began a search for new investment partners with no clear outcome and as yet no new partners. The major reason was the great costs. Originally estimated at $12 billion to $14 billion, they quickly grew to $30 billion. Increased costs had to do with concern over technical parameters of the project, while the partners also worried about Gazprom's monopolistic position. For example, the French saw the transport of gas and condensate to the land by pipeline, while the Russians logically saw use of ships for loading the gas and offloading it at terminals, given their extensive, if aging, Arctic fleet. In April 2011 the directors of Shtokman Development AG approved a project close to that of Total, but further agreement did not follow, and directors now estimate production only in 2016–17, but no further decisions have been made about establishing a new schedule for the project or putting it on mothballs. Another concern has been the falling price of gas since 2008, especially in the United States, but generally worldwide.[51]

Yet because of oil and gas, the Arctic will become a zone of extensive transportation, infrastructural development, and innovation. Even Amderma may live again. In 2011, Igor Fedorov, the head of the Nenets Autonomous Region, sent a letter to Arkhangelsk district court in opposition to bankruptcy proceedings for the Amderma port. He suggested that Amderma would make a great site for search-and-rescue and disaster cleanup units in the region. The closest such facility for oil spills was 1,000 kilometers away in Murmansk, while several potentially major offshore oil and gas projects were established in the Pechora Sea. Fedorov called for both private and state investors in rejuvenating the Amderma project. The Russian railroad also announced plans to complete—at long last—a line to Amderma, a project begun with slave labor in the 1930s. Amderma has the decaying but existing infrastructure, housing, and airport to handle such projects as the Prirazlomnoe, Dolginskoe, Medynskoe, and Varandey sites.[52] Nenets region officials welcome economic development but worry about the environmental consequences from drilling—especially in light of the Deep Horizon accident in the Gulf of Mexico; a similar spill in the Pechora Sea would have far worse consequences because of shallow waters in the region and strong river flows that enter the sea to carry oil far and wide.[53]

From the Abyss to IPY

Scientific research mirrored these changes and developments in the views of the Russian president, innovation in such Arctic technology as icebreakers, and fascination with the extent that resources may be easier to exploit because of climate change. Scientific research over the course of three International Polar Years provided the foundation for Arctic assimilation, and the third IPY (2007–8) crucially demonstrated both the recovery of Russian science from the economic crises of the 1990s and Russian leadership in Arctic science.

During the first International Polar Year (1882–83), expeditions focused primarily on geographical discovery, while scientific research was a secondary problem.[54] But the Year was founded on the notion that only joint research of the leading countries with financial support of their governments could open the Arctic and Antarctic regions to further discovery and understanding. In Russia the Pavlovsk Magnetic-Meteorological Observatory played a significant role in preparation for the year. Fourteen polar stations were opened, two each in Russia and the United States, and one in ten other countries; flask-sampling, meteorological, and other stations were opened as well. The Russian polar station Sagastyr' Island was in the delta of the Lena River under the direction of N. G. Iurgens, from which Russian researchers conducted meteorological and magnetic observations and explored the Lena delta. K. P. Andreev headed the second station at Malye Karmakuly, on the western shore of South Island of Novaia Zemlia, a kind of administrative center for Novaia Zemlia until 1926 when Belushia Bay was designated. Russian initiative and activity in the IPY were extensive.[55] With the permission of the Tsarist government, the Dutch established their station on Dikson Island. As Russian specialists write, "Unfortunately, all stations were closed after completing the program of IPY because the results of observations for such a short period did not indicate value for climatology."[56]

The second International Polar Year (1932–33) moved scientific knowledge of polar regions forward substantially on the decks of modern technology (icebreakers and airplanes) and science (advances in meteorology, oceanography, and so on). Fourteen countries organized polar stations, while a total of forty-four participated. In the USSR, scientists organized a committee headed by the director of Gidrometsluzhba A. F. Vangengeim. Such scholars as Otto Shmidt and Vladimir Vize participated. The USSR contributed ninety-two stations to the second IPY, thirty-three of them new—near Tiksi (on Guker Island), Cape Cheliuskin, on Rudolf Island, and so forth. The observatory on Guker had a radio station, electrical generator, meteorological observatory, and geophysical pavilion. On top of this the USSR carried out twenty-six maritime expeditions, primarily in the Barents and Kara seas and including that of the *Sibiriakov* that completed the Northern Sea Route in one navigation.[57]

The International Geophysical Year (1957–58), usually connected with the Soviet achievement of launching the first artificial satellite, Sputnik, in 1957, simultaneously grew on and triggered rapid development of Arctic regions. World War II had interrupted scientific cooperation, and the Cold War did little to encourage resumption of efforts. But in an effort to ease tensions and in response to Khrushchev's doctrine of "peaceful coexistence," Soviet, American, and other specialists began to cooperate in the IGY. The autarky that prevailed in Soviet science suddenly came to an end within six months of Stalin's death. Institutes began to stress and report international scientific contacts, albeit largely with specialists in the socialist countries of Eastern Europe. The first two Geneva conferences on peaceful uses of atomic energy (1955 and 1958) and the IGY contributed greatly to the recreation of international communities of research.[58]

The main goal of the IGY was the accumulation of geophysical data about the entire planet. For the first time using rockets and satellites, this research included the upper layers of the atmosphere. More than 10,000 scientists who represented sixty-seven countries participated. In 1956 the Presidium of the Academy of Sciences established a committee to carry out the IGY that had fourteen subcommittees. Soviet scientists operated thirty-three of fifty IGY Arctic stations including on Dikson, Kotel'nyi, Wrangel, the Preobrazheniia islands and Capes Cheliuskin and Uelen, and launched eighty-eight meteorological rockets, including thirty-five in high latitudes. The USSR operated three drift stations (SP-6 through SP-8, see Chapter 4) and conducted extensive oceanographic research of the entire Arctic basin.[59]

Political stability, the interest of political leaders in the Arctic as a key to Russia's future and as a symbol of legitimacy and superpower status, and economic recovery led to a revival of funding for basic and applied scientific research on the Arctic and Antarctica leading up to the third IPY in the early 2000s. Research included investigation of the hydrophysical parameters of deep sea, near-bottom layers in the Arctic Ocean, the influence of Atlantic waters inflow on vertical distribution of nutrients in the Arctic Ocean, meteorological results from the SP-32 through SP-34 stations, and study of the dynamics and deformation mechanics of Arctic basin ice. AARI's floating research institute, the *Akademik Fedorov,* carried out extensive research in August–October 2004 and July–September 2005, the latter the ship's twenty-third expedition, which Russian specialists characterize as having "decisive significance" for the stabilization of research programs and that served as the foundation of "the study, monitoring and assimilation of the ocean," with the Arctic basin of the Arctic ocean a key focus of activities. The research supported the "economic interests of Russia on the shelf" in the Pacific and Arctic oceans connected with the development of the oil and gas industries. A government resolution of June 2, 2005, addressed the importance of R and D targeted tightly on economic issues through technoparks, innovation centers, and even research vessels. Toward these ends, during the IPY researchers on the *Akademik Fedorov* carried out research on hydrometeorological processes, hydrological and hydrochemical conditions, radiational processes in the atmosphere, changes in Arctic ecosystems, the nature of the ice regime, icebergs, glaciers, wave dynamics (with changes in ice regime), and so on.[60]

The preparation for Russian participation in the third IPY took five years and built on the work of committees and resolutions of government reminiscent of Soviet political culture. Under the leadership of Chilingarov and others, in November 2002 the government's Council on Problems of the Far North and Arctic passed a resolution whereby the Ministry of Foreign Affairs and Rosgidromet would consider how best to participate in the IPY. Specialists from Rosgidromet and AARI (first of all Viktor Dmitriev, Ivan Frolov, Alexander Danilov, Alexander Klepikov, and others) and the Russian Academy of Sciences including Vladimir Kotliakov and Maxim

Moskalevskii worked with European specialists and the World Meteorological Organization to advance a conception for the IPY based on each country's potential scientific contribution.

Several government resolutions followed to permit Rosgidromet chief A. I. Berditskii to set about convening an organizational committee and appointing him chair of the committee and Chilingarov vice chair. Regular meetings began in April 2005. An important early step was to create within AARI a center to facilitate scientific and informational-analytical components of Russian participation in the IPY including the creation of a website. Only one member of the twenty-nine-member organizational committee, Larisa Abriutina, Vice President of the Association of Indigenous Peoples of the North, Siberia and Far East, was a woman,[61] and like many others she occupied a nominal position.

Ten main directions comprised Soviet participation in the IPY: (1) hydrometeorologocal and geophysical conditions in polar regions; (2) soils, the cryosphere, and nature of permafrost; (3) geological history and the lithosphere of polar regions; (4) land and marine ecosystems of the Arctic and Antarctic; (5) the socioeconomic development of polar regions; (6) the development of an observational system; (7) data management; (8) geodesic and cartographic work; (9) efforts to improve the educational and scientific potential of qualified polar specialists and others; and (10) popularization and publication of knowledge among the broad public.[62] It is nearly impossible to do justice to any nation's two-year participation in the third IPY, let alone such an important Arctic power as Russia. Therefore, I err on the side of brevity so as not to overwhelm the reader with discussion of the 159 different expeditions, twelve of which were posted on the Russian IPY website in great detail. Of 399 clusters that comprised the IPY, Russians participated in 147 of them or 37 percent. Of these 147 clusters, ninety-eight concerned the Arctic (67 percent), seventeen Antarctica (12 percent) and thirty-two (22 percent) were "bi-polar" (Arctic *and* Antarctic) clusters.[63]

Russian scientists were major players on all fronts. Drift stations, computer modeling, marine expeditions, and polar stations all were involved in the effort. The expedition "Arctic-2007" on the *Akademik Fedorov* had "no analogies in the scale of research in the high latitude Arctic using a ship." In light of the increasing intensity of Arctic cyclones, temperatures, and precipitation, one goal was to develop hydrometeorological, meteorological, and hydrological data bases to understand the condition of the environment and processes going on in it. In July 2007 the ship left Petersburg for Baltiisk in Kaliningrad, under expedition director Chilingarov, where it was loaded with machinery, equipment, and instruments. At the end of July the *Akademik Fedorov* departed for the Barents Sea and to the northeast of Franz Josef Land. In August the ship began complex measurements and data collection in the Kara and Laptev seas and ornithological and geological research on Severnaia Zemlia and Ushakov Island and prepared for "SP-35." In the fourth stage (essentially September), the expedition searched for an ice field for SP-35, left the Laptsev Sea, and entered the northern part of the East Siberia and Chukhotka seas. The ship returned to Petersburg on October 3, having sailed 14,447 nautical miles, including nearly 5,000 miles in heavy Arctic ice. Researchers undertook hundreds of hydrochemical measurements, investigated the thermal and radiational characteristics of the ice, and conducted atmospheric analysis.[64] Under Viktor Dmitriev, the academic secretary of AARI and a specialist in high-latitude climate and other studies, specialists have published seven volumes of *Vklad Rossii v Mezhdunarodnyi Poliarnyi god 2007–2008* (The Contribution of Russia to the International Polar Year) that reveal the significant achievements of Russian Arctic science broadly defined.

Modeling has occupied a central place in Arctic research for the past ten to fifteen years. One system modeled oils spills in ice-filled areas of the Arctic that considered dynamic-thermodynamic, wind, wave, oil transport, and other variables that would, apparently, make it possible to deal with ever more likely spills as Russia developed rich Arctic oil fields.[65] Other modeling research considered projected changes in the permafrost;[66] Arctic Ocean water and sea ice modeling;[67] use of explosives to destroy icebergs and grounded hummocks to ensure safe operation of submarine constructions in freezing seas;[68] and the use of explosives to deice "engineering structures."[69] Other mid-decade studies considered the tornado-like structure under icebergs in the Barents Sea involving upward fluxes of melt water and air bubbles and downward fluxes of released sediments and cooled ambient water.[70]

While the AARI was a major player in the IPY, many other institutes were active participants. The Research Institute of Polar Medicine (within the Northern State Medical University in Arkhangelsk) was created on the foundation of a research laboratory that dealt with problems of acclimatization to life in northern latitudes, including research into new methods and means of prophylactics, earlier diagnosis of pathologies, and correction of the state of health of indigenous people and long-term settlers. Regarding indigenous people, the institute worked closely with the NGO "Iasavei,"[71] an association of the Nenets people in a project called Kaninskii Red Teepee. The Kanin Peninsula divides the White and Barents seas[72] and is a region of the Nenets Autonomous Region whose administrative center is the town of Nes, about 100 kilometers from the Mezen River. As part of this project, all families of nomadic herders in the region were given pressing medical help. Specialists carried out analysis of the population and demographic and ethnic structure of these people with an eye toward improvement the health of tundra residents. They conducted similar work on Kolguev Island to provide treatments based on modern medicine and technology; it is unclear from the published sources what the nature of illnesses or treatments was. Another project concerned investigation of the widespread "shift method" in the oil and gas industries. Researchers sought to work with employers to maximize the health needs of employees working under such difficult work regimes. Other researchers investigated food and nutrition with the financial support of enterprises of the northern food industry. Finally, institute personnel have given special attention to children, children's pathologies, and diseases that afflict the very young.[73]

An energetic and talented scholar with wide-ranging interests, the director of the Research Institute of Polar Medicine, Galina Nikolaievna Degteva, has been connected with Arkhangelsk and the Nenets region since her early years and, after considering geography, chose biology and medicine for her career, along the way studying dentistry at the Arkhangelsk State Medical Institute. Upon graduation she worked in a hospital of the forestry industry in the Upper Toemskii region of Arkhangelsk province, sometimes engaging in difficult dental surgery, for example on a lumberjack after a horrific accident. She joined local party and government organizations and as an activist popularized medical knowledge among the regional residents. After three years she returned to Arkhangelsk to study hematology and better understand blood diseases among northern residents in cities and towns, including specialists and "aborigines," on the Northern Dvina River and its sources. Her major interests included uptake of iron in pregnant women, alcoholism, respiratory diseases, and erythropoiesis, the latter of which led to her candidate of science dissertation in the 1980s. Upon defending the thesis, Degteva began study of oil workers that involved legendary delays in travel from Arkhangelsk to Naryan-Mar by airplane and further by helicopter because of weather. This became the foundation of her doctoral work on ecological and physiological aspects of oil workers in the north in 1996. Degteva turned increasingly to issues of children's medicine. Given this background and training, her involvement in the IPY as a leading researcher on public health and medicine in the north surprised none of her colleagues in Russia or abroad.[74]

Other such institutes and laboratories as the Center of Polar Medicine of AARI, the St. Petersburg State Pediatric Medical Academy, the Institute of Geography, the NII of System Analysis (Moscow), and the Polar Academy (SPB) were closely involved in IPY medical studies. Among the more crucial findings were evidence of an increase in heart disease in connection with global warming and extreme weather patterns in the Arctic, expansion of disease vectors for anthrax, smallpox, and tularemia along with the expansion of habitat of rodents, and a finding that the border for encephalitis had shifted northward.[75]

Several researchers considered the peculiarities of polar stress syndrome—with its metabolic disorders, immune insufficiencies, northern tissue hypoxia, blood hypercoagulation, regenerative and plastic failure, and so on, with this stress leading to a number of adaption disorders and pathological states.[76] Specialists studied the impact of fluctuations of temperature, humidity, wind speed, and air pollution that contributed to increased risk of respiratory diseases. In one company pulp-and-paper town, they identified that the highest toxic effect of pollutants on children's respiratory systems was in February–April and June–August.[77] Other researchers determined that arterial hypertension—with its impact on cell membranes, vessel endothelium, blood cells, and organs—occurs at higher rates in Arctic regions than elsewhere in the Russian Federation.[78] Researchers studied the presence of ^{90}Sr and other radionuclides in the subarctic in the face of new plans to expand electricity and heat generation from nuclear power stations, including floating nuclear power stations. Shockingly, this article did not mention the presence of the Krasnoiarsk-26 plutonium production facility on the Enisei River, which was established in 1950, while providing data that the levels of

^{90}Sr discharged in that river exceeded that of the Northern Dvina, Pechora, and Lena rivers from 1961–90.[79] Residents of the Yamal-Nenets Autonomous Region have blood levels of such heavy metals as nickel and chromium that in some cases are up to ten times higher that the norm, likely from ingestion of such local food items as wild berries, mushrooms, grasses, and fish—and from local smelters and mines.[80]

Specialists at the Murmansk Marine Biological Institute, desolately located on the Solovetski Islands in the late nineteenth century as a research station, transferred in 1935 to the even more desolate Dalnye Zelentsy, roughly 150 kilometers from Murmansk on the northern Kola Peninsula Coast, and accessible after hours and hours by bus; in 1989 it was moved to Murmansk. The institute maintained research stations at Dal'nye Zelentsy and in the Franz Josef Land and Svalbard archipelagos. The biologists conducted research on the current state and dynamics of ecosystems and patterns of biological productivity processes in the Nordic and Russian seas; evaluated the state of the benthos, plankton, marine fish, bird, and mammal biology and ecology in the northern seas; created methodologies for environmental assessments; and supported the development of biotechnology polar mariculture based on "scientific principles" of rational usage.[81] In a recent article, they discussed the impact of climatic changes on marine ecosystems of the European Arctic that influenced biological productivity and bioresources of the seas, including changes in ice cover that determined biodiversity of organisms at all trophic levels. The research, conducted as part of the IPY, built on the well-known fact that the north Atlantic current, which entered the Barents Sea, kept the southern reaches usually ice-free, that is with floating ice, and only in several extremely cold years did the ice approach the northern coast of the Kola Peninsula. The White Sea is iced in from December to May, although ice sets as early as November from Kanin Nose to the Kara Gates, while the Arctic island archipelagoes are usually surrounded by ice from October to June. The researchers noted that climatic changes of the marine environment had a decisive influence on the migration of fish and therefore on the geography of the fish industry; cod for example in very cold periods stay largely to the coast, while in warm periods may reach Kolguev Island to the south and the coast of Spitsbergen to the North. As the ice cover has shrunk, cod fisheries have expanded throughout the Barents Sea. They conclude, however, that the organisms in the seas of the European Arctic adapt well to changes, although benthos has been much more sensitive in terms of population and makeup.[82]

Caught between Oil and an Icy Place

The indigenous people of the Russian northwest have gotten less attention than icebreakers, oil and gas, and the environment. Small in number, more prone to illness and disease from the processes of modernization, pushed to the outskirts of society, seemingly beyond the tundra, and often treated as backward, uneducated, and foolish, the Nentsy, Saami, and other aborigines faced great uncertainty in the early twenty-first century. While the rhetoric of national leaders called attention to the economic, cultural, and demographic plights of these people, their presence in the middle of increasingly valuable lands and shorelines makes it likely that they will be pushed aside again. In the 1990s they faced great hardships in the transition to a market economy as the regular payments they received through reindeer herding and fishing activities suddenly dried up. Younger people more willing to take risks have fled the region. The dangers of land grabs, oil spills, and loss of voice affect all those who remain behind.

The joys of cultural construction! In the 1920s and 1930s emissaries of Marxism to the Bolshezemelskaia tundra discovered illiterate reindeer herders whom they judged to be backward, and concluded that, in their shamanism and lack of understanding of the five-year plans, they were ultimately hostile to Bolshevism. They built an outpost of Soviet power on the Pechora River, Naryan-Mar (Red City), from which to bring the Nentsy into society and tie their reindeer to the economy. It grew from a sleepy port in 1931 to a town of 8,000 by 1934, when its first streets were named, the fisheries trust was established, and the first teachers arrived. The first building appeared in the flat emptiness of the Arctic desert in the outpost of Beloshchel'e, Naryan-Mar's predecessor, in 1923, the first book in Nenets was published in January 1933, the first "graduates" to receive certification to operate and work on motorboats—1934—important given the appearance of fishing collectives, a canning factory in late 1934, bread began to be sold freely in stores and coops in January 1935, and the first reindeer herder-shock workers were recognized soon thereafter. A weather bureau and drama theater followed. An airport of sorts opened with weekly flights to Arkhangelsk commencing in 1936 with baggage limited to five kilograms. The building of a real port with passenger terminal and docks was carried out in the mid-1930s; the workers crucially lengthened the coal dock, important because in 1936 alone 110,000 tons of coal were to pass through the port and in 1937 220,000 tons. Wide boulevards, mostly mud in spring and summer, were in place by the late 1930s. But while there was a library in 1938, with twenty to fifty readers each evening to take in fresh newspapers and magazines, there was no electricity, only kerosene lamps, even though the city theater on the second floor of the building had electricity. "Olympiads" to attract talented indigenous youth to perform and to celebrate their culture were held, and the first opened with a chorus of songs about Stalin, followed by dances—but not about Stalin; the Naryan-Mar school for the liquidation of illiteracy still had a great deal to accomplish. A nursing school was transferred from Shchel'iaiur, hundreds of kilometers upstream on the Pechora River, to Naryan-Mar in 1939, but the dormitory for students had no beds, mattresses, linens, or lamps.[83] The first brick building (1960s), multistory brick building (1975, the Communist Party headquarters, of course), and natural gas to apartments (1978) were long in rising on the tundra. But Naryan-Mar was the capital of the Nenets Autonomous Region, and the town fathers, distant as they were from Moscow, were in a great position in the twenty-first century to make money on oil and gas. The small town, accessible only by jet or helicopter, was as expensive as an oil city is expected to be.

I discovered Naryan-Mar fifteen years after the collapse of the USSR in 2007. It is doubtful I would have been permitted to visit in the 1980s during my first trips to the USSR had I even known of its existence. But I had no doubt Naryan-Mar had changed markedly since the breakup of the USSR. No longer a gray, distant outpost of Soviet civilization, it was a booming oil and gas town with overly expensive hotels and restaurants. Boeing, not Illushin, jets regularly flew in delicacies, cherries and kiwis and even flowers from Holland. With friends who drove me through the tundra I saw a huge radar, to my untrained eye long unmanned, although there was a shaman nearby. I discovered a new home in Naryan-Mar whose confusing contradictions made it welcoming. I have been in the summer and the winter, during blowing snows and in swarms of mosquitoes. Because of its isolation yet cosmopolitanism, its seeming quiet yet dynamic civic and political

Sengeiskii Island Meteorological Station, Nenets Autonomous Region. (Photo by Paul R. Josephson)

culture, its cold, and its mosquitoes, its 20,000 inhabitants approach life in the center of the tundra with a matter-of-fact attitude toward the challenges they face running a city above the Arctic Circle, only 140 kilometers from the Pechora Bay.

To get a better sense of aviation in the Arctic, in June 2010, as part of an international conference, I organized a helicopter trip to Kolguev Island and the Arctic Ocean coastline. The scores of polar stations that appeared on coastlines and islands of the Arctic Ocean provided constant and accurate information regarding meteorological conditions. The stations consisted of a residence/barrack, several outbuildings, scientific equipment, and radio transmitters and antennae. Thousands of rotting and rusting carcasses of civilization—trucks that did not survive, fuel drums, decrepit snowmobiles, construction materials—had been dumped nearby.[84] One or two people staffed the stations, in the short summer preparing for the long winter and in the long winter waiting for the short summer. They tuned snow vehicles;

An Arctic graveyard on Kolguev Island. (Photo by Paul R. Josephson)

gathered and piled driftwood on the shore; and wait for deliveries from civilization by helicopter or boat. I visited the Sengeiskii Island Polar Station, established in December 1954, on the southeast shore of the Barents Sea on a sandy, largely flat island with occasional hillocks. We dropped in by helicopter, spent about forty minutes saying hello to the two weather-burned inhabitants of the station, and then flew on to Naryan-Mar.[85]

On Christmas Day 2011 I heard the governor of the Nenets Autonomous Region (NAO), Igor Gennadievich Fedorov, address the town fathers and mothers of the Nenets region on the achievements of 2011. His address opened with the startling words, "Friends, associates, coworkers . . . comrades." That some officials still used the word *comrade* indicated that a striking continuity in attitudes and even personnel of the core leadership persisted on Russian soil. Many of the people in positions of power in Russia today teethed on Communist Party positions. Yet when Fedorov spoke in May 2012 about the importance of international programs and organizations in the future of the NAO, he had perhaps said *comrades* out of respect for real Soviet achievements, icebreakers, and the certainties of its stagnant daily life, not out of nostalgia for the xenophobia and surveillance endemic in the USSR.

Fedorov noted the growing interest of foreign researchers and organizations in the Nenets region, ranging from representatives of NGOs to social scientists, biologists, journalists, and others. Unfortunately, the Putin administration decided to destroy NGOs operating in Russia, fearing the seeds of political opposition and targeting in particular those who received even a pittance of foreign funds to register as "foreign agents," a xenophobic term that suggests that they operate as spies. The law includes huge fines and has been used to criminalize opponents of the state. The law has also been used to hamper the activities of such groups based entirely in Russia as Memorial to honor the memory of the victims of Stalinism. Any foreign–Russian partnership involving not-for-profit organizations thus has been threatened, even those in the sciences and humanities.[86]

When he spoke, perhaps Fedorov had no inkling of the growing determination of the Putin administration strictly to surveil, limit, or prevent NGOs to operate. Not only Arctic council nations, but Germany, the Netherlands, China, and South Korea had made their presence known. Relations were especially active with Norway since 1996 when Russia became a regular member of the Barents Euro-Arctic Council; the council's interests centered on climate change, youth, and indigenous people.[87] Since 2007 the Nenets region has had an agreement with Rogaland County in western Norway, the center of Norwegian North Sea oil activities with programs on logistics, oil industry safety, and exchanges in fisheries, culture, and ecology, and from 1991 cooperation and from 2010 a Sister City relationship with Kautokeino, Norway, in Finnmark County, an area of a great many Saami people. Statoil has indicated interest in development of the Khariaginskoe Oil Deposits in the Bolshezemelskaia tundra[88] and a readiness for long-term cooperation in the socioeconomic development of the region through support of youth programs, culture, sport, and schools. Through the efforts of the Intergovernmental Russo-Canadian Economic Commission, wind–diesel electrical power stations appropriate for thinly settled regions appeared on the tundra, and with the assistance of a Canadian telecommunications broadcast radio and TV in native languages expanded. The Finnish Ministry of Economic Development and Innovation also fostered improvements in wind energy, transport, energy in the housing sector, reindeer husbandry, emergency services, cultural ties, and tourism. Fedorov concluded that his administration had positioned the region as an "international space of expertise for the study of problems of the Arctic."[89]

In spite of these efforts, the life of the Nentsy—and of all circumpolar peoples—remained at a critical juncture. Access to medical care, housing, formal education, and social welfare programs had by many measures improved the quality of life of such people as the Nentsy. Yet acculturation, growing contacts with the outside world, and being tied to a wage economy have led to stress and alienation among indigenous peoples. Resistance to a dominant culture led many individuals to lose self-esteem and to identity confusion. Diseases increasingly served as causes of death versus accidents, infanticide, starvation, and suicide that were major causes centuries ago. Contact with bacteria and viruses from developed societies brought devastating epidemics of populations that lacked immunities to the diseases in question. Researchers identified rising (if often exaggerated) substance abuse, especially among the young, as well as rising suicide and homicide rates and more woman and child battering. The "modern life" seemed easier. Yet the shift from "country" to "market" food led to more obesity, hyperglycemia, and hypercholesterolemia. Previously, hunting for "country food" meant a high level of physical activity, a diet of animal flesh with low fat content, and low sugar levels. Ever-encroaching industrialization contaminated local foods with PCBs, heavy metals, and other pollutants. Sedentary life fostered the need for community exercise and sports programs. Evidence indicated aging of lung volumes because of smoking, use of snowmobiles, and lack of exercise, as well as increased death rates of lung cancers in many Arctic settlements, as well as more bronchitis and TB.[90] And then there was the matter of the precarious position of traditional culture—art, music, dress, language, and literature. As one solution to the risks to Chukchi, Saami, and Nenets culture of redoubled efforts to industrialize the Arctic in the twenty-first century, two researchers involved in the IPY called for the creation of a united circumpolar population based on a sociodemographic synthesis, what they call an "ethno-biospheric society—a unitary circulated population—and the basis of socio-demographic synthesis and guarantee of ecologically balanced . . . activity of man."[91] It is difficult to imagine what this would entail, how it would work, who found fund it and administer it, and how it could possibly slow acculturation with its attendant dangers.

Several indigenous writers have dealt with the tension between the need to preserve their cultures and lifestyles, the pressure of oil and gas to change, and the difficulty in seeing such technology as reactors as a gift of modern civilization. The Soviet Ministry of Middle Machine Building, the very ministry responsible for industrial manufacture of nuclear warheads, celebrated the construction of the Bilibino Nuclear Power Station, the northernmost nuclear power plant in the world, well above the Arctic circle, as part of its effort to create a nuclear-powered Arctic and to bring a special kind of "fire" to the simple Chukchi people. It built four graphite-moderated EPG-6 reactors, each producing 12 MW electric and 62 MW thermal power. To this day, Bilibino benefits Russians; of 5,000 people in the town, only 300 are Chukchi, mostly in service positions, and the rest are Russians. And the 2013 "Miss Atom" came from Bilibino.

The Chukchi writer Yuri Rytkheu (1930–2008), the son of a hunter and whaler himself, perhaps the most well-known indigenous author in Russia, addressed these incongruities in his prose. After graduating from university in Leningrad, he turned to writing. At first Rytkheu celebrated the way Soviet power brought progress to the Chukchi and other nomadic northern peoples. He was elected in 1954 to the Union of Soviet Writers as a symbol of the patronage of small peoples in the USSR. His early writings from the 1950s seemed to some readers to be a sellout to Soviet power, but from the 1970s he focused on the Chukchi life. He came to criticize "civilization" as dangerous to indigenous people as some king of "silent genocide." He turned to Chukchi legends and stories, becoming himself a wonderful storyteller. Rytkheu's *Chukchi Bible* published recently in English celebrated the little people's way of life.

In *Dorozhnyi Leksikon* (Traveling Lexicon), his last book, in a series of riotously funny "dictionary entries" Rytkheu frames the differences between the supposed superior way of life of the Russian acculturators who are oblivious to their condescension and the Chukchi who understand fully what they have gained from Russianization. The entries range from "automobile" and "arbuz (melon)" to "Evrei (Jew)," "kit (whale)," and "kniga (book)" and from "led (ice)" to "potselui (kiss)," "seks (sex)," "sol'ntse (sun)," and "khleb (bread)." Rytkheu demonstrated in fact how deeply the Russians changed their world, not only the meaning of their words. Regarding "melon," Rytkheu wrote, "There is no corresponding word in Chukchi and never was." He knew very little about this wonderful fruit that came from warm parts of the earth and only from a textbook on botany. But he first encountered one in 1948 in Vladivostok waiting a few days for a flight to Moscow when he and a friend saw these magnificent fruits and vegetables being unloaded from freighters. And then they saw it: a melon, which a stevedore sold them. They quickly ran away, lest someone steal their prized acquisition. They found a quiet place, cut off a small piece of the skin, and tried a taste: It tasted like a piece of typical tundra grass. Disappointed, they thought—maybe it was necessary to boil it first? They decided rather than risk illness from improper preparation to toss the melon into the ocean.[92] About ice, familiar and ubiquitous in his consciousness, Rytkheu writes, "Salted ice that accompanies the coastal hunter year round, is quite different in all of its manifestations." And the hunter must know each form, for otherwise he risks his life. Yet ice can be beautiful and form fairy-tale-like structures. In Uellen, the ice rarely disappeared, and if it did you only had to wait for a light northern wind. The Chukchis took dog sleds to the river to cut out slippery pieces of fresh water to cart back to the igloo. Ice had many, many forms, and the Chukchi knew them better than the Russians coming to settle Chukotka with their science.[93]

After a bumpy childhood education, Vasilii Ledkov (1933–2002), a Nenets writer, graduated from the Herzen Pedagogical Institute in Leningrad with a degree in philology, returned to the Nenets region, and worked for the *Naryan-Mar Vynder*, publishing poems and stories in that newspaper and in *Pravda Severa*. Without an alphabet, of course, the Nentsy had no written literature. They had various forms of oral art: fairy tales and epic legends that were gathered and published in 1936 and then periodically over the next decades. With literacy, the first Nenets authors began to publish in the journal *Taiga i Tundra*. The first number of *Taiga and Tundra*, filled with articles by students of the "Northern Department" of the Leningrad Eastern Institute, appeared in 1928. This was, apparently, the first time that indigenous, "uncultured, illiterate" individuals had produced such writing, whereas before that time travelers, explorers, ethnographers, police bureaucrats, and other Europeans had provided impressions. The journal was published in all for five years.[94] V. N. Ledkov belonged to the second generation of Nenets writers.[95]

Ledkov's writings revealed the difficult transition to Soviet life in the tundra among reindeer herders. In "Month of Little Darkness," he described the civil war between the Reds and Whites and how the Nenets did not understand the war or why people killed each other.[96] How quickly things changed. From 1929 the tundra was not simply land of the Nenets but the Soviet Nenets Autonomous Region, filled with collective farms,[97] yet this transformation brought changes that the Nenets found difficult to bear. In another story, a shaman laments the changes in the tundra. Wherever he goes he sees sharp saws and axes working like woodpeckers and at the sources of rivers—no water and no fish. The sky was filled with smoke. The "teepees" were made incongruously of wood.[98] How painful to look at the dead lakes; they used dynamite to kill fish that rose to the surface. The Nentsy did not fully understand the process, but apparently it was good for the seagulls.[99]

In almost all Soviet literature technology was a force of good, a solution to problems that stood before managers, collective farm workers, and electrical engineers. Even if a story or novel realistically raised concerns about the economy, productivity of laborers, the production of consumer goods, the quality of life in the countryside, or eternal problems with food, the authors suggested that more technology used efficiently would solve the concern. In another Ledkov story, two herders talk. One explains his love of the tractor. "You were at the front [during the war] . . . we say powerful machines, and they don't just crush people. Machines, there will come a time, when they will begun to work for us . . . Already not only in the south but here on the Pechora collective farms are buying tractors, acquiring technology. And the tractor—is also a tank. What a powerful thing! And then they are putting motors on sleds, so that they can go anywhere. You know what kind of machine I thought up? A winter machine to break up ice-crusted snow and black ice. The reindeer can't get to the lichen and moss with black ice, and without lichen—they will die. There you have it! What a machine!"[100]

As the World Warms

Evaluation of the extent of human impact on climate change is another aspect of the competition for Arctic resources. Scientific expeditions have the goal of sustainable methods of resource exploitation. Scientists tend to see the data they produce, being "objective science," as the only foundation for policy decisions, not to mention drawing of accurate boundaries among the Arctic countries and how best to exploit resources. The Kyoto agreement, drafted in 1997, required industrialized countries to cut their greenhouse-gas emissions to 5 percent below 1990 levels by 2012. At the World Climate Change Conference in Moscow in 2003, Russia demurred to ratify the treaty, which would have meant its final ratification even without U.S. participation. In his welcoming speech to the conference President Putin offered ambiguous and ironic comments. Andrei Illarionov, Putin's top economic adviser and a staunch opponent of any government interference in the economy, claimed that climate change was nonexistent. The chairman of the forum, Yuri Izrael, the seventy-four-year-old director of the Moscow-based Institute of Global Climate and Ecology, who served as the head of Soviet delegations in all international environmental forums since the early Brezhnev era, sought to put aside any discussion of the ratification of the Kyoto Protocol by Russia during the forum. His efforts have gained him the name "the communist fossil fighting for fossil fuel."[101] What made things worse is that in Russia very few members of the Russian scientific community acknowledged the central place of the human factor in climate change. Among developed countries Russia, perhaps, had the smallest portion of scientists who evince concern about any global environmental issues. In the end, the story was a happy one as Russia surprisingly ratified the protocol in November 2004. Journalists suggested that Russia agreed to ratify the treaty only in exchange for agreement of the European Union on terms for Moscow's admission to the World Trade Organization. If global warming continues, perhaps the Nenets will not need tractors to free the lichen for reindeer?

While most Russian specialists acknowledged global warming, they saw it as part of natural cycles, downplayed its dangers to the environment, and intended to take advantage of warming to pursue the Northern Sea Route. On March 16, 2010, President Medvedev conducted a meeting of his Security Council to consider the economic and military dangers of climate change for Russia. He ordered the government to create a unified plan of scientific research on the climate with two parts. The first would include measures for struggling with climate change, and the second measures to adapt to these changes. The administration was certain that Russia could take advantage of such "natural cataclysms" as climate change in agriculture and in river transport. At the same time, Russia abandoned Kyoto as Medvedev adopted a position of a climate-change down-player if not naysayer. In joining Canada and Japan in dropping from the Kyoto Accords according to which they agreed to limit and reduce production of greenhouse gases, Russia of course had the example of the United States, which, through the obfuscation of the Republican Party, never agreed to the Kyoto Accords, with the United States far and away the greatest emitter, although China had gained ground.

Medvedev claimed that "not everything here is as clear as is presented by ecologists and people who with a high degree of interest follow this process. To this point there is not one universal forecast and exact scenario of these processes, and we are obligated to be prepared for any eventuality and use it in order to strengthen the power of our government." He worried that monitoring and forecasting of climate change in Russia was far behind that in other developed countries. He called for study in particular of the reasons and mechanics of changes in the Arctic. This problem would be solved through a space-based monitoring system.

Medvedev's other concern involved raising the quality of civil and military infrastructure in the Far North, in as much as global warming might call forth deterioration of structures, sewer systems, and so forth. In a discussion with the Russian Security Council, Medvedev referred to the fact that climate change might induce the Arctic to become a region for conflict between governments connected with the development of energy resources, use of shipping lanes, and bioresources and with regional shortages of food and water resources. He claimed, "Already today the polar countries are taking active steps to broaden their scientific research, economic and even military presence in the Arctic zone. In this, unfortunately, attempts to limit the access of Russia to the . . . assimilation of Arctic deposits have been observed that, of course, are allowable from a legal point of view yet unfair from the position of the geographical situation and very history of our country." He criticized "carbon protectionism" among several developed countries against countries like Russia, which, he claimed, had not taken sufficient action to curb greenhouse gas production.[102]

In its annual report released in September 2012, the University of Colorado's National Snow and Ice Data Center reported that Arctic sea ice melted over the summer to cover the third-smallest area on record and issued a warning that global warming could leave the region ice-free in September 2030.[103] All data since that time have indicated a worsening situation.

To what extent will Russian leaders and the Russian parliament ensure enforcement of existing lows to prevent extensive damage to fragile Arctic ecosystems? The Deputy Minister of Natural Resources and Ecology, Sergei Donskoi, by training a petroleum engineer, celebrated Russian geologists' discovery of the twenty-five new oil and gas fields in the western part of the Russian Arctic sea shelf alone, including the Shtokman field. Donskoi offered that exploitation of the fields would be "environmentally sustainable." He insisted that, as an outcome of the third International Polar Year, Russia would join other countries to establish databases of environmental hot spots in its Arctic territory, create new protected areas, and set up an automated monitoring system.[104] But as its recent withdrawal from the Kyoto Accords indicated, Russia's growing carbon production and consumption will instead contribute to global warming.

Rosgidromet head Berdinskii, Chilingarov, and their IPY planning committee produced a document concerning proposed Russian research activities in which they set forth the crucial place of the Arctic for Russia's future. They noted that the study of natural and anthropogenic processes in the Arctic had great significance for developing resources and realizing production that would ensure a high quality of life and steady-state development of the northern latitudes. They acknowledged that human activities had had an impact on ice conditions, especially in shallower regions of the Arctic shelf where there were significant reserves of fossil fuels and the Northern Sea Route operated. Changes in ice regime, the appearance of icebergs, the development of gas and oil fields, and the declining quality of fisheries were all related. The gradual warming of the earth over the past 100 years of $0.6^{\circ} \pm 0.2^{\circ}C$, and somewhat higher in the Arctic, was subject to natural cycles. Since the 1990s the warming trend had accelerated, which would influence both economic activity and the life of indigenous peoples and would affect the interests of many countries of the world. For example, increasingly open water would certainly stimulate increased shipping activities, which required the development of infrastructure to facilitate the increased traffic and consideration of a variety of international legal issues. On top of this, they reminded readers, specialists, and policy makers of the need to consider Russia's strategic and tactical interests carefully in light of all these phenomena.[105] They urged that the industrial potential of the Barents, Bering, and Norway seas had to be balanced with the fragility of ecosystems in the seas and estuaries, especially because air, land, and water pollution had had a negative impact on many ecosystems with impacts on human health and the quality of life. Research on the natural and anthropogenic changes in high latitudes thus had become a matter of international importance.[106]

Modest protection efforts had been carried out. Russian leaders promoted the establishment of a number of parks and zapovedniki (nature preserves) in the early twenty-first century in part to generate money through ecotourism. The park "Russian Arctic" to preserve Franz Josef Land islands was such a site. In 2011 roughly 900 tourists set foot on Franz Josef Land, paying between $12,000 and $25,000 each. But ecologists wanted more concrete results. Stanislav Fomin of the World Wildlife Federation in Murmansk called for the passage of special laws and environmental standards to go beyond cleanup to ensure that the larger presence of the Ministry of Defense in the Arctic did not lead to new problems, on top of which he noted the very small scale of the cleanup.[107]

The Ministry of the Environment approved a project to remediate the archipelago of discarded military technology, oil drums, scrap metal, and hydrocarbon waste with a focus on the three most polluted islands: Guker, Aleksandra, and Graham Bell. Unfortunately, with a budget in 2012 of only 740 million rubles ($23 million) to remove about 8,000 tons of hazardous waste, it is hard to imagine comprehensive results for all existing "superfund" sites. In all for cleanup in 2012–13, the Ministry of the Environment budgeted two billion rubles ($60 million), obviously small potatoes, and will finish the cleanup of Franz Josef Land only in 2020. Another challenge was the need for special equipment to handle the waste and a landing to handle barges to carry the stuff away. Ships of the type *Pioneer Moscow* normally used for lumber, bulk, and dangerous cargoes will be used after being fitted with special cranes and wenches.

Politicians of all stripes in all countries tend to be provocative about the disposition of resources and their critical importance for jobs especially in terms of economic uncertainty. They tend to ignore, downplay, or underestimate the extent of anthropogenic impacts on the environment of gas and oil wells, pipelines, smelters, and other factories. In the Russian case, the leaders have often repeated the right things. In May 2010 President Medvedev declared that "Russia is fully aware of its responsibility and today it takes every effort to carry out its works on the Arctic shelf, in the Caspian Sea, the Sea of Okhotsk and in other places in strict compliance with international law and international environmental standards. This is not the time to sit back."[108] Later that year during an Arctic tour, then Prime Minister Putin called for cleanup of the "gigantic garbage dump in the Arctic": hundreds of thousands of oil drums, a large amount of garbage, hazardous waste, and the like.[109] "We have to clean up the mess created over decades and left behind on islands, on airfields in the tundra region and in the waters of the Arctic," he said.[110] In July 2012 on a trip to Arkhangelsk, President Putin repeated his worries. He met with specialists who had participated in an expedition to eliminate the ecological disaster on polluted regions of the islands of Franz Josef Land. He said that "the importance of this theme includes the fact that at the beginning of the 1990s almost all facilities on the archipelago—the first Soviet hydrometeorological station was opened there in 1929—were shut down. On top of that because of the high transportation costs the facilities were not mothballed in the right way, and equipment and materials in most cases were not removed . . . the age of Arctic refuse can be calculated at twenty years, and in the worst case more than seventy years. In this year practical efforts are beginning on the cleanup of the Arctic." The government had allocated, however, a paltry sum for cleanup, while bureaucrats did not blink over $30 billion to develop the Shtokman field.

Given the primacy of resource development and economic growth over environmental concerns, the weakness of Russia's environmental ministry and law enforcement capabilities next to powerful economic ministries responsible for growth based on resource extraction, and the nature of Putinism, it was unlikely that remediation and circumspection occupied a central place in Arctic development. And the challenges of Arctic remediation were very great. In the postwar years, the authorities expanded mining and smelting operations in the Arctic regions from Nickel near the Norwegian border to Salekhard and Norilsk in the northern Arctic. Salekhard, on the Ob delta, came into existence to export nickel from Norilsk; the notorious Salkhard-Igalka gulag rail line served to tie together Arctic ports. Simultaneously, the Navy built bases from the Kola Peninsula to Vladivostok for submarine and other nuclear vessels. At scores of nuclear facilities the authorities handled waste, fissile materials, and other aspects of the nuclear fuel cycle with disregard for personnel and the environment, especially apparently in the Northern Fleet based in the northwest of Russia.[111] By all accounts, the conquest of Arctic resources had significant impacts that will last long into the twenty-first century.[112]

Many individuals sanguinely alluded to warming temperatures in Arctic and subarctic regions as a positive thing for facilitating economic development and transport and even spreading agriculture to regions previously inhospitable to such activities. Russian political leaders joined this crowd. They referred to the way warmer climate enabled the Russian icebreaker fleet to keep the Northern Sea Route from Murmansk to Arkhangelsk to Pevek to Vladivostok open year-round. But while this prospect may be of some short-term (and doubted) benefit, they ignored the great risks of global warming: more volatile weather patterns, rising sea level, the spread of microbes and diseases (as Russian and other medical researchers have reported), the release of methane long trapped in permafrost, and destruction of fragile Arctic ecosystems, which will lead to extinction of such species as polar bears. In 2008 two specialists at the Institute of Physics of the Atmosphere in Moscow reported on the significant shrinking of the ice cover from 7.5 million km^2 to 4.3 million km^2 in 2007 (and to 4.1 million km^2 in summer 2012) and the increase in the amount of *open* water between 1997 to 2006 by thirty-five days. Yet in their analysis they offered no commentary on the impact on the environment.[113] More to the point, the director of the AARI and several collaborators published an article in 2010 on climate change in the Arctic and Antarctic. They acknowledged global warming over the past 100 years, but argued that this resulted not from greenhouse gases but from "natural causes" and cycles. The authors admitted the need to deal with greenhouse gases. Yet they sited references from the Heartland Institute, a notorious Chicago-based conservative think tank funded by petroleum that rejects global warming entirely—and believes that if it exists, global warming is a good thing.[114] Other AARI scientists, however, presented data that the warming of the marine Arctic occurred from the beginning of the 1990s and sped up in the twenty-first century, especially in the summer months. They argued that more than 30 percent of the temperature change is the result of a sixty-year cycle.[115]

At the end of June 2012 specialists on permafrost held the quadrennial gathering of the International Association of Permafrost in Salekhard. Major topics of discussion included climate change, Arctic building techniques, and influence of harsh weather on those who move to the Arctic. Researchers considered technogenic impacts on the ecosystem in regions of permafrost—or 70 percent of Russian territory. Academician Vladimir Melnikov, director of the Institute of Cryosphere of the Earth, joked that with every charter plane of delegates that arrived the local temperature rose six degrees. The gorilla in the room was the fact that permafrost had begun to melt—although not all are convinced that this will lead to the release of significant amounts of methane, a very powerful greenhouse gas. The governor of the Yamal Region, Dmitrii Kobylkin, opened the meeting with the words, "Science is the decisive factor in the rational and careful assimilation of the Arctic" during a period when freight tonnage increases annually, stimulating a need for new infrastructure, and Arctic resources are being developed at a rapid pace. He called for the establishment of the International Innovation Center for the Study of the Arctic, called Ekograd, on Belyi Island in Yamal region to serve not only science, but the needs of the inhabitants of Yamal.[116] Whom does the Arctic serve?

The Russian Arctic at a Crucial Crossroads

The *Akademik Fedorov* continued to sail into the Arctic and Antarctic. It departed Petersburg in November 2011 for Antarctica with a scheduled return of May 2012—for the fifty-seventh season of Antarctic exploration. It was involved in deep drilling through the ice to Lake Vostok.[117] A summer 2011 expedition of the *Fedorov* carried roughly 200 people of whom seventy-four were crew and seventy-six members of the scientific personnel. The *Rossiia* atomic icebreaker with a crew of ninety-six people accompanied it. The goal was precise measurement of the depths through seismic sounding to determine the power of the bed deposits. Recall that the determination of their thickness is one of the criteria for establishing the border of the continental shelf and that if Russia proved that the Lomonosov and Mendeleev ridges extend to Greenland and are in fact the geographic continuation of its shelf, then it hoped to gain the right to explore resources in another

1.2 million square kilometers of Arctic region with likely oil and gas reserves in the triangle Chukhotka–Murmansk–North Pole.[118]

Can Russian science and technology ensure the successful assimilation of Arctic resources in the twenty-first century without the great human and environmental costs that characterized the heroic Soviet efforts to settle the region and exploit its oil and gas, nickel and copper, coal, and other resources? Surely, the eighty-year effort to chart resources, understand meteorological processes, gauge currents and water chemistry, and reveal the nature of the ice regime will enable measured development. Yet many obstacles stand in the way. The environmental legacy of Soviet industrial and military programs; new, extravagant plans to develop vast Arctic fossil fuel deposits; the persistent danger of pollution from exploitation of new oil, gas, and coal fields; climate change; the fragility of tundra, Arctic estuaries, seas, bays, and the Arctic Ocean; more powerful icebreakers; and rejuvenated state power have come together to threaten hopes and intentions to preserve and protect the Arctic and the health of the people who live there. McCannon accurately characterized the Arctic during the Cold War as an atomic-era battlefield with a destroyed ecosystem; the northern latitudes were contaminated with petrochemicals, PCBs, fuel, and other wastes. Not only heavy industry was to blame, but also the various military bases including the DEW (Distant Early Warning) Line, which haphazardly stored and disposed of wastes. Atmospheric tests included the largest nuclear device ever, the "Tsar-Bomba" over Novaia Zemlia in 1961 at fifty megatons.[119]

Keep in mind the military bases and radars and oil drums strewn along the Arctic Circle to protect the USSR against surprise attack over the Pole from the United States—that mirrored the military bases and radars and oil drums strewn along the Arctic Circle to protect the United States and Canada against surprise attack over the Pole. In Alaska, at various Distant Early Warning (DEW) sites, PCBs, thousands of barrels of hazardous waste, radioactive materials, solvents, pesticides, fuels, perhaps unexploded bombs, and other weapons await final disposition. The example of Canada revealed what Russia faced. The Canadian federal government has announced redoubled efforts to clean up the some of the remaining waste left by the Canadian DEW system. In spring 2012 the Federal Health Minister Leona Aglukkaq committed another $15 million to clean up two more sites on the Distant Early Warning line, the last two of twenty-one being "remediated" by the

Defense Department. The Aboriginal Affairs and Northern Development Department was responsible for another twenty-one sites. After spending billions of dollars on defense, the Defense Department budgeted a total of $575 million to clean up its sites. A recent article noted that the DEW "was built to warn against the possibility of a Soviet missile attack that never came—but its impact in the North was large and long-lasting nevertheless. DEW line construction crews were one of the first and largest forays of southern institutions into the Arctic at a time when Inuit people were making the transition from life on the land into settled communities." According to Frank Tester, a sociologist at the University of British Columbia, the construction of DEW stations also changed Inuit lifestyle and culture overnight. DEW gave Inuit "their first industrial-type jobs, introduced them to money for the first time and disrupted centuries-old patterns of family life. It accelerated their movement off the land. It created public health problems as Inuit scavenged from DEW line sites and built shanty towns, which one federal health worker compared to the worst slums of India."[120] Soviet and Russian programs had similar impacts on the environment. And Soviet and Russian programs consciously strived to incorporate indigenous people into an industrial world.

The plans to develop vast Arctic fossil fuel deposits presented new risks for the north, for they will draw in more engineers, shift workers, and support staff who need homes and offices. While many Soviet-era towns and cities decay and collapse onto the permafrost, still other new towns with their extensive infrastructure must be built. The massive projects will trigger a second wave of Arctic urbanization. Will planners and builders and officials remember the problems faced in the first wave of Arctic urbanization? The fossil fuel projects will involve massive multinational and state corporations to handle the identification and development of the rich deposits. Which institutions can ensure that they follow the highest standards of environmental and industrial safety?

In order to power the Arctic of the twenty-first century with its Northern Sea Route and extractive industries and cities, Russia will turn to nuclear power to the extent possible. Rosatom, the state ministry for things nuclear—icebreakers, reactors, and floating reactors—has recovered its hubris twenty-five years after Chernobyl. Its officials dismissed Fukushima as irrelevant to the Russian experience. They gained the full support of the Russian president to pursue rejuvenation of superpower status, even as cleanup of the Soviet Arctic nuclear legacy was an open—and expensive—question and apparently will remain so.[121]

One of the most troubling technologies to come out of the ministry's Arctic reaches was floating nuclear power stations. Based on submarine reactors and developed in Severodvinsk and other facilities to maintain high levels of employment in the nuclear shipbuilding industry, floating reactors (and floating nuclear-powered oil platforms still in the design stage) will provide electric energy and heat above the Arctic Circle. To achieve this maritime Bilibino, Russia has announced as a first step plans to build twelve floating nuclear reactors. The reactors will produce ninety megawatts of electricity for Arctic assimilation, but could also be designed for desalinization and industrial heat production. Rosatom intends to sell them for $335 million each: China, Algeria, Indonesia, Brazil, and other countries have indicated an interest in the plants. The first floating reactor, scheduled for 2014, will be anchored in a closed military bay off Russia's far eastern Kamchatka Peninsula. The region is seismically active, but the head of Rosatom, Sergei Kirienko, hubristically dismissed concerns. He said, "I know Fukushima has sparked many inflammatory rumors and gossip, including on the floating nuclear plant. Some people say that if a ground plant could not withstand a tsunami, what would then happen with a waterborne nuclear plant. But nothing will happen. Everything will be just fine." By 2020, therefore, we will see floating reactors on barges docked close to shore and plugged into local grids.[122] And everything will be fine.

The danger of pollution from exploitation of new oil, gas, and coal fields has only grown and was only one of the reasons why Total and Statoil have backed away from working on the Shtokman gas fields. A brief survey of scientific literature reveals the impossibility of taming Arctic resources without pollution.[123] What has changed in the last decade to give scientists, engineers, and policy makers any certainty that they have mastered the issue of the fragility of tundra, Arctic estuaries, seas, bays, and the Arctic Ocean? In the Russian system, given the subjugation of environmental concerns into the Ministry of Natural Resources and the Environment, a kind of interior department devoted to extracting value from nature, it is difficult to imagine circumstances where environmental issues will trump break-neck resource development.

After decades of trial, error, and experience; even with powerful technologies for supply, transport, and power generation; even with access to modern medicine and housing; and even with unquestioned support from the Kremlin, Arctic research remains dangerous. In May 2013 the sixteen scientists of the drift station North Pole-40 sent out an SOS to be evacuated from ice floe that was melting and breaking up under them. Launched in October 2012, scientists anticipated working until September 2013, but from the start—owing to global warming—they could not find an ice floe large enough to begin the operation properly. On top of this "a near record-breaking high pressure system" pushed the floe more than 300 kilometers in a month toward the Canadian Basin of the Arctic where even some of the thickest ice of the Arctic was cracking up, too. President Putin himself approved a sixty-five-million-ruble (more than $2 million) emergency mission by the *Yamal* nuclear icebreaker to rescue the research team and its equipment; airplanes could not be used because it was impossible to build an ice runway in ever-changing and melting conditions, while the ice floe was out of the reach of helicopters.[124]

If Russian leaders could find funding for symbolic technology—and its rescue—they were less inclined to pursue sustainable development programs, remediation, or cleanup. Under Putin, the environment played second fiddle to the lucrative business of the oligarchs of sucking oil and gas out of the Arctic and smelting heavy metals. The city of "Nickel," a medal in the cap of Norilsk Metals, seven kilometers from Norway and less than 100 kilometers from Finland, continued to spew sulfur dioxide and other noxious gasses into the atmosphere at five to ten times questionable norms, while European leaders ignored the damage to their countries and their peoples, perhaps for fear of antagonizing Russia. According to the *Barents Observer*, the Nickel plant "alone produces about 90,000 tonnes of sulfur dioxide every year, which is about five times the total amount of Norway's emissions." Norilsk Metals has invested only a few million dollars into pollution control, far less than needed or can be afforded. According to the Bellona Foundation, "Nickel" has special permission for higher emissions and in good Soviet fashion finds it cheaper to pay fines than to clean up.[125]

Rejuvenated state power threatened any hope to preserve and protect the Arctic and the health of the people who live there. Since the time of Peter the Great, the Russian state has assumed a major role as innovator and engine of modern technology, in the Tsarist period because of the nascent private sector, in the Soviet period because of state ownership of the means of production, and at present in still not entirely clear ways, but through a mix of public and private initiatives to develop resources. In jumps and starts, similarly, the state sought military technology after defeat and industry in the face of technological lag.[126] This held for Russia in the twenty-first century. The Russian president became leader of a kind of oligarchy of resource ministers, and he used state power to push economic development programs that hearken to, if are not fully beholden to, programs from the Stalin era. There can be no doubt of the central role the state will have in pushing "modernization" of Arctic regions through a twenty-first-century Glavsevmorput.

Climate change should be a call to be careful, go slowly, and recognize the fragility of Arctic ecosystems rather than embrace global warming for opening the Northern Sea Route. Instead, judging by worldwide experience in the New World, the Brazilian rainforest, and the Soviet tundra, Arctic development, as presently constituted, will threaten what remains of indigenous communities and lifestyles. While I do not suggest abandoning all efforts—or abandoning all hope—I do call for learning from the Soviet experience to move more slowly, to adopt measured policies, to forego impatience for circumspection, and in the face of uncertainties about industrial impacts on the Arctic circle to see the region as an important international heritage, not a pie to be cut up rapidly and recklessly by contiguous Arctic nations. The alternative, the one being pursued now by Arctic nations, will lead to ruin as we know already from the history of the heroic period of Arctic conquest in the Soviet era.

NOTES

ACKNOWLEDGMENTS

INDEX

Notes

Introduction

1. http://eng.kremlin.ru/news/399
2. http://eng.kremlin.ru/news/3777
3. For a global history, see John McCannon, *A History of the Arctic* (London: Reaktion Books, 2012).
4. Gerald Cohen, *Marx's Theory of History: A Defense* (Princeton: Princeton University Press, 2000).
5. Stephen Kotkin, *Magnetic Mountain: Stalinism as Civilization* (Berkeley: University of California Press, 1995).
6. James Scott, *Seeing Like a State* (New Haven: Yale University Press, 1994).
7. Anthony Sutton, *Western Technology and Soviet Economic Development, 1917 to 1930* (Stanford: Stanford University Press, 1968), and *Western Technology and Soviet Economic Development, 1930 to 1945* (Stanford: Stanford University Press, 1971).
8. John Scott, *Behind the Urals* (Bloomington: Indiana University Press, 1973). First published in 1942.
9. On the case of pulp, paper, and electricity, see Karl-Erik Michelsen, "An Uneasy Alliance: Negotiating Transnational Infrastructures at the Finnish-Soviet Border," in a forthcoming volume edited by Arne Kaijser.
10. Michael Suprun, editor, *Lend-Liz i Rossiia* (Arkhangelsk: Pravda Severa, 2006) and idem., editor, *Severnye Konvoi* (Arkhangelsk: Arkhangelskii Filial Geograficheskogo Obshchestva SSSR, 1991).
11. See Yuri Slezkine, *Arctic Mirrors* (Ithaca: Cornell University Press, 1994), Anna Kerttula, *Antler on the Sea* (Ithaca: Cornell University Press, 2000), and John McCannon, *Red Arctic* (New York: Oxford University Press, 1998) for discussion of native peoples in Soviet Arctic regions.
12. Gunnar Seidenfaden, *Modern Arctic Exploration* (Boston: Hale, Cushman and Flint Publishers, 1939 [?]), p. 18.
13. Ibid., pp. 30–32.

14. Ibid., pp. 38–39.

15. Ibid., pp. 56–57.

16. E. S. Kokovin, *My Podnimaem Iakoria* (Arkhangelsk: Severo-Zapadnoe Knizhnoe Izdatelstvo, 1972), p. 99.

17. Ibid., pp. 123, 125–26.

18. http://strana.ru/routes/2366451

19. http://kirjasto.sci.fi/brodsky.htm. See also http://www.literaryreview.co.uk/mahon_02_11.html and http://www.nobelprize.org/nobel_prizes/literature/laureates/1987/brodsky.html.

20. IosifBrodsky,"PoltoryKomnaty," http://brodskij.clubleader.ru/?tag=1965%D0%B3.

1. Charting the Arctic Landscape

1. The Barents Sea stretches from the northern tip of Norway to Novaia Zemlia and is often covered with mist and fog. Most ships push through the Iugorskii Strait as the main thoroughfare to the Kara Sea; the Soviets marked it with lighthouses at its eastern and western ends, surveyed the channel carefully, and maintained the channel with buoys. Land beacons provided additional support, but most captains preferred bringing steamers through in daylight in any event. In the middle of October the narrow, shallow waters of the strait freeze and provide a natural bridge between the mainland and Vaigach Island and enable access to its rich populations of brown bears, wolves, reindeer, hares, foxes, and partridges; Russians join Nentsy and other "samoeds" with their dog sleds in hunting expeditions. Just inside the strait under a large bluff lies Varneka Bay where the Soviet military police established a large mining settlement; the administration of the Northern Sea Route built a meteorological station, a radio station, and barracks. Significant surface deposits of lead and zinc triggered the industrial interest in Vaigach—as did opportunities for shipping ore to Arkhangelsk. See Leonard Matters, *Through the Kara Sea* (London: Skeffington and Son, Ltd., 1929), pp. 68–74.

2. K. K. Sluchevskii, *Poezdki po Severu Rossii v 1885–1886 gg.* (Moscow: OGI, 2009), pp. 99–105.

3. Ruslan Aleksandrovich Davydov has written extensively on the history of Murmansk Gubernia including the development of the railroad.

4. On Middendorff, see Erki Tammiksaar and Ian R. Stone, "Alexander von Middendorff and His Expedition to Siberia (1842–1845)," *Polar Record,* vol. 43, no. 3 (July 2007), pp. 193–216. See also Konstantine Krypton, *The Northern Sea Route: Its Place in Russian Economic History Before 1917* (New York: Research Program on the USSR, 1953), pp. 88–89.

5. Sluchevskii, *Poezdki po Severu Rossii,* pp. 112–14.

6. Krypton, *The Northern Sea Route,* pp. 122–23.

7. Ian Stone, "Joseph Wiggins (1832–1905), *Arctic,* vol. 47, no. 4 (December 1994), pp. 408–10.

8. Krypton, *The Northern Sea Route,* pp. 81–87.

9. N. M. Knipovich, *Ekspeditsiia Nauchnopromyslovykh Issledovanii u Beregov Murmana,* vol. I (St. Petersburg: Khudozhestvennaia Pechat', 1902), pp. 1–6.

10. Krypton, *The Northern Sea Route,* pp. 88–89.

11. V. M. Shimkevich, N. P. Varner, and N. N. Polezhaev, "Iz Vospominanii Zoologa," *Zhurnal Ministerstva Narodnogo Prosveshcheniia: Nov. ser.,* 1908 (vol. 16), no. 7, part 4, pp. 1–18.

12. S. Makarov, *"Ermak" v L'dakh* (St. Petersburg: Yevdokimov, 1901).

13. Krypton, *The Northern Sea Route,* pp. 74–75. Dmitry Anuchin, the governor-general of eastern Siberia and a prolific author of historical analyses of transport technology, shared Witte's view, worrying that without the railroad—and a canal connecting the Ob' and Enesci rivers—that Russia's military could easily be cut off from supply during a war in the east.

14. A. Borisov, *Velikii Severno-Vostochnyi Morskoi Put'* (St. Petersburg, 1910), p. 35.

15. Krypton, *The Northern Sea Route,* pp. 89–91.

16. Ibid., pp. 77–78.

17. Ibid., pp. 79–81.

18. Ibid., pp. 93–95.

19. Ibid., pp. 102–9, 125–26, 129.

20. Ibid., pp. 128–30.

21. Ibid., pp. 99–101.

22. Ibid., pp. 88–89, 93.

23. M. I. Belov, *Istoriia Otkrytiia i Osvoeniia Severnogo Morskogo Puti,* vol. III (Leningrad: Morskoi Transport, 1959), pp. 58–62; http://www.hrono.ru/biograf/bio_we/vilkicky.html. See also N. I. Evgenov, V. N. Kupetskii, "Russkii Poliarnyi i Issledovatel' B. A. Vilkitskii," *Letopis' Severa,* 4 (1964), and Valerii Klaving, *Grazhdanskaia Voina v Rossii: Belye Armii* (Moscow: Voenno-Istoricheskaia Biblioteka, 2003).

24. Fridtjof Nansen, "Some Results of the Norwegian Arctic Expedition, 1893–1896," *Geographical Journal,* vol. 9, no. 5 (May 1897), pp. 473–505.

25. See http://hrono.ru/biograf/bio_s/sedov_gya.php for a biography of Georgii Sedov.

26. N. V. Pinegin, *V Ledianykh Prostorakh k Severnomu Poliusu (1912–1914)* (Moscow: OGI, 2009).

27. Ibid., pp. 11–18, 28–29.

28. Ibid., pp. 194–206.

29. Krypton, *The Northern Sea Route,* pp. 124–25.

30. Vice Admiral Makaroff, "The 'Yermak' Ice-Breaker," *The Geographical Journal,* vol. 15, no. 1 (January 1900), pp. 33–42.

31. Ibid., p. 40.

32. Much of the following discussion comes from these sources, which often repeat information in the other webpages: http://www.seaham.i12.com/myers/zamyatin.html, http://www.kirjasto.sci.fi/zamyatin.htm, and especially "Zamyatin in New Castle," at http://www.sclews.me.uk/zamyatin.html. According to one author, Zamiatin applied some form of the numbers of the ships or shipyard or tonnage to the characters of *We.* And "apart from icebreaker supervision in various towns, including Sunderland, South Shields and Glasgow, where he attended ship launches in 1916 and 1917, he endured Zeppelin and aircraft bombing, learned to drive a car and visited castle ruins. That is all he has to say about an episode which was crucial to his development as a writer."

33. *We,* it must be pointed out, was not directed precisely at Bolshevism but at authoritarianism. *We* appeared in an émigré journal in 1927. In 1929 the Soviet *Literary Gazette* attacked "bourgeois writers" operating in the USSR and mentioned Pilnyak by name. In fact, Pilnyak had idealized a Trotskyite official in a short story. Their expulsion led Boris Pasternak also to resign from the Union. Zamiatin wrote Stalin himself in 1931 with a request to leave the USSR; Stalin ordered him and his wife to go into exile. Zamiatin moved to Paris and never again achieved the same literary success; he died in poverty in 1937. Pilnyak was arrested and also died in 1937, no doubt at the hands of the secret police.

34. Krasin died in 1926 of a blood disease that his friend, Alexander Bogdanov, who had founded an Institute for the Study of Blood Transfusions, could not cure, Bogdanov himself dying of a gruesome experiment he performed on himself. See Nikolai Krementsov, *A Martian Stranded on Earth: Alexander Bogdanov, Blood Transfusions and Proletarian Science* (Chicago: University of Chicago Press, 2011).

35. Sergei Lar'kov, "Sud'by Uchastnikov Znamenitoi Ekspeditisii," in Lar'kov and Fedor Romanenko, *"Vragi Naroda" za Poliarnym Krugom* (Moscow: Paulsen, 2010), pp. 214–200.

36. http://en.wikipedia.org/wiki/Krasin_(1916_icebreaker)

37. John McCannon, *A History of the Arctic* (London: Reaktion Books, 2012), pp. 194–95.

38. Scientists in the Academy had founded KEPS in 1916 in an attempt to assist the Tsarist government in locating and harnessing resources for the flagging war effort, so badly managed by the Tsarist government, and in assisting national industry buffeted by its isolation from foreign technology, especially from Germany and from military industry, that made production of fuel, ordnance, and explosives difficult.

39. Archive of AARI (A AARI) R-1797 "Uchastie Akademicheskikh Uchrezhdeniiakh v Izuchenii Severa" (Moscow 1941), ll. 1–4.

40. A AARI, R-1797, ll. 6–7. L. L. Breitfus joined the Murmansk Scientific Trade Expedition of Knipovich in 1898, became director of the expedition in 1902, and kept that position until 1908, carrying out research in the Barents Sea on the *Andrei Pervozvannyi*, one of the first ships in the world designed for research and industrial ends, in the process creating the first hydrological map of the sea. From 1912 Breitfus headed the hydrometereological service of the Main Hydrographic Administration, from which position he organized four stations in the Kara Sea. He took the initiative to arrange to have airplanes look for G. Ia. Sedov, G. L. Brusilov, and V. A. Rusanov, who had disappeared in a polar expedition. In 1920 Breitfus went to Europe on business to serve research and development on the Arctic. But in 1921 Breitfus moved to Berlin and determined to remain in Germany to the end of his life. V. Volkov, "Breitfus," at http://www.ihst.ru/projects/emigrants/breitfus.htm; S. V. Popov, "L. L. Breitfus—Issledovatel' Arktiki," *Zemlia i Vselennaia,* 1990, no. x; and GARV, F. 5446, op. 37, d. 48, ll. 211–12.

41. A AARI, R-1797, ll. 8–9.

42. A AARI, R-1876 "Sostoianie Arkticheskogo Porta v 1918–1920 gg. i Pechorskaia Ekspeditsiia," l. 3–4.

43. A AARI, R-1876, ll. 1–2.

44. S. V. Mikhailov, T. S. Rass, "Razvitie Rybnykh Promyslov v Barentsovom More i na Murmanskom Poberezh'e," (Moscow, 1941) in A AARI, R-1840, ll. 12–14.

45. Belov, *Istoriia Otkrytiia i Osvoeniia Severnogo Morskogo Puti,* pp. 58–62.

46. Ibid. The first group of ships consisted of *Ivan Susanin* and *Poliarnyi* and headed toward Dikson Island, with the *Arktur* and *Altair* going to Vaigach Island. The second consisted of the smaller hydrographic ships *Inei, Orlik, Shuia, Poleznyi,* and *Anna* under the command of the hydrologist K. K. Neukoev. These ships quickly fell into Red hands. The third group consisted of the transport/icebreakers *Solombala, Kildin, Pakhtusov,* and *Kolguev* and the steamship icebreakers *Taimyr* and *Solovei Budimirovich,* directed also toward Dikson and whose task was to escort through the ice and ensure safe passage through uncharted deltas of Siberian rivers.

47. Ibid.

48. Ibid., pp. 182–84.

49. A AARI, R1795 "Proekt Vodnykh Putei na Severe Evropeiskoi chasti Rossii, 1920 g.," ll. 1–9.

50. A AARI, R1440, "O Predostavlenii Kontsessii na Velikii Severnii Zh-D Put'," ll. 8–10.

51. TsGAOR, Arkhiv VSNKh, F. NTO, d. 419/55, 1920), as cited in A. F. Treshnikov, "Ordena Lenin Arkticheskomu i Antarkticheskomu Nauchno-Issledovatel'skomu Institutu—50 let," *Problemy Arktiki i Antarktiki,* vol. 36–37 (1970), pp. 5–7. See also A AARI, R-1876, ll. 6–7.

52. Bertrand M. Patenaude, *The Big Show in Bololand* (Stanford: Stanford University Press, 2002).

53. A. F. Treshnikov, "Ordena Lenin Arkticheskomu i Antarkticheskomu Nauchno-Issledovatel'skomu Institutu—50 let," *Problemy Arktiki i Antarktiki*, vol. 36–37 (1970), pp. 7–8.

54. A AARI, R-1735, "Materialy po Istorii Severonogo Morskogo Puti: Rybnye i Zverinye Promysla Krainego Severa," pp. 1–8.

55. Ibid., pp. 71–72.

56. "Vtoraia Karskaia Ekspeditisa (1922 g.)," in A AARI, R. 1776, ll. 1–19.

57. A. S. Zhirmunskii, "Severnaia Nauchno-promyslovaia Ekspeditisiia VSNKh (1920–1924 gg.)," Leningrad, ANII, 1941, in A AARI, R. 1719. See also V. Kh. Buinitskii, "25 let Arkticheskogo Instituta," in V. Kh. Buinitskii, editor, *XXV Let Nauchnoi Deiatel'nosti Arkticheskogo Instituta* (Leningrad-Moscow: Izdatel'stvo Glavsevmorputi, 1945), pp. 10–11. See also his V. Kh. Buinitskii, "25 Let Arkticheskogo Instituta" *Problemy Arktiki*, vol. 1 (1946), pp. 11–19, for a speech at the celebratory meeting of the institute on January 13, 1946, in which director Buinitskii covers essentially the same topics.

58. A AARI, R-1834, "Plan Raboty Novozemelskikh Otriadov Severnoi Nauchno-Promyslovoi Ekspeditsii NTO VSNKh, 1922–1924," ll. 1–26. A number of efforts resulted in publications: V. V. Faas on exploitation of northern resources, Fersman on the Khibiny region, V. K. Soldatov and A. A. Zernov on ichthyology of the Pechora River, N. Ia. Volens on the economy of the Pechora region.

59. Ibid., ll. 27–32.

60. Ibid.

61. A. F. Treshnikov, "Ordena Lenin Arkticheskomu i Antarkticheskomu Nauchno-Issledovatel'skomu Institutu—50 let," pp. 8–10.

62. A AARI, R-1870, "Lenin o Razvitii Proizvoditel'nykh Sil Krainego Severa i Komissii VSNKh po Izucheniiu i Prakticheskomu Ispol'zovniiu Russkogo Severa," ll. 6–8.

63. A AARI, R-1887, "Materialy o Radiostantsii na Ostrovakh Novoi Semli i Zemli Frantsa Iosifa za 1922/29g." (Leningrad: ANII, 1929).

64. Ibid.

65. Ibid.

66. Ibid.

67. Krypton, *The Northern Sea Route*, pp. 89–90.

68. On Komseveroput, see http://www.emaproject.com/north_article.html?id=1571.

69. Savva Morozov, http://www.memorial.krsk.ru/public/80/19870424.htm. See also http://www.polarpost.ru/forum/viewtopic.php?f=8&t=2337. Dikson's guest book includes the signatures of such visitors as Fridtjof Nansen and Otto Smidt, head of Glavsevmorput.

70. Morozov, http://www.memorial.krsk.ru/public/80/19870424.htm. See also http://www.polarpost.ru/forum/viewtopic.php?f=8&t=2337.

71. Ibid.

72. Ibid.

73. Morozov, http://www.memorial.krsk.ru/public/80/19870424.htm.

74. Ibid. See also http://www.polarpost.ru/forum/viewtopic.php?f=8&t=2337.

75. See for example Anna Kerttula, *Antler on the Sea* (Ithaca: Cornell University Press, 2000) and Yuri Slezkine, *Arctic Mirrors* (Ithaca; Cornell University Press, 1994).

76. S. V. Kertselli, *Izbennoe Olenovodstvo i ego Znachenie v Sel'skom Khoziaistve* (Leningrad: Institut Narodov Severa, 1933).

77. S. V. Kertselli, *Po Bol'shezemel'skoi Tundre s Kochevnikami* (Arkhangelsk: Gubernaia Tipografiia, 1911).

78. Kertselli, *Po Bol'shezemel'skoi Tundre,* p. 64.

79. Ibid., pp. 5–7, 10, 13–14.

80. Ibid., p. 16.

81. Ibid., p. 20.

82. Ibid., pp. 5–7, 10, 13–14.

83. Ibid., pp. 5–7.

84. Ibid., pp. 10–11.

85. Ibid., p. 78. See pp. 77–87 on herding, seasonal requirements and activities, grazing habits, biological changes in the reindeer over the summer and early fall months, and other human and animal aspects of herding.

86. Ibid., p. 88.

87. Ibid., pp. 29–30.

88. Ibid.,, pp. 54–55.

89. Ibid., pp. 56–57.

90. Ibid., pp. 11, 62–64.

91. Ibid., pp. 64–67.

92. Ibid., p. 76.

93. A AARI, R-1449 "Narodnyi Komissariat po Delam Natsional'nostei (1917–1924 gg) i Malye Narody Severa RSFSR) (M: ANII, 1946), pp. 1–22. On Sibnats, see http://www.ipdn.ru/rics/va/_private/a8/C-109-114-Chumak.pdf.

94. http://finugor.ru/leaders/candidate/21124. From 1927 Smidovich also chaired the Central Bureau for the Study of Local Lore. In the Tsentral'noe Biuro Kraevedeniia (TsBK or the Central Bureau for the study of Local Lore), established in 1922 under the auspices of the Academy of Sciences, a number of activists pushed conservation. TsBK was not an armchair scientific organization, but rather a uniquely mass organization. By the late 1920s, 2,270 branches of the TsBK embraced roughly 60,000 members including not only academically based naturalists, but representatives of the provincial intelligentsia as well. Some chapters of

TsBK also studied indigenous people. Yet like the Committee of the North, this important organization was short-lived. Already in 1929–30 some of the members of the association were arrested as being parts of a "monarchist counterrevolutionary organization," and it too was liquidated in 1937.

95. http://www.suvenirograd.ru/sights.php?id=1234&lang=2. See also http://v-k-arseniev.ru/ and http://v-k-arseniev.ru/books-and-papers/dersu-uzala/.

96. Sergey Krivshenko, "Vladimir Klavdievich Arsenyev and Its [sic] Creative Heritage," http://www.vld.ru/ppx/Krivsh/Arsenev.htm.

2. Neither Cod nor Coal

1. In *Seeing Like a State* (New Haven: Yale University Press, 1999), James Scott describes the rise of high modernism and its essential features including such case studies as the USSR.

2. Matthew Farish and P. Whitney Lackenbauer, "High Modernism in the Arctic: Planning Frobisher Bay and Inuvik," *Journal of Historical Geography*, vol. 35, no. 3 (July 2009), pp. 517–44.

3. Alvin Weinberg, *Reflections on Big Science* (Cambridge: MIT Press, 1967).

4. V. Kh. Buinitskii, "25 let Arkticheskogo Instituta," pp. 12–13.

5. Elena Poltavskaia, "Otto Iul'evich Smidt," at http://arctic-online.ru/history/heroes/1.html. After being replaced as director because of age and illness, and no doubt a bit because of the high-profile failure of the zimovka of the *Sedov*, Shmidt focused on research and administrative activities in various geophysical institutions of the Soviet Academy of Sciences. In his last years, even as his tuberculosis worsened, Shmidt focused on pedagogy and publication as editor of *Priroda* (Nature).

6. V. Kh. Buinitskii, "25 let Arkticheskogo Instituta," pp. 12–13.

7. Gosudarstvennyi Arkhiv Arkhangel'skoi Oblasti, hereafter GAAO, F. 1735, d. 181, 149–50.

8. Ibid., l. 166.

9. GAAO, F. 1735, op. 1, d. 305, ll. 1–9. Glavsevmorput was organized along geographic and administrative lines. It had five territorial administrations. The Arkhangelsk administration was one of the largest from the point of view of personnel and obligations, but small from the point of view of territory and budget. In 1936 the territories received almost ten million rubles as follows: Arkhangelsk (1.3 million rubles), Omsk (three million rubles), Krasnoiarsk (0.7 million rubles), Iakutsk (1.5 million rubles), and the Far East (3.2 million rubles).

10. GAAO, F. 1735, d. 181, ll. 148–49.

11. GAAO, F. 1735, op. 1, d. 98, ll. 118–22.

12. GAAO, F. 1735, d. 181, ll. 16–19.

13. Glavsevmorput', *O Partiinoi Rabote v Arktike* (Leningrad: Izdatel'stvo Glavsevmorputi, 1939), pp. 58–59.

14. Ibid., pp. 58–59.

15. Ibid., p. 60.

16. Bruce Hopper, "Soviet Conquest of the Far North," *Foreign Affairs,* vol. 14, no. 3 (April 1936), pp. 499–505.

17. GAAO, F. 1735, d. 181, ll. 12–15.

18. Ibid., op. 1, ed. khr.345, ll. 86, 93, 99.

19. Ibid., l. 95.

20. Ibid., ll. 99–101.

21. Ibid., l. 81.

22. Glavsevmorput', *O Partiinoi Rabote,* pp. 44–45.

23. Ibid., p. 91.

24. GAAO, F. 1735, op. 1, d. 98, ll. 161–63.

25. Ibid., l. 187

26. Ibid., ll. 69–70.

27. Ibid., l. 62.

28. Ibid., l. 42.

29. Ibid., ll. 64–66.

30. Ibid., ll. 30–33.

31. Ibid., ll. 47–50. In *The Ghost of the Executed Engineer* (Cambridge: Harvard University Press, 1993), Loren Graham discusses how Peter Palchinsky and other engineers associated with various construction projects lamented the dreadful treatment of workers during the industrialization campaigns.

32. GAAO, F. 1735, op. 1, ed. khr. 98, ll. 47–50.

33. Ibid., ll. 45–46.

34. Glavsevmorput', *O Partiinoi Rabote,* pp. 39–40.

35. Ibid., pp. 71–73.

36. Ibid., p. 73.

37. Ibid., p. 74:

> Est' u nas Vasil' Davydov –
> Preotlichnyi Kochegar
> I vsia vaxta, chto Vasilii,
> I vsegda na marke par.
> Ostal'nye kochegary,
> Podtianites' vsled za nim,
> Chotby ne bylo zatorov,—
> Par Upustiat I stoim . . .

38. *Pokhod 'Sibiriakova'* (Moscow-Leningrad: OGIZizdat, 1933). The volume includes sixteen plates that reproduce paintings by Lev Kantorovich. See also O. Iu. Shmidt, *Pokhod "Cheliiuskina"* (Moscow: Pravda, 1934), *Za Osvoenie Arktiki* (Leningrad: Glavsevmorput', 1935), *Osvoenie Severnogo Morskogo Puti i Zadachi Sel'skogo Khoziaïstva Krainego Severa* (Moscow: VASKhNIL, 1937), and V. Iu.

Vize, *Issledovaniia Sovetskoi Arktiki,* 2nd edition (Arkhangelsk: Sevkraigiz, 1934), pp. 174–87. Vize was the scientific director of the "Sibiriakov" expedition.

39. A AARI, R-1764 "Ekspeditsii Glavsevmorputi s 1935–1938gg." (Leningrad: ANII, 1935), ll. 1–6.

40. Hopper, "Soviet Conquest of the Far North," pp. 501–2.

41. Shmidt, *Pokhod 'Cheliuskina'*, p. 15.

42. GAAO, F. 57, op. 1, ed. khr. 1, ll. 1–4.

43. GAAO Otdel DSPI, F. 290, op. 2, d. 549.

44. GAAO, F. 2590, op. 3N, d. 27.

45. Ibid., F. 1735, op. 1, d. 98, ll. 160–63.

46. Ibid., F. 57, op. 1, ed. khr. 1, ll. 1–4. The annual report for the activities of the society in 1937 indicates achievements along the same lines with the only major difference being that the institutional members, in spite of two reminders, had not paid their dues. Ibid., l. 6.

47. Ibid., ll. 16–18.

48. Ibid., F. 1735, op. 1, d. 98, ll. 45–46, 52–55.

49. Ibid., l. 59.

50. V. K. Buinitskii, *812 Dnei v Dreifuiushchikh L'dakh* (Moscow: Glavsevmorput, 1945), pp. 15–23.

51. S. A. Spirikhin, *Suda Severnogo Morskogo Parokhodstva i Poliarnoi Gidrografii* (Arkhangelsk: Pravda Severa, 2003), pp. 5–6.

52. Vladimir Zenzinov, "The Soviet Arctic," *Russian Review,* vol. 3, no. 2 (Spring 1944), pp. 68–71.

53. http://www.globalsecurity.org/military/world/russia/icebreaker-1.htm

54. GAAO, F. 1735, op. 1, ed. khr. 181, ll. 1–3.

55. Ibid., l. 31.

56. Ibid., d.423, ll. 1–16.

57. Ibid., ll. 98–99.

58. Ibid., ll. 103–7.

59. Among the many studies that consider the ideological and social aspects of large-scale technologies, see Walter McDougall, *The Heavens and the Earth . . .* (New York: Basic Books, 1985); Gabrielle Hecht, *The Radiance of France* (Cambridge: MIT Press, 1998); Stephen Kotkin, *Magnetic Mountain* (Berkeley: University of California Press, 1995); Thomas Zeller, *Driving Germany* (New York: Berghahn Books, 2007); Alf Nilsen, *Dispossession and Resistance in India* (London, New York: Routledge, 2010); and Mark Reisner, *Cadillac Desert* (New York: Viking, 1986).

60. On the importance of aviation for Stalinist legitimacy and the technological celebrations of aviation, see Kendall Bailes, "Technology and Legitimacy: Soviet Aviation and Stalinism in the 1930s," *Technology and Culture,* vol. 17, no. 1: pp. 55–81, and Scott Palmer, *Dictatorship of the Air: Aviation Culture and the Fate*

of Modern Russia (Cambridge: Cambridge University Press, 2006). Because these issues have been covered elsewhere, I will not stop for a long time on Arctic aviation.

61. I discuss these ideas in "'Projects of the Century' in Soviet History: Large Scale Technologies from Lenin to Gorbachev," *Technology and Culture*, vol. 36, no. 3 (July 1995), pp. 519–59.

62. Jay Bergman, "Valerii Chkalov: Soviet Pilot as New Soviet Man," *Journal of Contemporary History*, Vol. 33, No. 1 (January 1998), p. 139.

63. GAAO, F. 1735, d. 181, ll. 21–23.

64. http://gulagmuseum.org/showObject.do?object=1666453

65. http://www.bvvaul.ru/articles/text/rasskazi/loyko_povest/43.php. See also V. I. Artamonov *Zemlia i Nebo Vodop'ianova* (Moscow: Politizdat, 1991).

66. Iurii Kanev on F. B. Farikh at http://www.nvinder.ru/archive/2006/may/27/08.shtml. See also http://www.polarmuseum.ru/bio/pilots/bio_frh/bio_frh.htm.

67. Iurii Kanev on F. B. Farikh.

68. During the Finno-Soviet war, Farikh trained young pilots for blind, seat-of-the-pants flights, and he also carried out eighteen bombing raids. After the war he flew the familiar Bolshzemelskaia tundra—carrying postage, freight, and passengers. The secret police arrested him in 1948 and gave him twenty-five years in a labor camp, eight of which he served. But he was freed in 1956 and fully rehabilitated. He could no longer fly, but rather than write memoirs or play chess, he took a position as a worker at the "Red Metallist Factory"—from 1962 until 1975 until he was almost eighty years old. See Iurii Kanev on F. B. Farikh.

69. Ivan Doikov, "Ivan Khramov—Nachal'nik 'OLP Amderma'," at http://www.google.ru/url?sa=t&rct=j&q=khramov%20amderma&source=web&cd=1&ved=0CCUQFjAA&url=http%3A%2F%2Fwww.doykov.1mcg.ru%2Fdata%2F1%2FChramov.doc&ei=fWA7UMKuOKuu0AHqnYBo&usg=AFQjCNHIRKB-65cZKUp6UobT8VfDhfwy-Q&cad=rjt

70. A. Evsiugin, *Sud'ba Keimennaia Gulagom* (Nar'ian-Mar: Nar'ian-Marskaia Tipografiia, 1993), pp. 31–40.

71. Ibid. On I. Ia. Prourzin, see http://www.hrono.ru/biograf/bio_p/prourzin.html.

72. Evsiugin, *Sud'ba Keimennaia Gulagom*, pp. 33–34 and http://www.hrono.ru/biograf/bio_p/prourzin.html.

73. http://en.wikipedia.org/wiki/Sigizmund_Levanevsky

74. Evsiugin, *Sud'ba Keimennaia Gulagom*, pp. 36–38.

75. Chukhnovskii grew up in Gatchina, a city south of Leningrad connected with the history of the heavens through an observatory dating to the Tsarist era and an airport and flying school founded in 1917. (Looking not to the heavens, but to the basement of the Lubianka prison in Moscow, the NKVD purged Gatchina's

astrophysicists en masse during the Great Terror, 1936–37, decimating astrophysics for years. See Robert McCutcheon, "The 1936–1937 Purge of Soviet Astronomers," *Slavic Review,* vol. 50 (Spring 1991), pp. 100–17.) In 1917 Chukhnovskii completed a flight school in Oranienbaum and joined the navy, serving for the Reds in the aviation wing of the Volga-Caspian flotilla. He first flew to the Arctic in 1924 and in 1928 participated in the search for Nobile's *Italy* dirigible. His *Junker* was loaded onto the *Krasin* and carried out reconnaissance from the ice. He observed Mariano and Zapin on the ice, who had left the others in the hopes somehow of making Spitsbergen. On the return flight, Chukhnovskii crashed on the ice and had to wait five days for the *Krasin* to rescue him, for the ship first had to rescue the Italian survivors. In 1937–38, Chukhnovskii participated in the search for Levanevskii. During the war and after, Chukhnovskii served polar aviation as part of the northern fleet. Stalinism ruined these people—and Stalinism made their accomplishments possible in the first place.

76. http://www.bvvaul.ru/articles/text/rasskazi/loyko_povest/43.php. See also Artamonov *Zemlia i Nebo Vodop'ianova.*

77. M. Vodop'ianov, *Poliarnyi Letchik. Rasskazy* (Leningrad: Leningradskoe Gazetno-zhurnal'noe I Knizhnoe Izdatel'stvo, 1954).

78. http://dic.academic.ru/dic.nsf/enc_tech/2149/%D0%94%D0%BE%D0%B1%D1%80%D0%BE%D0%BB%D0%B5%D1%82

79. Elmer Plischke, "Trans-Arctic Aviation," *Economic Geography,* vol. 19, no. 3 (July 1943), pp. 283–91.

80. In *Thank you, Comrade Stalin!* (Princeton: Princeton University Press, 2000), Jeffrey Brooks explores how the central Soviet media developed a cult of Stalin in the 1930s, as well as cults of other heroes and villains, production, and the like.

81. In the 1930s, Wrangel Island fell under the arbitrary rule of its governor Konstantin Semenchuk, who controlled the local populace and his own staff through open extortion and murder. He forbade the local Eskimos to hunt walruses, which put them in danger of starvation, while collecting food for himself. He was then implicated in the mysterious deaths of some of his opponents, including the local doctor. At trail, Semenchuk was sentenced to death for "banditry," as charges of murder carried only a ten-year sentence. See McCannon, *Red Arctic,* pp. 56–57. See also http://www.time.com/time/magazine/article/0,9171,756168,00.html

82. Buinitskii, *812 Dnei,* pp. 115–16.

83. B. A. Krutskikh et al., "70 Let Ordena Lenina Arkticheskomu i Antarkticheskomu Nauchno-issledovatel'skomu Institut," *Problemy Arktiki i Antarktiki,* vol. 66 (Leningrad Gidrometeoizdat 1991), p. 10.

84. V. K. Buinitskii, "25 let Arkticheskogo Instituta," pp. 13–14.

85. Buinitskii, *812 Dnei,* p. 118.

86. Ibid., pp. 34, 36, 40.

87. Ibid., p. 39.

88. Ibid., p. 117.

89. Ibid., pp. 3, 12–15.

90. Ibid., pp. 8–9, 119–22.

91. Ibid., p. 16.

92. Ibid., p. 52. Russian divers claim to have identified the wreck of the *Cheliuskin*. Artifacts from the wreck are to be sent to Denmark, where the ship was built, to confirm its identity. http://memory.loc.gov/intldl/mtfhtml/mfdigcol/lists/mtfnarTitles2.html

93. Ibid., pp. 20–21, 26–30. See also McCannon, *Red Arctic*, pp. 66–67.

94. Ibid., p. 1.

95. Ibid., pp. 4–5.

96. Ibid., pp. 3–4.

97. Krutskikh et al., "70 Let Ordena Lenina Arkticheskomu i Antarkticheskomu Nauchno-issledovatel'skomu Institut," p. 8.

3. The Role of the Gulag in Arctic Conquest

The epigraph is from M. S. Kriukova, *Noviny* (Arkhangelsk: OGIZ, 1939). The "falcons" to which Kriukova refers were the heroic pilots of Stalinist aviation.

1. For a thorough, engaging study of the history of the Stalinist labor camps, see Anne Applebaum, *Gulag: A History* (New York: Anchor Books, 2004). See also Paul R. Gregory and Valery Lazarev, editors, *The Economics of Forced Labor: The Soviet Gulag* (Stanford: Hoover Institution Press, 2003).

2. GAAO Otdel DSPI, F. 296, op. 2, d. 1391, ll. 7–14. See Territorial'noe Otdelenie Arkhiva Nenetskogo Avtonomogo Okrug (hereafter TO ANAO), F. 289 (entire), on Glavsevmorput's "Amderma Expedition," 1929–53, and F. 263 (entire) on the Amderma district Central Executive Committee.

3. Sergei Lar'kov, Fedor Romanenko, *"Vragi Naroda" za Poliarnym Krugom* (Moscow: Paulsen, 2010), pp. 26–30.

4. In *Death and Deliverance* (Princeton: Princeton University Press, 2011), Stephen Barnes argues that the gulag was not intended to exterminate prisoners, but was primarily a penal institution where prisoners were given a final opportunity to reintegrate with Soviet society. Many of these individuals were allowed to leave. But many individuals had the sentences arbitrarily extended and were treated as pariahs to the ends of their lives even if they were released. Among other fine recent books on the gulag and on prisoners, see Golfo Alexopoulos, *Stalin's Outcasts: Aliens, Citizens and the Soviet State, 1926–1936* (Ithaca: Cornell University Press, 2003) and Miriam Dobson, *Khrushchev's Cold Summer* (Ithaca and London: Cornell University Press, 2009). See also http://russianhistoryblog

.org/ for ongoing discussions of the historiography of the gulag and other important issues.

5. On large-scale technological systems, see Thomas Hughes, *Networks of Power: Electrification in Western Society, 1880–1930* (Baltimore: Johns Hopkins University Press, 1983), and *American Genesis: A Century of Invention and Technological Enthusiasm, 1870–1970* (New York: Viking, 1989).

6. http://www.kraeved74.ru/pages_print_742.html

7. GARF, F. 9401, op. 1a, d. 98, ll. 39–43, as published at http://alexanderyakovlev.org/fond/issues-doc/1010448 and http://www.kraeved74.ru/pages_print_742.html

8. On Frenkel, as described here and below, see http://d-v-sokolov.livejournal.com/90266.html; K. A. Zalesskii, *Imperiia Stalina. Biograficheskii Entsiklopedicheskii Slovar'* (Moscow: Veche, 2000) at http://hrono.ru/biograf/frenkel.html

9. Frenkel worked at BBK, BAM, and other gulag infrastructure until 1942, then retired in 1947 because of health and died in the 1960s in Moscow.

10. Konstantin Gnetnev, *Belomor Kanal: Vremena i Sud'by* (Petrozavodsk: Ostrova, 2008), pp. 62–66.

11. Ibid., pp. 69–70.

12. Ibid., pp. 67–68.

13. As part of reeducation of those who managed to survive, the gulag administration of BBK established a series of clubs (fifty-six in the labor camp, twenty-one at villages), "red corners" for agitprop, libraries (twenty-five plus 400 book mobiles), movie theaters (five with sound, twenty-five without). In 1935 in the workers' villages there were eighteen elementary and four middle schools in which 5,274 of 6,040 children studied: 2,120 Russian, 1,083 Ukrainian, 731 German, 556 Polish, 275 Mordvinians, 264 Moldavian, 163 Chuvash, 38 Tatars, 15 Bulgarians, and 6 Jews. At a central theater with 338 seats, which opened in 1936, actors performed 152 plays and musicals. Such well-known individuals as the philosopher A. F. Losev (1893–1988) and the author Mir-Iakub Dulatov (1885–1935) contributed to the "arts" at BBK, the latter imprisoned for Kazakh "nationalism" and dying in the camp. See ibid., pp. 106–12.

14. Ibid., p. 332.

15. Ibid., pp. 187–98.

16. Ibid., pp. 207, 219–21.

17. Ibid., pp. 207–21.

18. BAMLag was dismantled and BAM was lost to memory until 1972 when Brezhnev and his oligarchs recast the railroad not only as an all-union (Komsomol) construction project, but a modest "project of the century." BAM was finished in name only, at great environmental cost, by official proclamation in 1983, but in many ways it remains unfinished. See Christopher Ward, *Brezhnev's Folly: The Building of BAM and Late Soviet Socialism* (University of Pittsburgh Press, 2009; paperback, 2010).

19. N. V. Protas'ev, *Proekt Soedineniia Ekaterininskoi Gavni na Murmane s set'iu Russkhikh Zheleznykh Dorog* (Petrozavodsk: Olonetskaia Gubernskaia Tipografiia, 1910).

20. Gnetnev, *Belomor Kanal,* pp. 240–44.

21. Ibid., pp. 240–44.

22. P. Korel'skii, *Na Moem Veku* (Arkhangelsk: Pravda Severa, 1996) at http://www.sakharov-center.ru/asfcd/auth/?t=page&num=5953

23. http://nworker.ru/2011/05/17/yagrinlag-vklad-v-pobedu.html

24. From M. M. Ostrochenko and A. M. Ostrochenko, *Severodvinsk—Gorod Korabelov* (1993) at http://www.sevska.net/index.php?option=com_content&task=view&id=2&Itemid=24

25. "Filial OTB v Molotovske—Dvadtsatyi Otdel Zavoda. Izvestnye Spetsialisty—Zakliuchennye v Gody Voiny (1941–1945 gg.) at http://scepsis.ru/library/id_1644.html

26. Ibid.

27. Cathy Frierson, *Children of the Gulag* (New Haven: Yale University Press, 2010).

28. Ostrochenko and Ostrochenko, *Severodvinsk—Gorod Korabelov* (1993).

29. GAAO Otdel DSPI, F 296, op. 1, d. 121, ll. 10, 24.

30. Ostrochenko and Ostrochenko, *Severodvinsk—Gorod Korabelov* (1993).

31. http://www.urokiistorii.ru/node/274

32. Ibid.

33. Ibid.

34. http://loshchilov.ucoz.ru/index/jagrinlag_pervye_gody/0-125

35. Ibid.

36. Ibid.

37. http://www.urokiistorii.ru/node/274

38. B. Simanovskii, "Piat' Minut Opozdaniia—Surovoe Nakazanie: Podrostki i Uzniki GULAGa Koval Pobedu v Tsekhan Vel'skogo RMZ No. 3," *Vel'skie Vesti,* January 21, 1995, and Boris Aleks. Nakhapetov, *Ocherki Istorii Sanitarnoi Sluzhby GULAGa* (Moscow: ROSSPEN, 2009).

39. GAAO Otdel DSPI, F. 296, op. 2, d. 191, ll. 1–16, 27, 30–33.

40. Ibid., ll. 15–15 ob.

41. Ibid., ll. 19–21 ob., 27. The Vaga, a left tributary of the Northern Dvina River, itself fed by the Kuloi, Ustia, and Vel rivers, runs through the Totemsky, Siamzhensky, and Verkhovazhsky districts of Vologda Oblast and the Velsky, Shenkursky, and Vinogradovsky districts of Arkhnagelsk oblast. It freezes in mid-November and stays under the ice until late April, so until the bridge was built, a snow-and-ice bridge on the river surface sufficed.

42. Ibid., ll. 16, 28 ob.

43. Ibid., ll. 2–5. Local officials sent summary reports of their political education activities to the secretary of the regional party committee and to one of the

heads of the political department of the NKVD in Moscow. See Ibid. DSPI, F. 296, op. 2, d. 191, ll. 13–15.

44. Ibid., F. 296, op. 2, d. 106, ll. 24, 24ob, and 25.

45. Ibid., ll. 26–26ob., 27–28.

46. Ibid., l. 11.

47. Ibid, 5–5ob. He could have been an honorary member of the National Rifle Association in the United States, however.

48. www.memo.ru/history/NKVD/GULAG/r3/r3–49.htm

49. Loren Graham, *The Communist Party and the Soviet Academy of Sciences* (New York: Columbia University Press, 1967).

50. http://www.gpavet.narod.ru/vittenburg.htm

51. www.stalin.memo.ru/spiski/pg09246.php and http://www.solovki.ca /camp_20/butcher_eihmans.php. Eikhmans's name is among the 134 NKVD agents recorded on Stalin's so-called third execution list that Stalin and Molotov signed.

52. Ibid.

53. www.bvvaul.ru/articles/text/rasskazi/loyko_povest/42.ph, www.pravda severa.ru/print.html?article=1650, www.naomuseum.ru / is_zak.php, and "Vaigach-skaia Ekspeditsiia OGPU" at http://wikimapia.org/7123916/Vaygachskaya-Expedition-OGPU-Vaygachsky-OLP-GULAG

54. http://amderma.livejournal.com/51300.html. The camp published such newspapers as *Northern Miner, Watchtower, On Guard, On the Job,* and *At the Dock* and the journal *The Bowels of the Soviet Union* that essentially glorified prison labor.

55. http://www.yaregaruda.ru/ru/node/51

56. http://www.ourbaku.com/index.php5/Кремс_Андрей_Яковлевич_-_профессор

57. http://www.yaregaruda.ru/ru/?q=book/export/html/40

58. Ibid.

59. V. Ia. Shashkov, *Spetspereselentsy v Istorii Murmanskoi Oblasti* (Murmansk: Maksimum, 2004), pp. 90–94.

60. Ibid., pp. 65, 97–98. See also Gennady Luzin, Michael Pretes, and Vladmir Vasiliev, "The Kola Peninsula: Geography, History and Resources," *Arctic,* vol. 47, no. 1 (March 1994), p. 5.

61. Shashkov, *Spetspereselentsy,* p. 8.

62. Ibid., pp. 9–10.

63. Ibid., pp. 11–12.

64. E. S. Kokovin, *My Podnimaem Iakoria* (Arkhangelsk: Severo-Zapadnoe Knizhnoe Izdatelstvo, 1972), pp. 32–33. In *Poslednii Son Bol'shogo Taleia,* published in the collection *We Raise the Anchor* (1972), Kokovin described the problems that faced animal husbandry even after the kulak has left the scene. He notes

that "for a long time there have been no kulaks who owned so many reindeer, nor has the bell of the Shaman rung, and forever the vile Siberian carbuncle [a kind of anthrax] has been eradicated. And yet enemies of the collective farmer reindeer herders still remain—a big wolf. But the medical student Anya who has returned to her homeland to work in the tundra puts out poisoned meat to sedate the wolf, and it remains only to kill the sleeping animal with an axe." Ibid., p. 63.

65. Shashkov, *Spetspereselentsy*, pp. 59–72.

66. Ibid., pp. 78–81.

67. Lar'kov, "Sud'by Uchastnikov Znamenitoi Ekspeditisii," in Lar'kov and Romanenko, *"Vragi Naroda,"* pp. 212–14.

68. Ibid., pp. 214–200.

69. Ibid., pp. 223–35. According to Lar'kov's analysis, the social makeup of the expedition was a rough approximation of Soviet society with workers, sailors, and white-collar employees, except that the *Krasin* had of course leading specialists, pilots, journalists, and party staff on board. Of the entire crew, twenty-four individuals were intelligentsia, forty were mid-level specialists, and seventy were sailors, mechanics, and stokers. Among the "intelligentsia," every fourth person was repressed, among the technical staff every seventh, among the proletariat every twelfth, and among the entire group every sixth. Ten of the repressed were members of the leading staff, those responsible for the success of the *Krasin*.

70. http://www.polarpost.ru/Library/Popov-avtograph/text-avtograf_na_karte-12.htm

71. Ibid.

72. Z. M. Kanevskii, "I Stanut vozvrashchat'sia Imena . . . " in *Zemlia i Vselennaia*, 1989 (no. 1), pp. 60–66 at http://www.ihst.ru/projects/sohist/papers/kan89zv.htm

73. Lar'kov, "Sud'by Uchastnikov Znamenitoi Ekspeditisii," pp. 220–22.

74. http://www.polarpost.ru/forum/viewtopic.php?t=646

75. M. M. Ermolaev, "Moi Lagernye Gody. Vospominaniia" at http://russcience.euro.ru/memory/erm99rg.htm

76. Ibid.

77. Ibid.

78. Ibid.

79. Ibid.

80. Ibid.

81. S. Popov, "Poliarnyi Gidrograf," published originally in *Krasnoiarskii Rabochii*, August 25, 1990, at http://www.memorial.krsk.ru/public/90/199008251.htm

82. Ibid.

83. Kanevskii, "I Stanut vozvrashchat'sia Imena . . . "

84. http://tungus-bolid.krasu.ru/node/93 and http://www.travelsmania.ru/study-185-6.html

85. http://www.polarpost.ru/Library/Popov-avtograph/text-avtograf_na_karte-12.html and http://tunguska.tsc.ru/ru/archive/suslov/. Suslov later founded the Museum of the Arctic and Antarctic where he worked until 1950, became deputy chair of the Yakut Branch of the Academy of Science, and finally served as director of the Museum of Ethnography in Leningrad.

86. Lar'kov, Romanenko, *"Vragi Naroda,"* p. 32.

87. http://www.polarpost.ru/Library/Popov-avtograph/text-avtograf_na_karte-12.html as accessed February 2, 2012.

88. Michael Suprun, *Severnye Konvoi: Issledovanie, Vospominaniia, Dokumenty* (Arkhangelsk: Arkhangelsk Branch of the Geographical Society of the USSR, 1991), and *Len-Liz i Rossii* (Arkhangelsk: Pravda Severa, 2006).

89. Nikolai Vasil'evich Ufarkin, "Buinitskii," http://www.polarpost.ru/forum/viewtopic.php?f=8&t=1599 and http://oceanology.narod.ru/rus/r_buynitskiy.htm

90. Lar'kov, Romanenko, *"Vragi Naroda"*, pp. 45–57.

91. Ibid., pp. 58–131.

92. Ibid., pp. 33–43.

93. Glavsevmorput', *O Partiinoi Rabote v Arktike* (Leningrad: Izdatel'stvo Glavsevmorputi, 1939), pp. 3, 7.

94. Ibid., pp. 7–8.

95. Ibid., pp. 8–9.

96. Ibid.,, p. 26.

97. Lar'kov, Romanenko, *"Vragi Naroda,"* p. 132.

98. Ibid., p. 133–34. See pp. 138–61 for a list of those individuals who were repressed.

99. Kanevskii, "I Stanut vozvrashchat'sia Imena."

100. Ibid.

101. Ibid.

102. GAAO Otdel DSPI, F. 296, op. 2, d. 1887, ll. 1–8.

103. Ibid., ll. 12–14.

104. Ibid., ll. 20, 22.

105. Ibid., ll. 12–14.

106. Ibid., ll. 15–17.

107. Ibid., l. 21.

108. Ibid., ll. 25–26.

109. Ibid., l. 27–28.

110. Ibid., l. 23.

111. Ibid., F. 296, op. 2, d. 191, l.16.

112. Ibid., F. 296, d. 1894, op. 2, l. 1.

113. Ibid., l. 3.

114. Ibid., ll. 5–6. On December 31, 1954, the chair of the military tribunal of White Sea Military District, G. Primak, sent a report to the secretary of the provincial party committee, Latunov, that indicated of the 154 persons convicted by a military tribunal, sixty-six had their terms reduced from twenty-five to ten years; twenty-five others to the term actually served; fifty-five to five years with amnesty; and eight were simply exonerated. One of those eight, Nikolai Fedorovich Prudnikov (b. 1910), was sentenced to ten years in the camps because he was trapped in occupied territory near Kursk, failed to engage the Germans from behind the lines, and in March 1942 voluntarily joined the German army. But it turned out that the investigator of transportation department of the KGB of North Pechora Railroad, a Lt. Tarakanov, whose last name means "cockroach," appropriately in this case, had fabricated everything. See ibid., l. 8.

115. Ibid., F. 296, op. 2, d. 1887, ll. 18–19.

116. Judith Pallot, "Forced Labour for Forestry: The Twentieth Century History of Colonisation and Settlement in the North of Perm' Oblast'." *Europe-Asia Studies* 54, no. 7 (2002), pp. 1068–69. On this issue see also Judith Pallot, "Russia's Penal Peripheries: Space, Place and Penalty in Soviet and Post-Soviet Russia," *Transactions of the Institute of British Geographers* 30, no. 1 (2005); Judith Pallot, Laura Piacentini, and Dominique Moran, "Patriotic Discourses in Russia's Penal Peripheries: Remembering the Mordovan Gulag," *Europe-Asia Studies* 62, no. 1 (2010).

117. Kanevskii, "I Stanut vozvrashchat'sia Imena . . . "

118. Dobson, *Khrushchev's Cold Summer*, pp. 2–6, 13–15.

119. Ibid., pp. 190–96.

4. The Arctic Sciences of Places and People

1. On the history and politics of Soviet science, see Nikolai Krementsov, *Stalinist Science* (Princeton: Princeton University Press, 1997), Loren Graham, *Science in Russia and the Soviet Union* (Cambridge, New York: Cambridge University Press, 1993), and *Science, Philosophy, and Human Behavior in the Soviet Union* (New York: Columbia University Press, 1987); Slava Gerovitch, *From Newspeak to Cyberspeak* (Cambridge: MIT Press, 2002); and Paul Josephson, *Lenin's Laureate: Zhores Alferov's Life in Communist Science* (Cambridge: MIT Press, 2010).

2. V. Kh. Buinitskii, "25 let Arkticheskogo Instituta," pp. 14–16.

3. Ibid., pp. 13–14.

4. B. A. Krutskikh et al., "70 Let Ordena Lenina Arkticheskomu i Antarkticheskomu Nauchno-issledovatel'skomu Institut," *Problemy Arktiki i Antarktiki,* vol. 66 (Leningrad Gidrometeoizdat 1991), p. 11.

5. Ibid., pp. 11–15.

6. A AARI, 0–537 "Predvaritel'nyi Otchet o Rabotakh Vostochnoi Vysokoshirotnoi Ekspeditsii na Ledokole 'Severnyi Polius' v 1946 g.," pp. 18–26.

7. Krutskikh et al., "70 Let," p. 8. Also, Alexander Girs developed a theory of atmospheric circulation in the northern hemisphere and related it to river discharges. See Mario Lucertini, Ana Millàn Gasca, and Fernando Nicolò, editors, *Technological Concepts and Mathematical Models in the Evolution of Modern Engineering Systems: Controlling Managing Organizing* (Basel: Springer, 2004), p. 167.

8. Krutskikh et al., "70 Let," pp. 16–17, 21.

9. A. O. Andreev, M. V. Dukal'skaia, and S. V. Frolov, "Stranitsy Istorii AANII," *Problemy Arktiki i Antarktiki,* vol. 84 (2010), pp. 14–16.

10. Glavsevmorput, *O Partiinoi Rabote v Arktike* (Leningrad: Izdatel'stvo Glavsevmorputi, 1939), p. 61.

11. Krutskikh et al., "70 Let," p. 23.

12. Andreev et al., "Stranitsy," pp. 16–18.

13. Krutskikh et al., "70 Let," p. 26.

14. Andreev et al., "Stranitsy," pp. 18–19. In 1970 in the Antarctic station Vostok specialists began drilling of the first super-deep shaft for sampling of ice core for paleogeographic and climatic research; two generations later Russian specialists reached Lake Vostok in 2012.

15. Ibid.

16. Ibid., pp. 19–21.

17. B. A. Krutskikh et al., "70 Let," pp. 13–14.

18. A. F. Treshnikov, "Ordena Lenin Arkticheskomu i Antarkticheskomu Nauchno-Issledovatel'skomu Institutu—50 let," *Problemy Arktiki i Antarktiki,* vol. 36–37 (1970), p. 15.

19. I. E. Frolov, editor, *Rossiiskie Issledovaniia na Dreifuiushchikh L'dakh Artiki* (St. Petersburg, AANII, 2010), p. 7.

20. Ibid., pp. 7–8.

21. Ibid., p. 9.

22. Sergei Chennyk, "Papanin," in *Pervaia Krymskaia,* no. 94 (October 7–13, 2005) at http://www.google.ru/url?sa=t&rct=j&q=papanin&source=web&cd=18&ved=0CJcBEBYwEQ&url=http://www.hrono.ru/biograf/bio_p/papanin_id.php&ei=_DwNUL3dD6XZ0QGq06GOBA&usg=AFQjCNHhQmBBtuH_Nemz0iPEtdNFEhIbFQ&cad=rjt and http://www.hrono.ru/biograf/bio_p/papanin_id.php

23. Discovered by Nils Nordenskjöld in 1893 and explored by Nansen, Cape Cheliuskin fell to Soviet explorer Samoilovich in 1932; Nordenskjöld was the first to sail the Northern Sea Route in 1878, sailing from the Bering Strait.

24. Ivan Papanin, *Zhizn' na L'dine. Dnevnik* (Moscow: Pravda, 1938), pp. 17, 18, 60, 150.

25. http://www.franz-josef-land.info/index.php?id=647&L=5. Rudol'f Island, the northernmost part of Franz Josef Land at 81°51′, has a diameter of 27 kilometers and is almost entirely covered by an ice cap.

26. Papanin, *Zhizn' na L'dine,* pp. 32, 76. On one occasion the men opened a crate to find food soaked in kerosene.

27. Ibid., pp. 159, 161, 165, 167, 183.

28. Ibid., p. 19.

29. Ibid., pp. 31, 47, 58, 60, 81, 135, 189–90, 207.

30. Ibid., pp. 159, 161, 165.

31. Ibid., pp. 213–20.

32. http://www.whoi.edu/page.do?pid=66677

33. Frolov, *Rossiiskie Issledovaniia,* pp. 19–28.

34. A AARI, 0–537 "Predvaritel'nyi Otchet," pp. 2–7.

35. Ibid., pp. 8–10.

36. Ibid., pp. 10–12.

37. Frolov, *Rossiiskie Issledovaniia,* pp. 27–40.

38. See for example Josephson, *Physics and Politics in Revolutionary Russia* (Berkeley: University of California Press, 1991), and Graham, *Science, Philosophy and Human Behavior in the USSR.* On Akademgorodok, see Josephson, *New Atlantis Revisited* (Princeton: Princeton University Press, 1997).

39. Frolov, *Rossiiskie Issledovaniia,* pp. 41–74.

40. Ibid., pp. 75–78.

41. Krutskikh et al., "70 Let Ordena," p. 21.

42. Ibid., p. 10.

43. Andreev et al., "Stranitsy Istorii AANII," pp. 14–16.

44. Frolov, *Rossiiskie Issledovaniia,* p. 92.

45. Ibid., pp. 127–34.

46. Ibid., pp. 254–61.

47. Ibid., pp. 267–70.

48. Ibid., p. 296.

49. Ibid., pp. 294–306.

50. Ibid., pp. 307–29.

51. http://www.sevmorgeo.com/rus/index_e.htm

52. Frolov, *Rossiiskie Issledovaniia,* pp. 531–38.

53. Treshnikov, "Ordena Lenin," pp. 16–17.

54. Frolov, *Rossiiskie Issledovaniia,* pp. 7–9.

55. On Fersman, see http://www.pandia.ru/296607/, http://iomn.net/?p=85, http://www.ladoshki.com/?books&author=2184&mode=a, and http://dic.academic.ru/dic.nsf/enc_geolog/5286/%D0%A4%D0%B5%D1%80%D1%81%D0%BC%D0%B0%D0%BD. Among his major publications are *Novyi Promyshlennyi Tsentr SSSR za Poliarnym Krugom (Khibinskii Apatit)* (Leningrad: Izdatel'stvo Akademiia

Nauk, 1931), Fersman and B. M. Kupletskii, editors, *Khibinskaia Gornaia Stantsiia* (Leningrad: Izdatel'stvo Akademiia Nauk, 1934), *Dostizheniia Sovetskoi Mineralogii i Geokhimii za Poslednie Gody* (Leningrad: Izdatel'stvo Akademiia Nauk, 1935), *Vospominaniia o Kamne,* 2nd edition (Moscow: Izdatel'stvo Akademiia Nauk, 1945), and *Ocherki po Istorii Kamnia* (Moscow: Izdatel'stvo Akademiia Nauk, 1954–61).

56. A. E. Fersman, *Tri Goda za Poliarnym Krugom* (Moscow: Molodaia Gvardiia, 1924) and *Puteshestvie za Kamnem* (Leningrad: Detgiz, 1956).

57. S. M. Kirov, *Izbrannye Stat'i i Rechi, 1912–1934* (Leningrad: OGIZ RSFSR, 1939), p. 82.

58. Archive of the Kola Science Center (hereafter A KNTs), F. 1, op. 6, d. 26, ll. 1–10.

59. Ibid., F. 1, op. 6, d. 56, ll. 5–16.

60. Ibid., F. 1, op. 6, d. 1a (Rukopis' E. P. Kesslera, 1930).

61. Ibid., F.1, op. 6, ed. khr. 6. For a history of the Kola Science Center written by an eyewitness, see A. M. Oranzhireev, *Rabota Akademii Nauk SSSR i Sotsialisticheskoe Stroitel'stvo na Kol'skom Poluostrove (1920–1935)* (Apatity: Kol'skii Nauchnyi Tsentr, 2008). Written in 1936.

62. A KNTs, F. 1, op. 6, d. 17, ll. 12–14.

63. Ibid, ll. 18–23.

64. Ibid., d. 20, ll. 2–4.

65. Ibid., d. 7, ll. 3–4.

66. Ibid., ll. 26–27.

67. Ibid., d. 20, l. 28.

68. Ibid., d. 7, l. 1.

69. Ibid.,, l. 2.

70. Ibid., ll. 2–3.

71. Ibid., d. 397, ll. 5–7.

72. Ibid., d. 16, ll. 1–5. Talks included A. I. Liakhov on "Agronomical Characteristics of Soils of the Kola Peninsula," T. V. Aristovskaia on "Microflora of Soils of Kola Peninsula," S. N. Ignat'evskaia on the "Introduction of Feed Bean Culture," and S. A. Kasparova on "Biochemical and Physiological Preconditions of Advancement of Cultured Plants in the Far North" and "Preservation of Potatoes Against Disease During Vegetation and Storage."

73. http://www.kolasc.net.ru/napr.html

74. See Roger D. Launius, James Rodger Fleming, and David H. DeVorkin, editors, *Globalizing Polar Science: Reconsidering the International Polar and Geophysical Years* (New York: Palgrave, 2010) for a study of the role of the IPYs in historical and scientific studies.

75. Viktor Dmitriev, "The Russian View of the International Polar Decade," paper presented at "Artic Perspectives" Eco-Pechora Conference, June 2010, in Naryan-Mar, Russia.

76. GAAO, F. 5577, op. 1, d. 56, l. 35.

77. Ibid., d. 549 a, l. 37.

78. Ibid., ed. khr. 26, ll. 1–6.

79. Ibid., op. 1, d. 549 a, l. 37. In October 1943, in honor of the twenty-sixth anniversary of the October Revolution, researchers of the Maritime Observatory took on socialist obligations (which meant at least sixty-three of them filled out small strips of paper) to finish various research papers or write up observations on flows, currents, level of water, seas, and the Onega and Mezen' rivers and to constitute maps and other instruments of research. They also pledged "to attend a study circle for the book of Stalin, *O Sovremennoi Otechestvennoi Voine* (On the Present Fatherland War)." See GAAO, F. 5577, op. 1, d. 263, ll. 1–11.

80. Ibid., d. 549 a, ll. 29–30.

81. Ibid., l. 31.

82. Ibid., ll. 33–35.

83. On top of this, few staff had experience: 240 had been employed only one year and 649 from two to five years. Programs to improve qualifications of staff lagged in particular at isolated radio stations. High turnover did not help matters. As for many Arctic organizations, one reason for turnover was the inability of Gidrometsluzhba to provide housing for employees. Ibid., ll. 6–9, 12. In 1950 the staff of the Syktyvkar office in the Komi Republic was increased by twelve people with the formation of a long-term weather forecasting group. In Vologda region the office was larger and therefore able to serve industry and agriculture more efficiently. By nationality, they were exclusively Russian: 1,270 were Russians, twenty-three Urainians, four Belorusian, ten Jews, and only one each Nenets, Komi, Chuvash, and Tatar.

84. Ibid., ll. 13–15.

85. Ibid., ll. 37–38.

86. Ibid., ll. 20–23.

87. Ibid., d. 720, ll. 1–2, 7–9.

88. A. L. Matusov, *Iz Istorii Meditsinskogo Obespecheniia Russkikh Morskikh i Poliarnykh Exspeditsii* (Leningrad: Gidrometeoizdat, 1972), *Trudy AANII,* vol. 305, p. 63.

89. Ibid., pp. 64–69.

90. On Soviet medicine, see Susan Solomon, editor, *Health and Society in Revolutionary Russia* (Bloomington: Indiana University Press, 1990); Frances L. Bernstein, Christopher Burton, Dan Healey, editors, *Soviet Medicine: Culture, Practice, and Science* (DeKalb: Northern Illinois University Press, 2010); Bernstein, *The Dictatorship of Sex* (DeKalb: Northern Illinois University Press, 2007); and Mark Field, *The Soviet Doctor: A Case Study of the Professional in Soviet Society* (Maxwell Air Force Base, Ala., Air University, Human Resources Research Institute, 1952).

91. Alfred Crosby, *Ecological Imperialism* (Cambridge: Cambridge University Press, 1993).

92. GAAO, F. 1735, op. 1, ed. khr. 480, l. 36.

93. Ibid., ed. khr. 483 l. 2.

94. Ibid., d. 293, l. 26.

95. Ibid., ed. khr. 642, ll. 46ob–47.

96. V. N. Ledkov, *Mesiats Maloi Temnoty,* trans. by V. Leont'ev (Moscow: Sovetskaia Rossia, 1989), p. 117.

97. Ibid., p. 155.

98. Ibid., p. 243.

99. GAAO, F. 1735, op. 1, ed. khr. 295, ll. 16–19.

100. A AARI, 0–183 Otchet o Deiatel'nosti Kompleksnoi Ekspeditsii AMN SSSR i AANII po Izucheniiu Voprosov Akklimatizatsii Cheloveka v Usloviiakh Krainego Severa v 1948–1949 gg.," pp. 1–2.

101. Ibid., pp. 2–3.

102. Ibid., pp. 3–5. Lectures on such pseudoscientific subjects as I. S. Kandror's on the "Victory of Michurinist Biology in the Struggle with Reaactionary Weissmanism-Morganism" did not slow medical research.

103. Ibid., pp. 6–8.

104. Ibid., pp. 9–11.

105. Ibid., p. 12.

106. Ibid., p. 23. Sysin (1879–1956), active in the student revolutionary movement, a specialist in public health, helped the Bolsheviks organize an epidemiological service after the revolution and founded *Gigiena i Epidemiologiia* (Hygiene and Epidemiology) in 1922.

107. Krutskikh et al., "70 Let Ordena," pp. 22–23.

108. http://lib.bioinfo.pl/auid:6378574 and http://www.ntis.gov/search/product.aspx?ABBR=JPRS47626

109. L. P. Roshchevskaia et al., "The Founder of Soviet Ecological Physiology G. M. Danishevskii in Komi Autonomous Republic," *Voprosy Kurortologii, Fizioterapii, i Lechebnoi Fizicheskoi Kultury* no. 2 (2011 Mar–Apr):, pp. 46–49.

110. A. L. Matusov, "Mediko-Sanitarnoe Obespechenie i Nauchnaia Deiatel'nost' Vrachei v Arkticheskikh i Antarktichesikh Ekspeditsiiakh," in A. L. Matusov, editor, *Meditsinskie Issledovaniia v Arkticheskikh i Antarkticheskikh Ekspeditsiiakh* in *Trudy AANII,* vol. 299 (Leningrad: Gidrometeorologichekoe Izdatel'stvo, 1971), pp. 10–14.

111. On the cultural and economic significance of reindeer, see Piers Vitebsky, *The Reindeer People: Living with Animals and Spirits in Siberia* (Boston: Houghton Mifflin, 2005). See also Gerald T. Conaty, *The Reindeer Herders of the Mackenzie Delta* (Toronto: Key Porter Books, 2003); David Anderson, *Identity and Ecology in Arctic Siberia: The Number One Reindeer Brigade* (Oxford: Oxford University Press, 2000); and Andrei Golovnev and Gail Osherenko, *Siberian Survival: The Nenets and Their Story* (Ithaca: Cornell University Press, 1999).

112. GAAO Otdel DSPI, F. 296, op. 2, d. 2024, ll. 72–75, 104–7, 160–64.

113. GAMO, F. 640, op. 1, d. 1, ll. 29–31.

114. On passportization, *David Shearer,* "Elements Near and Alien: *Passportization,* Policy and Identity in the Stalinist State 1932–1952," *Journal of Modern History,* vol. 76 (2004), pp. 835–81, and Shearer, *Policing Stalin's Socialism: Repression and Social Order in the Soviet Union,1924–1953* (New Haven: Yale University Press, 2009). It also bears recalling how the Russian Federation, beginning in 2002, distributed thousands of passports in Abkhazia and South Ossetia to create Russian citizens within the borders of Georgia and then use the presence of these "Russians" as one of several pretexts to carry out a war with Georgia—to protect Russian citizens.

115. GAAO, F. 1735, op. 1, no. 632, ll. 1–4, 11–15.

116. Ibid., d. 482, l. 1.

117. Ibid., no. 632, ll. 1–4, 11–15.

118. Ibid., d. 495, l. 11.

119. F. Ia. Gul'chak, *Severonoe Olenovodstvo* (Moscow: Sel'khozizdat, 1954).

120. Constantine Krypton, "Soviet Policy in the Northern National Regions After World War II," *American Slavic and East European Review,* vol. 13, no. 3 (October 1954), pp. 348–51.

121. GAAO, F. 1735, op. 1, ed. khr. 642, ll. 1–2.

122. Ledkov, *Mesiats Maloi Temnoty,* p. 172.

123. Ibid., p. 191.

124. See for example F. Ia. Gul'chak et al, *Razvedenie Severnykh Olenei* (Leningrad: Izdat Glavsevmorputi, 1939), E. I. Gorbunov, et al., *Infektsionnye i Invazionnye Zabolevaniia Severnogo Olenia* (Moscow-Leningrad: Izdat Glavsevmorputi, 1940), and V. N. Andreev et al., *Obsledovanie Pastbishchnykh i Promyslovo-oxotnich'ykh Ugodii Krainego Severa s Pomoshch'iu Samoleta* (Moscow-Leningrad: Izdat Glavsevmorputi, 1940).

125. GAMO, F. 640, op. 1, d. 1, ll. 2, 6, 35–37.

126. Ibid., ll. 8–9.

127. Michael David-Fox, *Revolution of the Mind* (Ithaca: Cornell University Press, 1997).

128. John McCannon, *A History of the Arctic* (London: Reaktion Books, 2012), pp. 207–9.

129. Ibid., pp. 203–5.

130. The following brief discussion draws heavily on Yuri Slezkine's excellent book on the subject, *Arctic Mirrors.*

131. Krypton, "Soviet Policy in the Northern National Regions," pp. 339–42.

132. N. M. Koviazin, V. M. Krylov, and A. G. Podekrat, *Ocherki po Promyslovomu Khoziaistvu i Olenevodstvu Krainego Severa* (Leningrad: Izdatel'stvo Instituta Narodov Severa TsIK SSSR im. P. G. Smidovich, 1936), pp. 74–83.

133. Ibid., pp. 106–7.

134. Ibid., p. 115.

135. GAAO, F. 1735, op. 1, ed. khr. 480, 13–15, 17–18.
136. Ibid., F.194, op. 1, d. 10, l. 99.
137. Ibid., d. 10, ll. 62, 74.
138. Ibid., d. 50, ll. 4–5.
139. Ibid., d. 23 (1934–35), ll. 25–27.
140. Ibid., d. 10, l. 70. In mid-1934, the committee forwarded to Moscow fifty copies of *Materialy po Razvitiiu Iazykov i Pis'mennosti Narodov Severa v Murmanskom Okruge* (Materials Concerning the Development of Written Language of the People of the North in Murmansk Region) to tout these achievements. Ibid., ll. 25–27.
141. Eikhfeld (1893–1989) specialized in agriculture from his early years and studied under leading Soviet botantist and geneticist Nikolai Vavilov, who encouraged him to move to the Khibiny Agricultural Experimental Station in the mid-1920s, which became a subdivision of the All-Union Institute of Plant Breeding under director Vavilov. Eikhfeld then rejected his mentor, adopted Lysenko's quack Lamarckianism and took over direction of Vavilov's institute in 1940, while Vavilov perished in the gulag as an opponent of Lysenko. For his ideological impurity Eikhfeld became president of the Estonian Academy of Sciences from 1950–68. See http://www.warheroes.ru/hero/hero.asp?Hero_id=10722.
142. GAMO, F. 194, op. 1, d. 23 (1934–35), l. 11.
143. Ibid., d. 1, l. 36.
144. Ibid., d. 23, ll. 3–6.
145. Ibid., d. 293, ll. 35–36.
146. Krypton, "Soviet Policy in the Northern National Regions," pp. 343, 346.
147. Sergei Lar'kov, Fedor Romanenko, "Iz Kamennogo Veka—za Koliuchuiu Provoloku . . . " in Lar'kov and Romanenko, *"Vragi Naroda" za Poliarnym Krugom* (Moscow: Paulsen, 2010), pp. 165–70.
148. Ibid., pp. 170–92.
149. Ibid., p. 193.
150. Ibid., pp. 194–97.
151. GAAO, F. 1935, op. 1, d. 305, ll. 47–50.
152. Ibid., ed. khr. 642, 42–42 ob.
153. Ibid., ll. 43ob, 44, 44 ob, and d. 305, l. 8.
154. Ibid., ll. 42 ob., 43, 43 ob.
155. Ibid., d. 305, l. 30.
156. Ibid., ed. khr. 295, ll.33
157. Ibid., d. 293, l. 2–5 ob. At least they addressed a few problems: The kultbaza acquired a motorboat and built a dock at Kungas. But the absence of motors for the boats—and the absence of bricks, blankets, and much else—meant lags everywhere.
158. GAAO, F. 1935, op. 1, d. 305, ll. 58–59.

159. Ibid.

160. Ibid., l. 43.

161. Ibid., ed. khr. 295, ll. 13–14

162. Ibid., ll. 13–16, 45–46.

163. Ibid., ed. khr. 642, ll. 45–46, and ibid., ed. khr. 295, ll. 13–16.

164. Ibid., F. 1735, op. 1, ed. khr. 480, ll. 47–49.

165. Ibid., ed. khr. 640.

166. Ibid., ed. khr. 483, l. 1. There were sixty-five boys and fifty-one girls; eighty-four Nenets, twenty-four Komi, eight Russian; by class origin seven workers, 103 peasants, and six white collar; by party affiliation five Komsomol and sixty-five Pioneers.

167. Ibid., ll. 16–17.

168. Ibid., ed. khr. 480, l. 1. (or 383–16/17?).

169. Ibid., d. 305, l. 31.

170. In one House: *Bolshevik, Partinyi Stroitel'* (Party Builder); *Severnaia Arktika* (The North Arctic); *Vestnik Znanie* (The Herald of Knowledge); *SSSR na Stroike* (The USSR Under Construction); *Krasnyi Bibliotekar'* (Red Librarian); *Bezbozhnik* (Athiest); *Ogonek* (Spark); *Krokodil; Bloknot Agitatora* (Notebook of the Agitator); *V Pomoshch' Partuchebe* (In Support of Party Education); and newspapers: *Pravda, Izvestiia, Komsolmolsk, Pionerskaia, Pravda Severa,* and *Arkticheskaia Zvezda* (Arctic Star). See ed. khr. 642, ll. 47.

171. Ibid., d. 293, ll. 39–40.

172. Lar'kov, *"Vragi Naroda" za Poliarnym Krugom,* pp. 30–31.

173. Pelia Punukh, *Iz-pod Piati Vekok* (Arkhangelsk: Severozapadnoe Knizhnoe Izdatel'stvo, 1985), p. 180.

174. Ibid., p. 198.

175. Ibid., p. 201.

176. Ibid., pp. 232–33.

177. Ibid., p. 254.

178. Ibid., p. 235.

179. Ibid., p. 289.

180. Ibid., p. 349.

181. GAAO, F. 1735, op. 1, ed. khr. 295, ll. 3–4, 10.

182. Ibid., l. 5.

183. Ibid., l. 6.

184. Ibid., l. 7.

185. Ibid., ed. khr. 642, ll. 49, 68. This might have had to do with the fact that most employees were young, ranging in age from twenty to forty-five years old. Only three had finished higher education. There were but four party members and six Komsomol among them. Three-quarters were Russian, 70 percent of peasant origin, and only 8 percent (five individuals) of working class roots.

186. Ibid., ed. khr. 295, ll. 13–14.

187. Ibid., d. 293, l. 1.

188. Natal'ia Khaimina, "Dom u Berezovoi Sopki," *Vybor Naroda,* no. 4 (January 22, 2005) at http://www.arhpress.ru/vibor/2005/1/22/12.shtml

189. GAAO, F. 1735, op. 1, d. 293, ll. 51–51 ob.

190. Krypton, "Soviet Policy in the Northern National Regions," p. 353. See also *Pravda,* August 6, 1948, and September 26, 1948.

191. Krypton, p. 344.

192. Ethel Dunn, "Education and the Native Intelligentsia in the Soviet North: Further Thoughts on the Limits of Culture Change," *Arctic Anthropology,* vol. 6, no. 2 (1970), pp. 112–22.

193. Email from Ekaterina Boikova to the author, July 6, 2013.

5. The Nickel That Broke the Reindeer's Back

Epigraph: Author translation of stanza from well-known poem of Aleksandr Pushkin, "Osen' [Autumn]".

1. Studies of urbanization in the former Soviet Union indicate how Lenin and the Bolsheviks reveled in the potential of cities to enhance the economic, political, and cultural life of the masses; like Marx they found rural life to be backward and unpromising for their programs of social and economic transformation. Of course, problems of crime, housing, transport, infrastructure, and so on were as central in the attention of socialist planners as they were of capitalist planners. No matter the great potential of officials to push the city in the direction of utopia in the USSR, the challenges were greater than they imagined. They encountered constant tension between the need to provide adequate shelter for the masses and the hope to give individuals and families privacy while still holding onto the collectivist ethos. Centralized control of power and finance frustrated attempts to create a clean, quiet, safe Soviet city. People sought any means to move to larger cities where elites lived where they, too, might have more comforts, better apartments, and higher salaries than in the provinces. Among several studies of the Soviet city, see Blair Ruble, *Leningrad: Shaping a Soviet City* (Berkeley: University of California Press, 1990); James Bater, *The Soviet City: Ideal and Reality* (London: Edward Arnold, 1980); and Henry W. Morton and Robert Stuart, editors, *The Cotemporary Soviet City* (Armonk: M. E. Sharpe, 1984).

2. V. Gladkov, *Monchegorsk* (Murmanskoe Knizhnoe Izdatel'stvo, 1961), pp. 1–15. See also A. E. Fersman, *Novyi Promyshlennyi Tsentr SSSR za Poliarnym Krugom (Khibinskii Apatit)* (Leningrad: AN SSSR, 1931), and Fersman and V. M. Kupletskii, editors, *Khibnskaia Gornaia Stantsiia: Sbornik Statei* (Leningrad: AN SSSR, 1934).

3. From O. Pisarzhevskii, *Aleksandr Evgen'evich Fersman* (Moscow: Molodaia Gvardiia, 1955) at http://www.detskiysad.ru/raznlit/fersman14.html

4. http://murmanarchiv.ru/index.php?option=com_content&view

=article&id=244:2011-03-25-08-53-43&catid=37:a-series-of-articles-murmansk-movie&Itemid=77

5. http://guides.rusarchives.ru/browse/guidebook.html?bid=79&sid=185080. In February 1934, Novaia Zemlia, Vaigach, and several smaller islands in the Barents and Kara seas were moved to the Nenets National Okrug. In March 1936, the Pechorskii Okrug, which was part of the Northern Krai, was established in response to rapidly developing coal mining in the area. Next the Northern Krai was simply abolished in keeping with the new Soviet Constitution of 1936, and its territory was divided between newly established Northern Province and the Komi Republic, which in 1937 was split into Arkhangelsk and Vologda provinces, the former which consisted of Arkhangelsk, the NAO, and twenty-seven regions.

6. Gladkov, *Monchegorsk,* pp. 19–23.

7. Ibid., p. 30.

8. Ibid., pp. 30, 33, 39.

9. Sergei Tararksin, "Sudeb Sgorevshikh Ochertan'e," http://helion-ltd.ru/red-dir-mistake/

10. "Iz Vospominanii L. Tartakovskoi," at http://www.arctic.org.ru/2001/fer_kon.htm. Kondrikov always tried to gain recruits to Khibinogorsk. Late one night on the streets of Leningrad Kondrikov encountered a familiar face. They began to talk about the north and its promise, and Kondrikov tried to convince the man to drop everything and move to Khibiny. The friend declined. Later that evening Kondrikov recalled it had been Ruvim Shapiro, the director of the renown Mariinskii Theater.

11. Tararksin, "Sudeb Sgorevshikh Ochertan'e."

12. "Iz Vospominanii L. Tartakovskoi."

13. V. Ia. Pozniakov, *Kombinat "Severonikel" im. V. I. Lenina* (Murmanskoe Knizhnoe Izdatel'stvo, 1979), pp. 18–19.

14. Tararksin, "Sudeb Sgorevshikh Ochertan'e."

15. "Zadachi stroitelei," *V Boi za Nikel,* no. 21, no. 1 (1938), p. 1.

16. "Iz Vospominanii L. Tartakovskoi."

17. Tararksin, "Sudeb Sgorevshikh Ochertan'e."

18. "Iz Vospominanii L. Tartakovskoi."

19. P. I. Negretov, "How Vorkuta Began," *Soviet Studies,* vol. 29, no. 4 (October 1977), pp. 565–75.

20. http://www.kolagmk.ru/about/history

21. GAMO, F. 459, op. 1, d. 1, ll. 2–3.

22. Ibid., ll. 16, 33.

23. Ibid., d. 60, l. 13. Excavators, tractors, and other equipment had been moved to Liinaxamari, a natural ocean port belonging to Finland on the Murmansk fjord that became territory of the USSR in the treaty ending the Winter War, so most work at Ianiskoski was carried out by hand. The equipping of a concrete factory also lagged.

24. Ibid., ll. 59–59 ob.
25. Ibid., ll. 53–54.
26. Ibid., d. 1, l. 28.
27. Ibid., d. 60, l. 3.
28. Ibid., ll. 53–54.
29. Ibid., ll. 10–12.
30. Ibid., op. 5, d. 39, ll. 1–18.
31. Ibid., ll. 31–32.
32. Ibid., ll. 38–40.
33. Ibid., ll. 103, 150, 162–67.
34. Ibid., l. 192.
35. Ibid., op. 18, d. 1, l. 1.
36. Ibid., op. 5, d. 39, ll. 168–79.
37. *Planovoe khoziaistvo,* no.7, 1970.
38. In *Killing for Coal: America's Deadliest Labor War* (Harvard University Press, 2008), Thomas G. Andrews describes many of these features of the Colorado (and other) coal industry.
39. See Timothy LeCain, *Mass Destruction* (New Brunswick: Rutgers University Press, 2009).
40. In the American west private agribusinesses, mines, timber, and other resource extraction industries came to control vast sections of the land, which they engineered with impunity. High-pressure water cannons that used mercury to trap gold washed away the earth and poisoned the land. This led, in California, to an active environmental movement in responsive the agro-industrial despoliation of the nineteenth century. For discussion of the impact of gold mining on the California environment, see Andrew Isenberg, *Mining California* (New York: Hill and Wang, 2005).
41. http://www.bellona.ru/positionpapers/nickel
42. The FSB declared in 2005 that the White Sea–Baltic Canal was a "strategic facility," although its outmoded construction, small locks, and multitudinous overpasses made it out of date as a strategic object in 1941.
43. On these cities and general themes of population migration control, Michael Gentile, "Former Closed Cities and Urbanisation in the FSU: An Exploration in Kazakhstan," *Europe-Asia Studies,* vol. 56, no. 2 (March 2004), pp. 263–78; Richard H. Rowland, "Russia's Secret Cities," *Post-Soviet Geography and Economics,* vol. 37, no. 7 (1996), pp. 426–62; and Mervyn Matthews, *The Passport Society-Controlling Movement in Russia and the USSR* (Boulder: Westview Press, 1993). See also Asif Siddiqi, "ZATOs In View," *Russian History Blog,* posted April 20, 2012.
44. GAAO Otdel DSPI, F. 296, op. 2, d. 1411, ll. 79–86.
45. Ibid., d. 1533, ll. 11–14, 19–27, 30, 31.
46. Ibid., ll. 19–30.
47. Ibid., op. 2, d. 1411, ll. 8–13, and op. 2, d. 2017, l. 31–32.

48. Ibid., op. 2, d. 1411, l. 40.

49. Ibid., d. 908, l. 50.

50. Ibid., d. 1572, ll. 21–24. No doubt the agitators did not explain the fourteen-year lag as a result of Stalin's murderous purges, including of high party officials, military officers, and other members of the Soviet elite.

51. Ibid., d. 1411, l. 56. For example, of twenty-four secretaries of primary party organizations of Factory 203 in 1952, four had served since before 1939, sixteen from the war years, and four since 1945. Only one of these men had higher education, four had middle technical education, and three only middle school experience, with the rest at most seven years of school.

52. Ibid., d. 1572, ll. 2–10.

53. Ibid., d. 1883, l. 23.

54. Ibid., l. 22.

55. http://vaga-land.livejournal.com/331049.html

56. http://stal-sever.ru/stroitelstvo-metiznogo-zavoda.html. In Vologda, Latunov encountered one major disappointment. In May 1956 the Council of Ministers passed a resolution about the construction in Vologda province of a hardware factory to come on line in 1959 and produce 300,000 tons of nuts, bolts, and other goods annually. A government commission examined potential sites near Cherepovets and Vologda and chose the former, which deeply upset the provincial leadership including Latunov. He and the chair of the provincial executive committee, I. Volkov, wrote the Central Committee and Council of Ministers a letter in which they accused the commission of bias for basing its decision not so much on economic consideration of expending production throughout Vologda province as on the bureaucratic interests of two ministries to locate two metallurgical plants close to each other (the Cherepovets Metallurgical Works—now "Severstal" [Northern Steel]—which commenced operation in the 1930s, and the machine tool factory. Cherepovets remained the second-most polluted city in all of Russia in 2012.) But, Latunov argued, Cherepovets did not have the labor force to support a second facility; it was necessary to choose Vologda to help in the improvement of the economic situation of the provincial capital. Moscow decided on Cherepovets.

57. GAAO Otdel DSPI, F. 296, op. 2, d. 2017, l. 57.

58. Ibid., ll. 62–63.

59. Ibid., l. 64.

60. Ibid., ll. 49–50. Work was required to build a special mooring to handle oil deliveries, but neither Glavsevmorput nor any other organization had the funds to underwrite the construction needed because of safety concerns; transport of oil and other highly flammable materials required special handling and emergency safety equipment. See ibid., d. 1389, ll. 8–10.

61. Ibid., d. 1411, ll. 79–86, 90.

62. Ibid., d. 1533, ll. 11–12.

63. Ibid., d. 1389, l. 1. The Limendskii Factory sadly closed in 2010, and workers continued legal action against it to receive back wages for several years. See http://news.nordportal.ru/novosti/ekonomika/limendskiy_sudostroitelnyy_zavod_otmetil_grustnyy_yubiley/.

64. For criticism of housing policies and practices in New York City, see Lewis Mumford, *From the Ground Up* (New York: Harcourt and Brace, 1956).

65. On housing and consumer goods under Khrushchev, Mark B. Smith, *Property of Communists: The Urban Housing Program from Stalin to Khrushchev* (DeKalb: Northern Illinois University Press, 2010). See also Henry Morton, "The Housing Game," *The Wilson Quarterly*, vol. 9, no. 4 (Autumn, 1985), pp. 61–74; Susan Reid, "Khrushchev Modern: Agency and Modernization in the Soviet Home," *Cahiers du Monde Russe*, vol. 47, no. 1/2 (January–June 2006), pp. 227–68; and Jan Prybyla, "The Soviet Consumer in Khrushchev's Russia," *Russian Review*, vol. 20, no. 3 (July 1961), pp. 194–205.

66. GAAO Otdel DSPI, F. 296, op. 2, d. 2017, ll. 31–32.

67. Ibid., ll. 41–42.

68. Ibid., ll. 51–52.

69. Ibid., d. 1883, ll. 10–12

70. Ibid., d. 2017 ll. 13–17.

71. Ibid., ll. 18–27.

72. Ibid., d. 1389, ll. 2–5.

73. Ibid., d. 1411, ll. 15–20.

74. Ibid., ll. 19–20.

75. Ibid., ll. 15–18, and d. 1533, l. 38.

76. Ibid., d. 1411, l. 39.

77. Ibid., d. 1883, ll. 2–5, 10–12, 19–22, 26–27, and ibid., d. 2017, ll. 13–17, 18–27.

78. Ibid., d. 2017, ll. 18–27.

79. Ibid., l. 54.

80. Savelii Prokhorovich Loginov was born into a peasant family in 1913. He finished a high party school. He was a member of the Communist Party from 1939 and was a teacher and instructor in party, union, and Komsomol organs. Second secretary of the Arkhangelsk obkom from 1951–55, he became secretary in November 1955 and was a delegate to the nineteenth, twentieth, and twenty-first party congresses. In the five years of his service the forestry, medical and pedagogical, and other higher educational institutions expanded as Arkhangelsk became more of a center of science, culture, and the forestry industries. He has a reputation to this day of being "democratic" in his attitudes toward others and helped many city residents with their infinite housing problems. He insisted on the construction of the railroad bridge over the Dvina. He also bought into the "corn" fantasy of Khrushchev and supported its planting, but it did not amount to anything in the Far North. See http://arh-necropol.narod.ru/obnovleniya

/loginov/ and http://www.edu.severodvinsk.ru/after_school/nit/2008/work/gener _ivan/podr.html.

81. GAAO Otdel DSPI, F. 296, op. 2, d. 2017, l. 53.
82. Ibid., ll. 1, 4, 13–17, 53.
83. Ibid., d. 1533, ll. 1–10.
84. Ibid., d. 2017, ll. 13–17, 21.
85. Ibid., d. 1883, l. 21.
86. Ibid., d. 2017, ll. 13–17, 21.
87. Ibid., ll. 55–56.
88. http://www.hrono.ru/biograf/bio_n/nosenkoii.php
89. http://www.domikorabl.ru/review/Zdravstvui_zdravstvui_amdermasvetlana_pavloval1.html
90. M. M. Kolovangina, *Nenetskii Okrug: Pervoe Desiatiletie* (Nar'ian-Mar: Krasnyi Gorod, 2009), pp. 54–82.
91. GAAO Otdel DSPI, F. 296, op. 2, d. 1391, ll. 7–14.
92. http://goroda-prizraki.narod.ru/goroda_amderma.html
93. Svetlana Zamlelova, "Iz Istorii Amdermy" at http://www.zamlelova.ru/index.php?id=995
94. GAAO Otdel DSPI, F. 296, op. 2, d. 1391, l. 16.
95. Ibid., l. 19.
96. Ibid., l. 21.
97. Ibid., l. 24.
98. Ibid., l. 25–26.
99. Ibid., ll. 28–29.
100. Ibid., ll. 30–33
101. Ibid., l. 33.
102. Ibid., ll. 2–4.
103. Ibid., d. 1389, l. 85.
104. Svetlana Zamlelova, "Iz Istorii Amdermy" at http://www.zamlelova.ru/index.php?id=995
105. http://goroda-prizraki.narod.ru/goroda_amderma.html
106. Ibid.
107. http://poilka.com/index.php?newsid=343 and http://ru.wikipedia.org/wiki/%D0%90%D0%BC%D0%B4%D0%B5%D1%80%D0%BC%D0%B0
108. http://arctic.ru/news/2011/12/tons-scrap-be-evacuated-amderma-2012
109. A. M. Gorky, *Sobranie Sochinenie,* vol. 30, pp. 421–24, at http://s-marshak.ru/works/prose/prose17.htm
110. H. P. Smolka, "Soviet Development of the Arctic New Industries and Strategical Possibilities," *International Affairs,* vol. 16, no. 4 (July 1937), 564–78.
111. Savva Morozov, http://www.memorial.krsk.ru/public/80/19870424.htm. See also http://www.polarpost.ru/forum/viewtopic.php?f=8&t=2337.
112. P. I. Negretov, "How Vorkuta Began."

113. On Vorkuta, see Alan Barenberg's forthcoming *Gulag Town, Company Town: Forced Labor and its Legacy in Vorkuta.*

6. Transformation of Taiga and Tundra

The epigraph is from Oleg Mitaev, "Mesta Glukhie," at http://www.mityaev.ru/song.cfm?text=Mesta_gluhie. My thanks to Dr. Tatiana Kasperski for translating this poem.

1. For a thorough and engaging examination of many of these programs in Murmansk region, see Andrew Bruno, "Making Nature Modern: Economic Transformation and the Environment in the Soviet North," Ph. D. Dissertation, University of Illinois, Champaign-Urbana, 2010, and the forthcoming book based on the dissertation.

2. Mark Reisner, *Cadillac Desert* (New York: Viking, 1986).

3. National Archive of the Republic of Karelia (hereafter "NARK"), R2546, op. 1, ed.khr. 1/1, ll. 11, 17.

4. Ibid., d. 2/27, ll. 1–3.

5. Ibid., ll. 7–8.

6. Ibid., l. 32.

7. Ibid., d. 1/6, l.32.

8. Ibid., edkhr. 5/85, ll. 1, 4–5.

9. Ibid., R626, op. 2, ed. khr. 1 / 2, l. 36, 41. In the Zaonezhskii Region, reclamation work in the Velikogubskii, Vygozerskii, Kazhemskii, Paianitskii, Tipinitskii, Tolvuiskii, Iandomozerskii, and Kosmozerskii village councils ranged across 24 kolkhozy from 20 to 142 hectares in 1950–51; in the Rugozerskii region at 10 kolkhozy with a total of 200 hectares; in the Rugozerskii region 4 kolkhozy and 50 hectares; in the Medvezhegorskii 14 kolkhozy with 192 hectares of arable land and 280 hectares of meadow; in the Sortovalskii region19 kolkhozy with a total of 1200 hectares; in the Segezhskii region with 9 kolkhozy with 748 hectares and 8 villahydropower station at 7,158 hectares; and in the Olonetskii Region 17 kolkhozy with a total of 1,700 hectares of arable land, 450 of swamp, and 120 of meadow. See ibid., ed. khr. 1/1, ll. 1–2, 8, 18, 28, 30, 31, 33, and R2546, op. 1, ed. khr. 5/85, ll. 16–18.

10. On the environmental and cultural costs of taming the Columbia River through a series of stepped dams, for example, see Richard White, *The Organic Machine* (New York: Harper Collins, 1995).

11. On Sardar Sarovar, see Philippe Cullet, "The *Sardar Sarovar Dam* Project: An Overview," at http://www.ielrc.org/content/a0704.pdf.

12. See Elena Gasikova, "Electric Power Infrastructure in the European North of Russia (in Russian)," paper given at "Eco-Pechora" Conference, Naryan-Mar, Russia, June 2010.

13. I discuss some of these issues in Josephson, "Technology and the Conquest of the Soviet Arctic," *Russian Review,* vol. 70, no. 1 (2011).

14. Loren Graham, *The Ghost of the Executed Engineer;* Paul Josephson, *Industrialized Nature;* Philip Micklin, "Water Diversion in the Soviet Union," *Science,* vol. 234 (October 24, 1986), p. 411. See also Robert Milko, "Potential Ecological Effects of the Proposed Grand Canal Diversion Project on Hudson and James Bays," *Arctic,* vol. 39, no. 4 (December 1986), pp. 316–26; Valentin Rasputin, *Vek Zhivi-Vek Liubi* (Moscow: Izvestiia, 1985), including "Proshchanie s Materoi" and "Baikal, Baikal." See *Petr Stepanovich Neporozhnii i Energetika Velikoi Strany* (Moscow: Energoatomizdat, 2010) for a sense of how Russian engineers nostalgically recall the glory years when their organizations willy-nilly dammed rivers and altered ecosystems.

15. On the Rural Electrification Administration, founded in the United States in 1935, see http://newdeal.feri.org/tva/tva10.htm.

16. http://gov.karelia.ru/News/2009/01/0128_05.html and http://www.kondopoga.ru/main/257-kondopozhskaja-gjes8201.html

17. Castro Alves, "A Cachoeira (The Falls)," http://www.revista.agulha.nom.br/calves13.html. My thanks to Dr. Antonio Botelho of Rio de Janeiro for the translation of Alves's poem.

18. http://petrozavodsk.rfn.ru/rnews.html?id=1492843&cid=6 and http://www.tgc1.ru/production/complex/korelia-branch/vyg-cascade/http://hydropowerstation.russgid.ru/belomorskaya.php

19. For example, construction on the Putkinskaia station (84 MW, three turbines each at 28 MW) began in 1962 and finished in 1970 with the first turbine operational in 1967. The dam is nearly 1,000 meters in length and 17.5 meters at its highest point. It includes a spillway for lumber.

20. NARK R3604, op. 1, d. 1/1, ll. 2–3, 6, 23–25, 33–36, 59–60, 63–64.

21. See for example Fekri Hassan, "The Aswan High Dam and the International Rescue Nubia Campaign," *The African Archaeological Review,* vol. 24, no. 3/4 (September/December 2007), pp. 73–94.

22. Panukrashev wrote, "In the case of your refusal to provide additional funds (for the third time), the institute cannot carry out the work necessary, and this will again lead to the destruction of historical monuments which are crucial to science (which, as you know, are protected by law)." See NARK R3604, op. 1, d. 1/1, ll. 109–11.

23. Gosudarstvennyi Arkhiv Murmanskoi Oblasti (GAMO), F. 959, op. 1, ed. khr. 1, ll. 3–6, 18. In fact, the hydrotechnology of prison labor found full expression in the design institute Giproproekt im. Zhuka, which grew out of and was named after the Gulag colonel and engineer who learned his skills at the Belomor-Baltic Canal and applied them elsewhere.

24. Ibid., ll. 8–12.

25. Ibid., d. 40, 1. 4, 21, 33, and d. 4, ll. 1–2. The earthen dam is 270 meters in length and 29 meters high, with concrete spillways and a fish ladder of 507 meters in length. The machine hall itself is 84 meters long and holds four 14.3-MW turbines from the Leningrad Metal Factory, while the generators, at 15 MW, are "Elektrosila" in origin.

26. GAMO, F. 731, op. 1, d. 8, ll. 18, 24, 28–29, 43, 59, 65–66. For modest plans to provide illumination for the "Dalnye Zelentsy" collective farm at 10 kW, for the "Belokenka" at 35 kW, the Krasnyi Sever kolkhoz on the Kolvitsa River, for the "Moriak" collective farm on the Northern Luvenga, another on the Kulonga River, and even one to power the Lapland Zapovednik, see ibid., ll. 3, 5, 11, 12, 35, and 92.

27. Ibid., d. 33. The Teriberka hydroelectric power station would support the burgeoning Murmansk fjord and nearby settlements of fish state farms, machine tractors, and other centers—a total of fifty inhabited points. See http://www.lhp.rushydro.ru/works/general.

28. NARK, R2725, op. 1, ed.khr. 4/64, 4–16.

29. http://www.kola-nature.org/node/7

30. GAMO, F. 731, op. 1, d. 8, l. 1.

31. Ibid., ll. 9, 18.

32. Ibid., d. 10, ll. 3, 5, 11, 12, 35.

33. Ibid., l. 48.

34. Ibid., d. 8, ll. 28–29.

35. Ibid., l. 92. In 1948 the "Dalnye Zelentsy" collective farm would get 10 kW of illumination; "Belokenka" 35 kW; "Moriak" a 6-kV power line and 150 kW of illumination; "Povir" a power line at 6kV and 250 kW of illumination, and a station was planned for "Krkino" at 35 kW.

36. Ibid., d. 10, l. 35.

37. Ibid., ll. 43–44.

38. Ibid., ll.17, 21, 21a, 23–26.

39. Ibid., l. 41.

40. GAAO, F. 1239, op. 1, d. 9, ll. 1–2, 5–9, 12.

41. Ibid., d. 199, ll. 1–51.

42. Ibid., op. 1, d. 6, ll. 1–5.

43. Ibid., ll. 6–7, 10, 14–16, 25–27, 43–50, 65–71. The persistent problems of interminably long construction, poor design, and faulty operation were repeated wherever Glavselektro tried to bring light and power to the countryside: the modest Apavazhskaia station on the Apavazh River at Karino in the Vilegodsk Region to serve the "Zavet-Lenina" collective farm and at the Veriugskaia, Topetskaia, and Tegrinskaia stations (on the Veriuga, Topetsk, and Tegra Rivers). See ibid., d. 21, ll. 1, 2, 8, 11. See also ibid., d. 37 concerning the Lopshen'gskaia hydropower station of the Belomorsk Region of Arkhangelsk.

44. GAMO, F. 731, op. 1, d. 8, l.43.

45. Ibid., ll. 58–59, 63, and http://energomuseum.ru/history/novostroyki/

46. http://www.kt70.com/~jamesjpn/photos/russia/teriberka/Teriberka.htm . For discussion from Bellona Foundation on the dangers of the exploitation of the Stokhmann gas deposits to Teriberka, see http://www.bellona.org/articles/articles_2010/shtokman_scepticism and http://barentsobserver.com/take-a-closer-look-at-teriberka.4615131.html.

47. Julia Lajus has written extensively on the environmental history of Soviet fisheries, 1920–40. Among her very many publications, see "Colonization of the Russian North: A Frozen Frontier," in Christina Folke Ax, Niels Brimnes, Niklas Thode Jensen, and Karen Oslund, editors, *Cultivating the Colony: Colonial States and their Environmental Legacies* (Athens: Ohio University Press, 2011), and "Science, Politics and Practice in the Fishery: Scientists, Industrialists and Fishers in the Russian North, 1898–1940," in Poul Holm and David J. Starkey, editors, *Technological Change in the North Atlantic Fisheries* (Esbjerg: Studia Atlantica, 1999), vol. 3, pp. 49–59.

48. NARK, R2170 op. 2, d. 4/21.

49. Ibid., d. 4/22, ll. 156–60.

50. Ibid., R2419, op. 1, d. 1/1, ll. 4–6, 17–19, 44–48.

51. NARK, R. 2419, op. 1, d. 1/9, ll. 2–15.

52. GAAO Otdel DSPI, F. 296, op. 1, d. 1082, ll. 2–6.

53. Ibid., ll. 12–16, 18, 30–31, 45–48, 60, 62.

54. Ibid., d. 582, ll. 33–34.

55. Ibid., ll. 116–19.

56. Ibid., d. 1389, ll. 24–25.

57. Ibid., ll. 6–8. Loginov requested a new vessel at 100 to 150 tons. On top of this the port needed six cranes, a storage facility for 15,000 tons of coal, and funds to repair four moorings.

58. Ibid., d. 1882, l. 5. The Pechora basin was quite isolated, and with essentially no roads transport was by air or boat during the short summer navigation season. The authorities worked bit by bit to improve communications, in 1954 building a telephone network from Oksino to Viska, upstream from Naryan-Mar and on opposite sides of the Pechora River, and acquiring two river motorboats. Ibid., l. 6.

59. Ibid., d. 1882, ll. 57–66.

60. Ibid., ll. 68–71.

61. Ibid.,, ll. 102–6. The authorities had made some progress, however, in establishing a series of schools to train the merchant marine.

62. Ibid., ll. 108–14.

63. Ibid., ll. 114–16.

64. Ibid., op. 2, ed. 192, ll. 1–2.

65. Ibid., l. 40.

66. Ibid., l. 41.

67. Ibid., l. 42.
68. GAAO, F. 1735, op. 1, d. 98, ll. 66–68.
69. Ibid., ll. 50–52.
70. Josephson, *Industrialized Nature* (Washington: Island Press, 2002).
71. GAAO Otdel DSPI, F. 296, op. 2, d. 908, l. 37.
72. Ibid., d. 1942, l. 19.
73. See, for example ibid., d. 908, l. 40.
74. Ibid., ll. 37–38.
75. Ibid., d. 1942, ll. 36–51.
76. Ibid., ll. 52–57.
77. Ibid., d. 908 1950, ll. 37–38.
78. Ibid., d. 978, l. 1. To give some idea of the isolation of the towns that grew up around the forestry operations, Shenkgursk is on the Vaga River, a left tributary of the Northern Dvina River 373 kilometers to the southeast of Arkhangelsk and 143 kilometers from the Arkhangelsk–Kotlas railroad. Lumber and woodworking operations are central to the town's economy, and there are also various factories connected with food (milk, bread, and so on). From more than 2,500 inhabitants in 1929 Shengursk grew to greater than 7,500 in 1989 and dropped gradually to 5,736 in 2010.
79. Ibid., ll. 2–3, 7–27.
80. Ibid., d. 1882, ll. 15–17.
81. Ibid., d. 582, ll. 3–5.
82. Ibid., ll. 22, 24–25.
83. GAAO Otdel DSPI, F. 296, op. 2, d. 582, l. 6.
84. Ibid., l. 9.
85. Paul Goble, "Corruption Keeping Russian Highways among Worst in the World," http://economie.moldova.org/news/corruption-keeping-russian-highways-among-worst-in-the-world-206976-eng.html
86. GAAO Otdel DSPI, F. 296, op. 1, d. 40, ll.4–8, ll. 10–11.
87. Ibid., 1, d. 1332, ll. 17.
88. GAAO F. 4369, op. 2, d. 56, ll. 3–4, 6–7.
89. Ibid., ll. 10–11, 19.
90. GAAO Otdel DSPI, F. 296, op. 2, d. 1539, ll.66–70, 75–76, 82–83.
91. Ibid., d. 1389, ll. 47–51.
92. Ibid., d. 908, ll. 31–36
93. Ibid., d. 1882, l. 12.
94. Ibid., d. 1942, l. 103–5.
95. Ibid., ll. 26–28, 68–69, 108–112.
96. Ibid., d. 1389, l. 114.
97. Ibid., d. 191, l. 33.
98. Ibid., d. 1389, ll. 57, 68.
99. Ibid., op. 1, d. 1332, l. 30.

100. Ibid., l. 49.

101. Ibid., op. 2, d. 319, l. 15. See also J. Patrick Lewis, "Communications Output in the USSR," *Soviet Studies,* vol. 28, no. 3 (July 1976), pp. 406–17.

102. Lenin, *Collected Works,* vol. 34, (Moscow: Progress Publishers, 1960–1970), p. 247.

103. Steven Solnick, "Revolution, Reform and the Soviet Telephone System, 1917–1927," *Soviet Studies,* vol. 43, no. 1 (1991), pp. 157–76.

104. GAMO, F. 360, op. 1, d. 2, l. 45, ll. 98, 99.

105. Ibid., d. 43, l. 4.

106. Ibid., ll. 39–46.

107. Ibid., ll. 47, 58.

108. Ibid., l. 51.

109. GAAO Otdel DSPI, F. 296, op. 2, d. 1884, l. 20.

110. Ibid., l. 36.

111. Ibid., op. 1, d. 1332, ll. 98.

112. Ibid., op. 2, d. 1563, l. 15 and d. 1884, l. 13. The Emetskii district (which existed from 1929 until 1959 when it was joined to the Kholmogorsk district) in the western part of Arkhangelsk province consisted of 290 inhabited points and a total of 21,000 people in 1929 and had several collective farms in the mid-1950s without radio communications. To give a sense of the continued isolation of this district, in 2009 the town center, Emetsk, had but 1,441 inhabitants, more than one-third (548) of whom were pensioners.

113. Ibid., ed. khr. 2024, ll. , 22–23, 173–75.

114. Ibid., op. 1, d. 1211, ll. 7–8, 28–29, and op. 2, ed. khr. 2024, l. 8. The same story of technological failure held for the airplane as well, with spotty service, a high frequency of accidents, and personnel not up to the task of sober operation. On the expansion of the Naryan-Mar civil aviation service, important given the absence of roads or a railroad to Naryan-Mar, see TO ANAO F. 157 (entire).

115. GAAO Otdel DSPI, F. 296, op. 2, d. 1563, l. 9.

116. Ibid., l. 12.

117. GAMO, F. 959, op. 1, d. 73, ll. 43–49.

118. GAAO Otdel DSPI, F. 296, op. 2, d. 908, ll. 31–36.

119. Ibid., d. 1389, l. 84.

120. Ibid., ll. 11–13.

121. Ibid., d. 630, ll. 97–98.

122. Ibid., d. 1389, ll. 31–36.

123. Ibid., d. 582, ll. 1–2.

124. Ibid., d. 1945, ll. 23, 27, 30, 39, 41, 45

125. Ibid., d. 908, ll. 4–5.

126. Ibid., d. 1942, l. 24.

127. Ibid., d. 1882, ll. 142, 183.

128. Ibid., d. 1942, ll. 61–65.

129. Ibid., d. 1882, l. 1.

130. N. K. Zhernakov, *Krutye Suvoi* (Arkhangelsk: Severo-Zapadnoe Knizhnoe Izdatel'stvo, 1975), p. 54.

131. Ibid., pp. 72–73.

132. http://publ.lib.ru/ARCHIVES/B/BOGDANOV_Evgeniy_Fedorovich/_Bogdanov_E.F.html

133. E. Bogdanov, *Visokosnyi God* (Moscow: Sovremennik, 1986), p. 33.

134. Ibid., p. 55.

135. Ibid., p. 204.

136. Ibid., p. 49.

137. Ibid., p. 87. Bogdanov referred in *Leap Year* to Khrushchev's crazy idea to plant corn—even in northern regions of the country—and build silos to hold it as feed. But the corn simply died in the northern regions, and those directors who fought the order faced reprisals. See p. 89.

138. Ibid., p. 223.

139. Ibid., p. 294.

140. GAAO Otdel DSPI, F. 296, op. 2, d. 1389, l. 67.

141. Ibid., d. 1882, ll. 143–46.

142. Ibid., d. 1389, l. 62.

143. *Pravda Severa,* 2 November 2007, p. 5.

7. Rediscovering the Arctic

The epigraphs are from A. F. Sukhanovskii, I. Iu. Slobodianiuk, *Arkticheskaia Rossiia* (Arkhangelsk: SK-Rossiia, 2007), pp. 17 and 243.

1. John McCannon, *Red Arctic* (New York: Oxford University Press, 1998).

2. I discuss the impact of the Putin presidency on science in *Lenin's Laureate* (Cambridge: MIT Press, 2010), pp. 215–46, from which these brief comments are taken.

3. Education and research have been tied to these efforts. The Northern (Arctic) State University, founded in 2010 in Arkhangelsk, brings together the humanities, social sciences, sciences, and engineering of the former Pomor State University, the Arkhangelsk State Technical University, the Arkhangelsk Forestry Institute, and the Severodvinsk Technical College, together with several other institutions across the province, to facilitate training of scores of young people to participate in Arctic assimilation and to expand basic and applied research across the region. See http://narfu.ru/en/university/about/his/.

4. A. M. Via'min et al., "Osobennosti Demograficheskikh Protsessov na Evropeiskom Severe Rossii na Rubezhe XX–XXI vekov," in Galina Degteva, editor, *Problemy Zdravookhraneniia i Sotsial'nogo Razvitiia Arkticheskoi Zony Rossii* (Moscow-St. Petersburg: Paulsen Editions, 2011), pp. 185–203.

5. http://www.prb.org/Articles/2002/RussiasDemographicDeclineContinues.aspx

6. T. Heleniak, "Out-migration and Depopulation of the Russian North During the 1990s," *Post-Soviet Geography and Economics,* vol. 40, no. 3 (1999), pp. 155–70. See also Ward Kingkade, "Population Trends: Russia," IB/96-2, (Washington: U.S. Department of Commerce, Bureau of the Census, 1997).

7. http://archive.kremlin.ru/eng/text/speeches/2008/09/17/1945_type82912type82913_206564.shtml

8. Loren Graham and Irina Dezhina, *Science in the New Russia* (Indiana University Press, 2008), p. 37. See also Graham and Dezhina, pp. 10–21, and E. A. Tropp, "Peterburgskaia Nauka: Devianostye Gody XX v.," in Zhores Alferov, editor, *Akademicheskaia Nauka v Sankt-Peterburge v XVIII–XX vekakh: Istoricheskie Ocherki,* ed. Zh. I. Alferov (Petersburg: Nauka, 2003). For background, see also Paul Josephson and Irina Dezhina, "The Slow Pace of Reform of Science in Russia and Ukraine," *Problems of Post-Communism,* vol. 45, no. 5 (1998), pp. 48–63; Paul Josephson and Igor Egorov, "Ukraine's Declining Scientific Establishment," *Problems of Post Communism,* vol. 49, no. 4 (2002), pp. 43–51.

9. http://rbth.ru/science_and_tech/2013/07/08/controversial_reform_of_russian_academy_of_sciences_postpone_27891.html

10. A. O. Andreev, M. V. Dukal'skaia, S. V. Frolov, "Stranitsy Istorii AANII," *Problemy Arktiki i Antarktiki,* no. 1 (84), (2010), pp. 21–23.

11. I. E. Frolov, editor, *Rossiiskie Issledovaniia na Dreifuiushchikh L'dakh Artiki* (St. Petersburg, AANII, 2010), pp. 557–74.

12. Ibid., p. 574.

13. http://www.meteorf.ru/default.aspx

14. Frolov, *Rossiiskie Issledovaniia na Dreifuiushchikh L'dakh Artiki,* pp. 578–600.

15. Lassi Heininen (with Heather Nicol), "The Importance of Northern Dimension Foreign Policies in the Geopolitics of the Circumpolar North," *Geopolitics,* vol. 12, no. 1 (February 2007), pp. 133–65.

16. Harley Balzer, "Vladimir Putin's Academic Writings and Russian Natural Resource Policy," *Problems of Post-Communism* (January/February 2006), pp. 48–54.

17. For a balanced view of Putin's role in the Russian state, see Allan Lynch, *Vladimir Putin and Russian Statecraft* (Dulles, VA: Potomac Books, 2011).

18. http://www.oceanlaw.org/downloads/arctic/Russian_Maritime_Policy_2020.pdf

19. For discussion of the relation between politics, investment, and oil, see Marshall Goldman, *Petrostate* (New York: Oxford University Press, 2009).

20. *The Basics of State Policy of the Russian Federation in the Arctic Region* adopted by the President of the Russian Federation D. Medvedev, September 18, 2008, at http://img9.custompublish.com/getfile.php/1042958.1529.avuqcurreq

/Russian+Strategy.pdf?return=www.arcticgovernance.org> and <http://www.arcticgovernance.org/russia-basics-of-the-state-policy-of-the-russian-federation-in-the-arctic-for-the-period-till-2020-and-for-a-further-perspective.4651232-142902.html

21. See http://www.scrf.gov.ru/documents/98.html.

22. http://izvestia.ru/news/511452#ixzz25myJTzVW

23. http://izvestia.ru/news/498517#ixzz25mygUo3h

24. Sukhanovskii and Slobodianiuk, *Arkticheskaia Rossiia*, p. 7. On Chilingarov, see http://lenta.ru/lib/14178687/, http://www.rg.ru/2012/01/27/arktika.html, and http://www.lobbying.ru/content/persons/id_337_linkid_2.html.

25. http://www.kp.ru/daily/23892/66464/

26. http://www.bbc.co.uk/news/business-11316430 and http://www.kp.ru/daily/23892/66464/.

27. http://www.kp.ru/daily/23892/66464/

28. http://www.km.ru/inosmi/2011/08/19/mezhdunarodnaya-politika/militarizatsiya-arktiki

29. http://www.cbc.ca/news/canada/story/2012/08/24/pol-harper-north-tour-friday.html

30. "Canada to Fortify Arctic Sovereignty With New Icebreakers," http://www.ens-newswire.com/ens/jul2007/2007-07-10-03.html

31. http://www.themoscowtimes.com/business/article/putin-sees-stalins-industrialization-as-model/467476.html

32. http://www.kp.ru/daily/25661.4/823178/

33. http://kp.ru/daily/25767/2751896/. Now a museum in Murmansk, school-children who visit it refer to the "Lenin" as the *Titanic* and ask who is the grandpa with the beard whose bust sits in front of the ship?

34. http://www.bellona.ru/weblog/andrey-ozharovsky/1241708613.5. For a map of other reactors dumped in the Arctic Ocean, see http://www.solovki.ca/danger/radiation_02.php.

35. http://www.kp.ru/online/news/1224536/ on August 16, 2012. AARI aerial research under director Sergei Frolov on the Arctic ice regime facilitated the North Pole journey of the *Arktika*.

36. http://www.barentsobserver.com/new-ice-breakers-for-the-arctic.4834701-116320.html and http://www.globalsecurity.org/military/world/russia/10521.htm and Yuri Golotuik, "Safeguarding the Arctic," in *Russia in Global Affairs*, August 9, 2008, at http://eng.globalaffairs.ru/person/p_1887

37. http://kp.ru/daily/25767/2751896/

38. http://kp.ru/daily/24568/741054/

39. http://izvestia.ru/news/496692#ixzz25myy1B4j

40. http://kommersant.ru/doc/2007338?isSearch=True (August 24, 2012).

41. http://kp.ru/daily/25767/2751896/

42. http://helion-ltd.ru/babich-rus-variant/

43. http://ria.ru/arctic_news/20120730/712578195.html#ixzz2WbacgHFe

44. http://pravdaurfo.ru/news/yamalo-neneckiy-okrug/putin-vozrodit-severnyy-morskoy-put-s-yamala and http://www.kremlin.ru/news/17173.

45. http://www.vesti.ru/doc.html?id=577224

46. Vladislav Larin, "'Nuclearification' of the Russian Offshore Oil and Gas Industry," *Bellona Foundation Report* (2008) at http://www.bellona.org/articles/articles_2007/sevmorput_drilling

47. http://en.rian.ru/analysis/20070409/63375060.html

48. http://in.reuters.com/article/2008/10/20/shtokman-budget-idINLK43254820081020

49. http://www.bellona.org/articles/articles_2012/Shtokman_freeze. See also see Goldman, *Petrostate.*

50. http://www.bellona.org/articles/articles_2012/Khanty_mansi_spills_statoil

51. Ol'ga Mordiushenko, August 30, 2012, at http://kommersant.ru/doc/2011009?isSearch=True

52. http://www.adm-nao.ru/?show=news&id=4744

53. Atle Staalesen, "New Life for Amderma," June 15, 2011, at http://www.barentsobserver.com/new-life-for-amderma.4933206-16149.html

54. William Barr, "Geographical Aspects of the First International Polar Year, 1882–1883," *Annals of the Association of American Geographers,* vol. 73, no. 4 (December 1983), pp. 463–84.

55. Erki Tammiksaar, Natalia G. Sukhova, and Ian R. Stone, "Russia and the International Polar Year, 1882–1883," *Polar Record,* vol. 45, no. 3 (July 2009), pp. 215–23.

56. A. O. Andreev, M. V. Dukal'skaia, and S. V. Frolov, "Iz Istorii Mezhdunarodnogo Poliarnogo goda," *Problemy Artiki i Antarktiki,* vol. 75 (2007), pp. 7–9.

57. Ibid., pp. 10–12.

58. "The Geneva Conference on Peaceful Uses of Atomic Energy," *American Journal of Public Health. Nations Health,* vol. 45, no. 11 (November 1955), pp. 1481–82, and *Proceedings of the Second International Conference on Peaceful Uses of Atomic Energy* (Vienna: IAEA, 2008).

59. Andreev et al., "Iz Istorii Mezhdunarodnogo Poliarnogo goda," pp. 13–16.

60. I. E. Fedorov, V. T. Sokolov, and I. M. Ashik, "Osnovye Itogi i Predvaritel'nyi Nauchni Rezul'taty . . . Reisov . . . 'Akademik Fedorov'," *Problemy Arktiki i Antarktiki,* vol. 75 (2007), pp. 18–34.

61. http://ipyrus.aari.ru/

62. Ibid.

63. http://ipyrus.aari.ru/russian_particip_inprojects.html. Total higher than 100 percent because of rounding.

64. I. M. Ashik and V. T. Sokolov, "Osnovnye Itogi i Predvaritel'nye Nauchnye Rezul'taty Ekspeditisii 'Arktika-2007'," *Problemy Arktiki i Antarktiki,* vol. 80 (2008), pp. 72–85.

65. V. V. Stanovoi, I. V. Lavrenov, and I. A. Neelov, "Sistema Modelirovaniia Razlivov Nefta v Ledovitykh Moriakh," in ibid., vol. 77 (2007), pp. 7–16.

66. Ibid., vol. 78 (2008), pp. 7–16.

67. Ibid., pp. 7–26.

68. Ibid., vol. 79 (2008), pp. 44–48.

69. Ibid., vol. 81 (2009), pp. 80–84.

70. Ibid., vol. 77 (2007), pp. 56–66.

71. http://www.yasavey.org

72. S. G. Grigorev, *Poluostrov Kanin* (Moscow: Trudy Geograficheskogo Nauchno-Issledovatel'skogo Instituta Moskovskogo Gosudarstvennogo Universiteta, 1929).

73. http://www.nsmu.ru/science/nii_pol_med/konf_MPG/istoria/

74. http://www.nsmu.ru/university/professors/degteva.php

75. B. A. Revich et al., "Kilmatciheskie Izmeneniia kak Faktor Riska Zdorov'iu Naseleniia Rossiiskoi Arktiki," in Degteva, editor, *Problemy Zdravookhraneniia,* pp. 9–68.

76. V. I. Khasnulin and P. V. Khasnulin, "Ekologicheski Obuslovlennyi Severnyi Stress (sindrom poliarnogo napriazhenia)," in ibid., pp. 69–82.

77. T. N. Ungurianu, "Klimaticheskie Faktory i Zagriaznenie Atmospfernogo Vozdukha Kak Faktory Riska Zabolevanii Neinfektsionnoi Prirody," in ibid., pp. 83–97.

78. Khasnulin and Khasnulina, "Arterial'naia Gipertenziia na Severae," in ibid., pp. 241–53.

79. *Problemy Arktiki i Antarktiki,* vol. 81 (2009), pp. 126–31.

80. A. N. Marinchuk et al., "Soderzhanie Tiazhelykh Metallov v Produktakh Pitanie i Plazme Krovi Naseleniia Priural'skogo Raina," in ibid.,, vol. 81 (2009), pp. 146–52.

81. http://www.mmbi.info/eng/

82. G. G. Matishov, "Kilmaticheskie Izmeneniia Morskikh Ekosistem Evropeiskoi Arktiki," *Problemy Arktiki i Antarktiki,* vol. 86 (2010), pp. 7–21.

83. Aleksandr Chuprov, *Nar'ian-Mar: Znakomyi i Neizvestnyi* (Naryan-Mar: OM-Mediia, 2007), pp. 1–65, 68, 75–76.

84. Sukhanovskii and Slobodianiuk write, "Millions of barrels of fuel have been disposed by civilization throughout the reaches of the Arctic." See Sukhanovskii and Slobodianiuk, *Arkticheskaia Rossiia,* p. 155.

85. http://sevprostor.ru/expeditions/karatayka-amderma-barents-sea-kara-sea/241-beliy-nos-yugorskiy-shar.html

86. http://www.guardian.co.uk/world/2013/mar/27/vladimir-putin-crackdown-ngo-russia

87. http://www.beac.st/in_English/Barents_Euro-Arctic_Council/Introduction

.iw3. According to the council's website, "Out of a total population in the Barents Region of approximately 5.3 million people, there are around 85,000 Sami inhabitants in Sápmi, the traditional area of the Sami people. 50–60,000 of those are living in Norway, some 20,000 in Sweden, some 7,000 in Finland and around 2,000 in Murmansk Oblast in Russia. In the Russian Nenets Autonomous Okrug there are 7,000 Nenets people. Approximately 6,000 Vepsians are living in the Republic of Karelia. These peoples are represented in the working group established by the Regional Council. The Komi people are represented by the representative of the Russian Republic of Komi, a member of the Barents Regional Council since 1 January 2002." See http://www.beac.st/in_English/Barents_Euro-Arctic_Council/Indigenous_Peoples.iw3.

88. On the Khar'iaginskoe Oil Deposits and their exploitation, see http://www.nies.ru/progects/projects_ecoizisk/item_36/.

89. Igor Gennadievich Fedorov, "Uchastie NAO v Realizatsii Programm i Proektov v Ramkakh Mezhdunarodnogo Arkticheskogo Sotrudnichestva, v tom Chisel po Linii Mezhdunarodnykh Organizatsii," May 30, 2012, at http://www.adm-nao.ru/?show=statics&id=303. Every two years the administration sponsors an international conference called "Eco-Pechora." In June 2010, in part with a National Science Foundation grant, the NAO sponsored a meeting on "Arctic Perspectives in the Twenty-first Century" with participation of twenty-five scholars, mostly beginning scholars, from around the world.

90. Roy Shephard and Andris Rode, *The Health Consequences of 'Modernization': Evidence from Circumpolar Peoples* (Cambridge: Cambridge University Press, 1996).

91. V. P. Khlopov et al., "Opredelenie Ekologo-Toksicheskikh Nagruzok na Naselenie Iamalonenetskogo Avtonomnogo Okruga," in Degteva, editor, *Problemy Zdravookhraneniia*, p. 266.

92. Iurii Rytkheu, *Dorozhnyi Leksikon* (St. Petersburg: "Zvezda," 2010), pp. 23–28.

93. Ibid., pp. 180–85.

94. Viacheslav Ogryzko, "Kak Zarozhdalas' Naniskaia Literatura," at http://www.litrossia.ru/archive/110/history/2593.php

95. M. G. Voskoboinikov at http://encyclopedia2.thefreedictionary.com/Nenets+National+Okrug

96. V. N. Ledkov, *Mesiats Maloi Temnoty. Dialogy*, trans. by V. Leont'ev (Moscow: Sovetskaia Rossia, 1989), p. 4.

97. Ibid., p. 32.

98. Ibid., p. 9.

99. Ibid., p. 25.

100. Ibid., p. 204.

101. Q. Schiermeier and B. MacWilliams, "Crunch Time for Kyoto," *Nature*, vol. 431 (September 2, 2004), pp. 12–13. For more discussion of Kyoto and Russia,

see Paul Josephson et al., *An Environmental History of Russia* (New York: Cambridge University Press, 2013).

102. http://kp.ru/daily/24457/619917/. The minister of natural resources, Yuri Trutnev, said that as a measure against "carbon protectionism" Russia might pursue limits on forest exports.

103. "Weather and Feedbacks Lead to Third-Lowest Extent," October 4, 2010, at http://nsidc.org/arcticseaicenews/

104. http://blogs.nature.com/news/blog/reports/earth_and_environment/arctic_frontiers/ January 2009

105. *Scientific Program of the Participation of the Russian Federation in Carrying Out the International Polar Year (2007–2008)* (in Russian) (Moscow: N. P. [State Duma?] 2006), p. 3.

106. Ibid., p. 4.

107. http://izvestia.ru/news/517624#ixzz25mxwWaaS

108. Speech at State Council Presidium Meeting on Improving State Regulation in the Area of Environmental Protection, May 27, 2010, http://archive.kremlin.ru/eng/speeches/2010/05/27/1525_type82913_226324.shtml

109. http://izvestia.ru/news/517624#ixzz25mxwWaaS

110. http://www.guardian.co.uk/world/2010/sep/23/putin-arctic-claims-international-law

111. Aleksandr Nikitin, Igor Kudrik, and Thomas Nilsen, "The Russian Northern Fleet: Sources of Radioactive Contamination," *Bellona Foundation Reports*, no. 2 (1996).

112. For a discussion of some of these issues from the U.S. perspective, see Ron Doel, "Constituting the Postwar Earth Sciences: The Military's Influence on the Environmental Sciences in the USA after 1945," *Social Studies of Science*, vol. 33, no. 5 (October 2003), pp. 635–66.

113. V. Ch. Khon and I. I. Mokhov, "Analiz Uslovii v Arkiticheskom Basseine i Perspektivy Razvitiia Severnogo Morskogo Puti v XXI Veke," *Problemy Arktiki i Antarktiki*, vol. 78 (2008), pp. 59–65.

114. I. E. Frolov et al., "Izmeneniia Klimata Arktiki i Antarktiki—Rezul'tat Desitviia Estestvennykh Prichin," in ibid., vol. 85 (2010), pp. 52–61. On obfuscation of such issues as global warming by the Heartland Institute as well as such cases as DDT, tobacco, and acid rain, see Naomi Oreskes and Eric Conway, *Merchants of Doubt* (New York: Penguin, 2012).

115. G. V. Alekseev et al., "Izmeneniia Kilmata v Morskoi Arktike v Nachale XXI Veka," *Problemy Arktiki i Antarktiki*, vol. 86 (2010), pp. 22–34.

116. http://izvestia.ru/news/529446#ixzz25mx91U6j

117. http://eco.ria.ru/nature/20111108/483779489.html

118. http://www.inosmi.ru/infographic/20110704/171577630.html

119. John McCannon, *A History of the Arctic* (London: Reaktion Books, 2012), pp. 236–45.

120. http://www.ipolitics.ca/2012/04/18/ottawa-to-clean-up-cold-war-arctic-relics/. Stacey Fritz has been working on a dissertation at the University of Alaska, Fairbanks, on the subject of the impact of DEW on ecosystems and Inuit lifestyle.

121. Among the dozens of excellent compilations and reports that indicate the extent of nuclear waste and other problems of the Soviet cold war legacy, see the Bellona Foundation reports: Nils Bøhmer, Aleksandr Nikitin, Igor Kudrik, Thomas Nilsen, Andrey Zolotkov, and Michael H. McGovern, *The Arctic Nuclear Challenge* (Bellona Foundation, 2001), and Igor Kudrik, Aleksandr Nikitin, Charles Digges, Nils Bøhmer, Vladislav Larin, and Vladimir Kuznetsov, *The Russian Nuclear Industry—The Need for Reform* (Bellona Foundation, 2004).

122. http://www.rferl.org/content/russia_pushes_ahead_with_controversial_floating_nuclear_reactors/9502474.html

123. W. G. Rees and A. P. Kapitsa, "Industrial Pollution in the Kol'skiy Poluostrov, Russia," *Polar Record*, vol. 30, no. 174 (1994), pp. 181–89; D. Kiselov, "Sereznost' Ekologicheskoi Situatsii v Rossiskii Federatsii," *Econord Magazin*, vol. 1 (1992), pp. 4–5; M. Kozlov, "Changes in Wind Regime Around a Nickel-Copper Smelter at Monchegorsk, Northwestern Russia," *International Journal of Biometeorology*, vol. 46, no. 2 (2002), pp. 76–80; L. Gytarskii et al. "Monitoring of Forest Ecosystems in the Russian Subarctic—Effects of Industrial Pollution," *Science of the Total Environment*, vol. 164, no. 1 (March 1995), pp. 57–68; B. A. Revich, "Public Health and Ambient Air Pollution in Arctic and Subarctic Cities of Russia," *Science of the Total Environment*, vol. 161 (1995), pp. 585–92; N. V. Vasiliev et al., "The Role of Migration Processes in Oncological Epidemiology of Siberia and the Far East," *Vestnik Rossiiskoi Akademii Meditsinskikh Nauk*, no. 7 (1994), pp. 34–39; A. R. Bond, "Air Pollution in Norilsk—A Soviet Worst Case," *Soviet Geography*, vol. 15, no., 9 (1984), pp. 665–69; A. A. Vinogradova, and V. A. Egorov, "Contributions of Industrial Areas of the Northern Hemisphere to Air Pollution in the Russian Arctic," *Izvestiia Akademii Nauk. Fizika Atmosfery i Okeana*, vol. 33, no. 6 (November/December 1997), pp. 750–57; D. R. Klein, "Arctic Grazing Systems and Industrial Development: Can We Minimize Conflicts?" *Polar Research*, vol. 19, no. 1 (2000), pp. 91–98; V. Shevchenko et al., "Heavy Metals in Aerosols Over the Seas of the Russian Arctic," *Science of the Total Environment*, vol. 306, nos. 1–3 (May 2003), pp. 11–25; and S. M. Allen-Gil et al., "Heavy Metal Contamination in the Taimyr Peninsula, Siberian Arctic," *Science of the Total Environment*, vol. 301, nos. 1–3 (January 2003), pp. 119–38.

124. http://www.dailykos.com/story/2013/06/05/1213506/-Russian-Arctic-Researchers-Await-Rescue-on-Shattered-Sea-Ice-Battered-by-Storms

125. Alex Boyd, "More Profitable to Keep Polluting the Region than Modernize the Production," http://www.nunatsiaqonline.ca/stories/article/65674arctic_governments_stand_by_as_deadly_pollution_spews_from_russias_nik/

126. Loren Graham, *Science in Russia and the Soviet Union* (Cambridge and New York: Cambridge University Press, 1993).

Acknowledgments

I am honored to have spent roughly seven months in the Russian northwest conducting field work and using libraries and state and Academy of Sciences archives in Murmansk and Apatity, Murmansk Province; Petrozavodsk, Karelia; Arkhangelsk and Naryan-Mar, Arkhangelsk Province; and the Arctic and Antarctic Research Institute, St. Petersburg. I am grateful to all of the archivists for their efforts on my behalf. I am also grateful to the National Science Foundation and program officers Drs. Anna Kerttula and Fred Kronz. Their support enabled my research and the holding of an international conference in Naryan-Mar in June 2010, "Eco-Pechora," which included a helicopter field trip from Naryan-Mar across the Arctic Ocean to Kolguev Island and then to a meteorological station for twenty-two conference participants.

I would like to acknowledge the assistance of a large number of people in the writing of this book. I apologize if I have missed anyone. Elena Nikiforova introduced me to the Russian Northwest when she invited me to be a member of a "moving research seminar" from St. Petersburg to the Kola Peninsula in 2007. Michael Suprun and Aleksei Feldt in the Department of History at Pomor State University, and other faculty and students in the department, kindly hosted me during a Fulbright Lectureship in autumn 2007. The late Aleksandr Beliaev, mayor of Severodvinsk, my friend, frequently supported my visits until, to our incredulity, the State Security Service (the "FSB") decided to deny my access to the city. Aleksandr's assistant, Olga Deriaeva, has professionally assisted my research needs. Sergei Borsky, a physician and onetime vice mayor of Severodvinsk who died suddenly in December 2012, offered the services of the municipality of the city—and the warmth of family gatherings over the past dozen years. Fortunately, my colleagues in the P/S Connection, a sister city organization to bring schoolchildren and ordinary citizens of Portsmouth, New Hampshire, and Severodvinsk together, have continued to provide a welcome to me—in Arkhangelsk. As always Robert Clemente, Willie Stargell, and Paavo Nurmi provided inspiration. Viktor Dmitriev, scientific secretary of the Arctic and Antarctic Research Institute, and the staff of the library and archive provided a warm environment in which to talk about the history of Russian Arctic science. Viktor taught me a great deal about Arctic science in conversation and in giving me access to recent AARI publications. Antonina Chetverova and Elena Morozova, graduate students at AARI, kindly discussed this project. Ekaterina Kosareva opened an understanding of the place of Soviet literature in Arctic assimilation. Dmitrii Kozlov generously supported my research efforts during my exploration of northern writers and poets. Allan Gamborg is the world's most generous host and warmly supported my research effort in Moscow, including at the White Horse. Galina and Dmitrii Medvedev ensured that all of my trips to Naryan-Mar were fruitful, well organized, and warm and escorted me to the tundra, where I met a shaman, saw reindeer, and stumbled into a radar. Tatiana Kasperski read the introduction to this book, offered suggestions for other chapters, and helped with several translations. Stephen Brain, Alan Barenberg, Michael Robinson, John McCannon, Nick Breyfogle, Andy Bruno, and Ron Doel each read one chapter and made crucial critical comments. Jim Fleming and Cathy Frierson discussed this project carefully with me and provided important insights. Jennifer Alexander asked me to lecture at the University of Minnesota on this book in February 2009, where she and her colleagues provided important insights to my ongoing writing. Julia Herzberg and Klaus Gestwa invited me to participate in a workshop in February 2012 on ice and snow in Russian culture in which I delivered preliminary results of this research; Julia's work on snow and ice in Russian culture is an example for us all. Elizabeth Knoll, my editor at Harvard University Press, provided excellent guidance on this book, as she did on my first book in 1991.

Manny Gimond, a GIS specialist at Colby College, produced the maps for this book. Victor Boyarsky and the Russian State Museum of the Arctic and Antarctica offered kind critical suggestions and also provided photographs. Katya Boikova, whose life in Severodvinsk began in the tundra with her parents, provided several photos of a Red Teepee. Katherine Gorunova, photographer extraordinaire, edited my photos of the Russian Arctic. My research assistants at Colby College, Jennifer Dahnke, Lauren Kerr, and Charlie Spatz, could not have been more superb and professional. Indiana Jones meticulously produced the index. Ukrainian football fans, including Sasha Popov, discussed the project with me during Eurocup 2012. Erik Seastead, Jon Chapin, Darren Matthews, and Steve Saunders offered their opinions on long runs. Last, the Fulbright program, the National Science Foundation, the Rachel Carson Center and its co-directors Helmuth Trischler and Christof Mauch, the Davis Center of Harvard University, and Colby College's internal grant program—all provided vital intellectual and financial support.

Index

www.ingramcontent.com/pod-product-compliance
Ingram Content Group UK Ltd.
Pitfield, Milton Keynes, MK11 3LW, UK
UKHW041421290926
475585UK00013B/27/J

9 780674 728905